The Fundamentals of Human

EMBRYOLOGY

Student Manual
Second Edition

John Allan and Beverley Kramer

Wits University Press
1 Jan Smuts Avenue
Johannesburg
2001
http://witspress.wits.ac.za

First published 2010

ISBN 978 1 86814 503 4

Cover design and layout by Hothouse South Africa
Printed and bound by Creda Communication, Cape Town

CONTENTS

CHAPTER THREE: **Organogenesis**

PREFACE

Few teachers of the basic medical sciences would dispute the value of a knowledge of embryonic development in the understanding of general anatomy and in the production of congenital defects.

Traditionally, the teaching of the basic aspects of human developmental anatomy is coupled with that of general anatomy and is rendered as part of the preclinical curriculum to students in the medical, dental and allied medical disciplines. Hopefully, when the student studies paedriatrics, some of the basics will have remained to assist in making sense of congenital anomalies. With the worldwide trend to reduce the preclinical sciences to a bare minimum, both anatomy and its essential partners – histology and embryology – will suffer accordingly.

While the authors feel that this trend is unfortunate and undesirable as well as of educational denegation, they feel, nevertheless, that it is necessary to provide students with a text in embryology, with the purpose of indicating, in a simplified way, the essentials of the subject. In this way a reasonable 'working knowledge' of embryonic development and its aberrations may be acquired.

Not only is this necessary from an educational point of view but also from a practical standpoint in the light of the increasing incidence of congenital abnormalities resulting from the industrial, chemical and radioactive pollution of the earth's surface. The practising doctor is very likely to encounter one or more of these abnormalities in his/her career.

In this text we have attempted to adhere to the 'fundamental' aspects of embryonic development, providing a progressive account of the processes which lead to the development of the human organism.

Our goal is to impart to students a comprehensive overview of how the human embryo forms, not only as a basis for the study of human anatomy, but also as a link to possible abnormalities that they will encounter in their clinical careers.

In the near future, genetic engineering will attempt to correct congenital abnormalities. Gene manipulation will challenge normal and abnormal development in an attempt to reduce the risk of, for example, a congenital heart abnormality or a cleft lip. Unfortunately, modern technology may also increase the incidence of certain abnormalities. A thorough grounding in the fundamentals of human development will prepare the professional-in-training for the 'progress' of the future. As we progress towards 'molecular medicine', we should not lose sight of the basic facts which make humans human.

The First Edition of the book was well received by both academics and students. One criticism voiced by students was that some of the diagrams were too small. In the interim too, the Federative International Committee on Anatomical Terminology substantially revised the *Terminologia Embryologica*.

The authors felt that these aspects of the book should be corrected. In addition, we believed that it was necessary to make alterations to the sequencing of some of the topics to allow the development to flow more logically.

The authors also felt that the addition, as an appendix, of coloured photographs of congenital abnormalities would help students to form a more realistic idea of developmental abnormalities. The authors would like to thank Professor S. Levin and Dr P. Evan for kindly allowing them to include in the appendix, photographs of specimens from their collections. In addition the School of Anatomical Sciences, University of the Witwatersrand, Johannesburg is acknowledged for photographs of specimens acquired under the Human Tissues Act of South Africa

INTRODUCTION

When we look at pictures of embryos in books we generally do not appreciate their 'size'. After the passage of 20 days from the time of fertilisation, the embryo is about 2mm in length and after 35 days it is about 8mm in length. From this it is evident that the earliest stages of development may only be studied effectively by viewing sections of embryos under a light microscope. Another way of studying early development is to project and trace enlarged images from microscope sections, on to sheets of cardboard or wax of known thickness. The images may then be cut out and mounted one upon the other to create an enlarged model of the embryo, organ or system which is being studied **(Fig. 1)**. More recently, the surface appearances of early embryos have been studied by the use of the 'scanning electron microscope' (SEM).

Obtaining very early human embryos for study has always presented a problem and in the past, these have been harvested mainly from hysterectomised uteri. Much of what is known of early embryonic development has been gleaned from these specimens. More recently, with the improved techniques in harvesting ova from human females and fertilising them *in vitro*, it has been possible to observe the very earliest stages of fertilisation and development. This is called the ***in vitro* fertilisation** technique.

Determination of the age and size of embryos is another problem. To overcome this, embryologists have developed a number of methods for determining age or size. One such method is to count the number of segments (somites) seen in the embryo. A somite is a body segment and the first of these appears when the embryo is about 1.5mm in length. The somites increase in number as the embryo increases in age and size. Clearly, because of the small size of the embryo, it is only possible to count the number of somites under the microscope. Later, when the embryo becomes larger, the length and therefore the age is assessed by measuring the greatest length from the crown of the head to the caudal curvature (rump) **(C-R length) (Fig. 2)**. By the end of 6 weeks (42 days), the C-R length is about 12mm and by the end of 8 weeks (56 days), the embryo has attained the length of about 30mm. By this time the embryo has developed the basics of all the organs and systems required by the adult. This is said to be the end of the

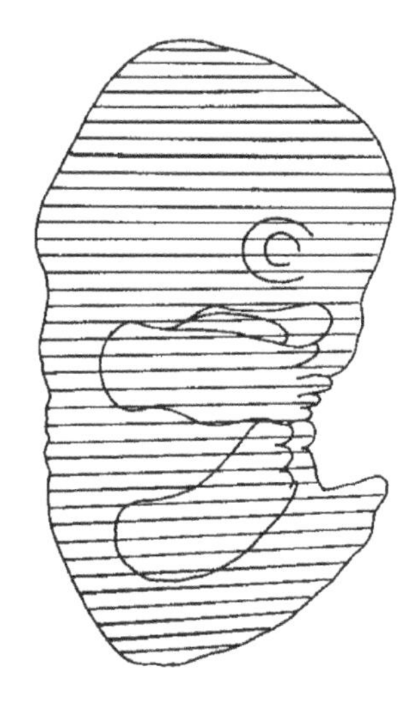

Figure 1: Reconstruction of an embryo by laminar wax cut-outs.

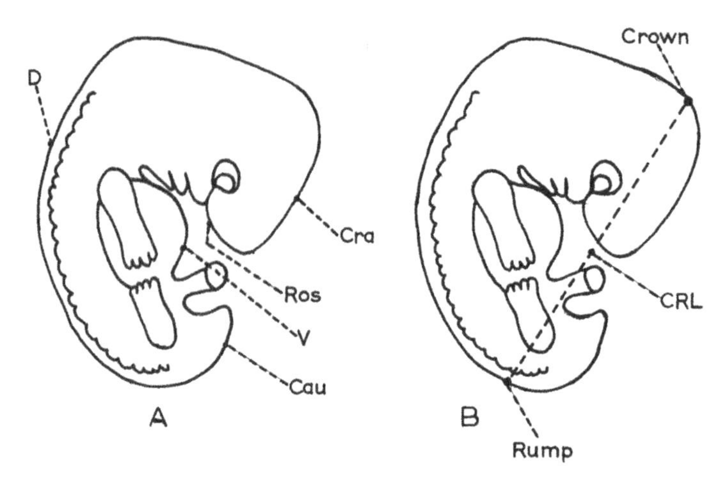

Figure 2: Crown-rump length (CRL) for determination of the age of the embryo. A: General aspects of the embryo. **B:** Measurement of crown-rump length. Cra = cranial; Cau = caudal; D = dorsal; Ros = rostral; V = ventral.

'embryonic phase' of development **(Fig. 3)**. After this, the **'fetal phase'** of development takes place, when **growth** of the existing organs and systems predominates. During the growth phase, the C-R length increases in an almost linear fashion, whereas the weight increases in an exponential manner **(Fig. 4)**.

In the early stages of development, the embryo is a relatively shapeless mass, but by the beginning of the fetal period, it has unmistakably human features. It must be obvious that when the embryo is very small and when many things are happening simultaneously in it, any infection in the mother such as German measles (rubella) will be transmitted to the embryo and may possibly result in several disturbances in the normal development of the embryo. Likewise, a mother who is addicted to alcohol or drugs or who smokes during pregnancy is likely to produce an abnormal or retarded or mentally deficient child.

From the end of the 8th week post-fertilisation (**pf**), the fetus grows until, at about 40 weeks after fertilisation, the time comes for it to see the light of day. By the 36th week, the fundus of the uterus has reached the level of the mother's 9th costo-chondral junctions (transpyloric plane). At about the 40th week, the head of the fetus descends (engages) into the true pelvis, a phenomenon which the mother regards as **lightening** and now the time for parturition is near **(Fig. 3)**.

Generally, a pregnancy is said to be 40 weeks in length. This is considered as ten lunar months or nine calendar months in length. Obstetricians calculate the starting point of pregnancy as the first day of the last menstrual flow.

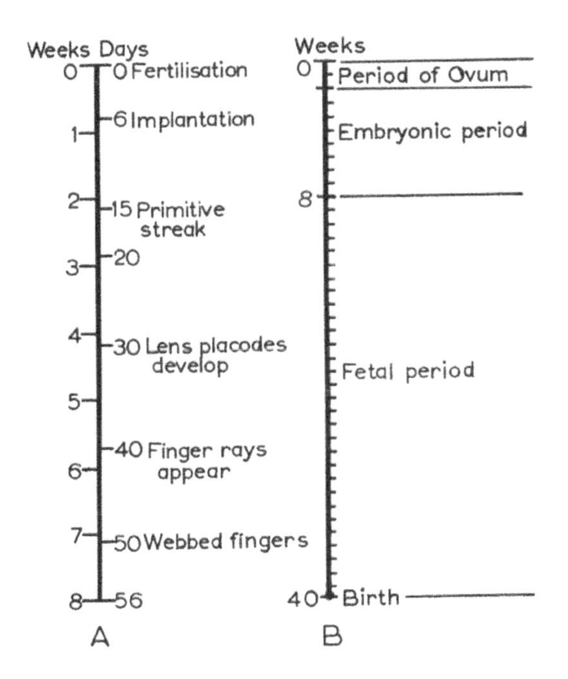

Figure 3: Table of embryonic and fetal events.

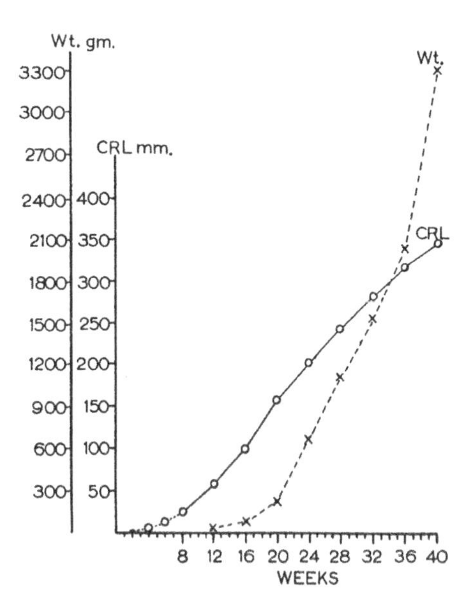

Figure 4: Relationship of weight to crown-rump length. CRL = crown-rump length; Wt = weight.

TERMINOLOGY

As mentioned previously, embryology is related to general anatomy. However, the terminology of the two states cannot readily be applied to each other. In 'adult' anatomy the terminology refers to the body being in the anatomical position (a fixed position or posture); in development the embryo may implant itself in various positions and, with the development of the umbilical cord, the conceptus may assume in a variety of positions in the uterine cavity. Consequently, one cannot apply the same terminology used in the adult to the embryo and fetus.

In this text we use the terminology relating to the embryo itself, whatever its intra-uterine position may be. Thus, the term **cepahalic** or **cranial** refers to the head end of the embryo, while **caudal** refers to that part of the embryo which is designated as the tail region **(Fig. 2)**. Similarly, the term **dorsal** is applied to the surface where the vertebral column exists, and the term **ventral** is applied to the opposite surface of the embryo. While in the cranial region of the embryo, the term **rostral** is used to refer to structures in relation to the nose. **Medial** and **lateral** refer to positions in relation to the central line of the embryo (median sagittal plane). The terms **proximal** and **distal** are used in relation to the origin of a structure (see midgut loop rotation).

Often an embryo is sectioned along a particular plane to view its organs. Thus the following terms are used. A **median sagittal plane** passes through the longitudinal, central axis of the embryo thus dividing the embryo into two longitudinal symmetrical halves. A **parasagittal plane** refers to a plane parallel to the median sagittal plane and will therefore divide the embryo into two unequal parts. A **coronal plane** passes through the embryo at right angles to a sagittal plane and will divide the embryo into ventral and dorsal parts. A **transverse plane** through the embryo passes at right angles to both the sagittal and the coronal planes and divides the embryo into equal or unequal upper or lower parts.

ADVICE TO THE STUDENT ABOUT EMBRYOLOGY

From the point of view of the student, two problems exist when studying embryology:

(a) Since development is a dynamic and ongoing process in which several 'things' are happening at the same time, the beginner is apt to become confused by the multiplicity of events. The confusion leads to frustration with a tendency to avoid or neglect this important subject. The solution to the problem is to take the development in small steps and to bring each to its final conclusion. Later, the single processes come together as a coherent whole.

(b) Many people have a problem in creating in the mind, a three-dimensional image from a two-dimensional picture. The authors are fully aware of this problem and have attempted to overcome it by providing suitable serial drawings of the developing parts of the embryo. By following these simple drawings and with simple explanations from the text and diagrams, the student should be able to study the ongoing processes of development with a minimum of difficulty. It should be obvious to any beginner that the study of embryonic development requires a modicum of imagination as well as a reasonable amount of concentration and study.

However, one of the most important aspects of the ability to understand human embryology is to have a sound knowledge of basic biology and the general anatomy of the human body. Without this, the study of embryology becomes a tiresome and lacklustre drudgery! To a person having a knowledge of human structure and function, the study of embryology is a fascinating, educational and explanatory pastime. Anatomy and embryology studied together result in a circular type of understanding, in that the embryology explains the intricacies of the anatomy while the anatomy gives a sound basis for the understanding of how structural form came about from a developmental point of view.

CHAPTER ONE

Preparation of Gametes for Embryonic Development

THE CELL

The study of embryonic development requires an understanding not only of how and when structures arise and develop but also of how they enlarge and grow. The human individual arises from the conjugation of two minute structures called **cells**, one from the mother (oocyte) and one from the father (spermatozoon). These are called **gametes**. Together, these gametes form a single cell, the **zygote**, from which the entire embryo, including its surrounding membranes, grows. The zygote undergoes successive cleavages and the cells thus produced multiply so that all the organs of the embryo are developed. Thereafter, the formed organs grow and enlarge in the fetus until the moment of birth. The newborn grows into the infant, the adolescent and the adult. **Cell growth** and **cell multiplication**, therefore, are *the* most fundamental aspects of embryonic development and of existence, and are present from 'conception' to the 'cessation' of life.

Before considering the actual process of cell multiplication or cell division, let us consider briefly those structures of a cell which are intimately associated with the process of cell division. The cell is regarded as the basic 'working unit' of all animals and plants. It is a minute living structure consisting of a mass of **cytoplasm** surrounded by a **cell membrane**. The cytoplasm contains a number of **organelles** and a **nucleus** (Fig. 1.1). The nucleus is also confined within a **nuclear membrane** and contains **chromatin**, a chemical substance composed of **deoxyribonucleic acid (DNA)**. This substance consists of simple proteins (histones) combined with the bases adenine, thymine, guanine and cytosine. These are bound randomly in pairs (A-T, G-C etc.) across two complementary strands of DNA set in the form of a double helix (Fig. 1.2). Combinations of these base pairs form **genes** which confer upon individuals their distinguishing characteristics. The chromatin is condensed into **chromosomes** which, when suitably stained, is seen as strands under the microscope. Although much emphasis is placed upon the activity of the nucleus in cell multiplication, the cytoplasmic contents also play an important role in the process. The **mitochondria** provide the energy for the division, while the presence and splitting of the **centrosome** is essential for the process of division.

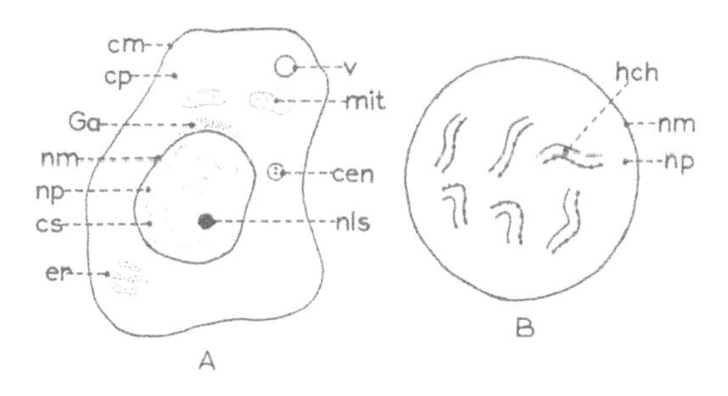

Figure 1.1: A diagram to illustrate the structure of a cell (A) and the formation of chromosomes in the nucleus (B). cen = centrosome; cm = cell membrane; cp = cytoplasm; cs = chromatic substance; er = endoplasmic reticulum; Ga = Golgi apparatus; hch = homologous chromosomes; mit = mitochondrion; nls = nucleolus; nm = nuclear membrane; np = nucleoplasm; v = vacuole.

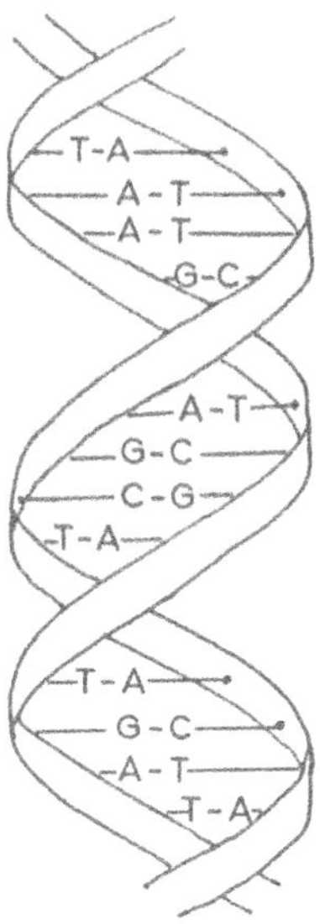

Figure 1.2: The double helix.

There are two types of cell division in the animal kingdom; the one is called **mitosis** and the other **meiosis**.

Mitosis

In this type of division, the two resulting cells have an identical composition and an identical genetic constitution. Established adult somatic cell lines such as fibroblasts, liver cells and the epidermal cells of the skin replicate by mitosis. The cell which is about to divide originally contains the normal number of chromosomes, which is 46 in the human. This is called the **diploid** (double) number and may be referred to as **2n**. By a process of doubling and separating, each of the 'daughter-cells' resulting from the division will still contain the same number of chromosomes as the original 'parent' cell. By contrast, in the process of meiotic division (described below), the number of chromosomes in the 'daughter cells' is halved. This is called the **haploid** (halved) number and is referred to as **1n** or **n**.

Since the process of mitosis occurs in a continuous way, it is convenient for proper understanding to divide it into four relatively simple phases. These are prophase, metaphase, anaphase and telophase, which are descriptive events in the process of cell division.

Most adult cells are found to exist in a 'resting' state and this may be called the **interphase** stage. At this time, the nucleus normally has the appearance of a rounded body surrounded by a stout nuclear membrane. The interior of the nucleus appears to be empty but by applying suitable staining techniques, it is possible to see fine threads of chromatin which are parts of chromosomes (**Fig. 1.1A**).

A situation which is frequently lost sight of, is that *all* people and, therefore, *all* cells have parents. In the cell nucleus, the parents are represented by **homologous chromosomes**

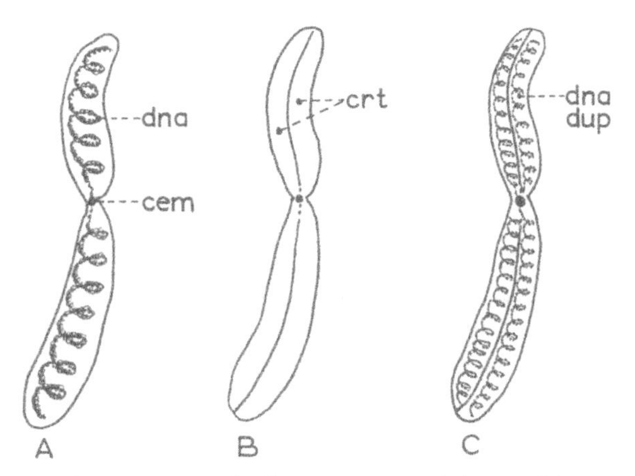

Figure 1.3: The structure of a chromosome and chromatid. A: Chromosome showing DNA content. **B:** Chromosome splitting to form chromatids. **C:** Chromatids containing duplicated DNA. cem = centromere; crt = chromatids; dna = deoxyribosenucleic acid.

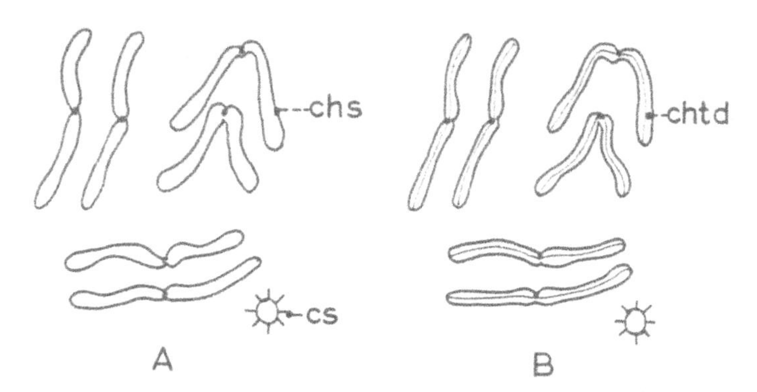

Figure 1.4: Appearance of condensed chromatin into chromosomes. chs = chromosomes; chtd = chromatids; cs = centrosome.

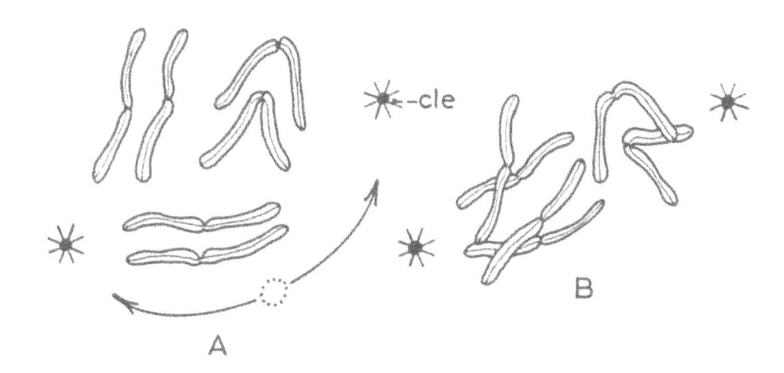

Figue 1.5: Diagram to illustrate division of centrosome into centrioles as well as formation of asters around centrioles. cle = centriole.

(**Fig. 1.1B**); some are from the mother and an equal number are from the father. The visible homologous chromosomes are identical in appearance and are set in pairs. It should be noted that the number of chromosomes is 'species specific' in that each species of animal has a fixed number of chromosomes, so that usually only animals of the same species may mate with one another to produce homologous pairs of chromosomes.

Each chromosome has a constriction somewhere along its length. This is called the **centromere** and is constant in position for any particular chromosome (**Fig. 1.3**). The centromere plays a crucial part in cell division by orientating the chromosomes in the correct position in the cytoplasm. When a cell receives a signal to divide, a series of events is set in motion which begin in interphase but soon pass into the stage of prophase.

Prophase (Initial or First Phase)

The chromatic material (coloured material of the nucleus) becomes condensed so that the chromosomes are clearly visible (under the microscope) (**Fig. 1.4**). The chromosomes are set in homologous pairs and soon each chromosome separates longitudinally into a pair of parallel **chromatids**. At this time the DNA undergoes duplication. The chromatids gradually separate from one another but remain attached at the centromere. At the same time, the **centrosome** in the cytoplasm divides into two **centrioles** which move to the opposite poles of the cytoplasm (**Fig. 1.5A**). When they reach these positions, cytoplasmic 'rays' form around them. These are called **asters** (stars). The nuclear membrane now disintegrates and disappears leaving the chromosomes as a tangled mass in the cytoplasm, between the centrioles (**Fig. 1.5B**). The rays of the asters around the centrioles extend across the cytoplasm passing between the chromosomes, to form a **spindle** (**Fig. 1.6A**). When this process is completed, the cell division enters the next descriptive stage – metaphase.

Metaphase (Second Phase)

The chromosomes become shorter and thicker and become arranged on the equator of the spindle. When this occurs, the rays of the spindle become attached to the centromeres of the chromosomes (**Fig. 1.6B**). It is clear from the configuration of the cell content that the rays of the spindle are ready to pull the mixed chromatids to the opposite ends of the cell. With completion of these events, the division passes to the next stage – anaphase.

Anaphase (Third Phase)

The centromere of each chromosome now splits and the chromosomes also split lengthwise. Being attached to the strands of the spindle, the 'daughter' chromosomes (chromatids) are separated from one another and are drawn towards the centrioles (**Fig. 1.7A,B**). As the chromosomes move towards the poles of the cell, the cell membrane develops a circumferential groove, the **cleavage furrow**, and at the same time the central part of the spindle becomes narrowed (**Fig. 1.8A**). With the completion of these events, the division process enters its final stage – telophase.

Telophase (Terminal Phase)

The chromosomes lose their crisp and orderly shape and revert to a disorderly mass of chromatin (similar to that of interphase). A nuclear membrane regenerates around each newly formed chromatin mass as the cleavage furrow deepens until the resulting two daughter cells are completely separated from one another. A small concentrated mass of chromatin appears among the chromosomes as the nucleolus and the centrioles revert to a centrosome (**Fig. 1.8B**).

Each of these cells will have 2n chromosomes as a result of the original chromosomes duplicating into identical chromatids and these will form the chromosomes of the new cells.

Meiosis

This type of cell division is confined to the **sex cells** and is characterised by the fact that the number of chromosomes in the 'daughter-cells' is reduced to half of the original number. Since the original number of chromosomes is 46 (2n), which is the diploid number, the 'daughter-cells' after division will contain the haploid number (1n) or 23 chromosomes.

This chromosomal reduction is necessary to form the gametes (sex cells responsible for **fertilisation**) because the mother and father will each contribute one gamete containing 1n chromosomes leading to the formation of the new individual. The *sum* of the chromosomes in the zygote, after fertilisation is, therefore, 2n, the normal diploid number.

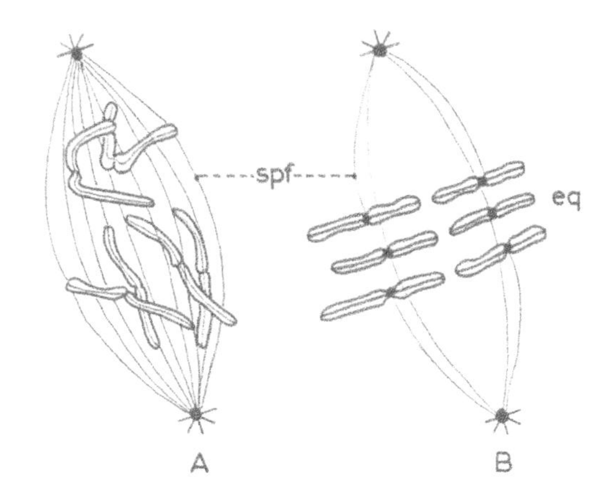

Figure 1.6: Diagram to illustrate formation of the mitotic spindle (A) and attachment of the centromeres to the spindles (B). spf = spindle fibres; eq = equator.

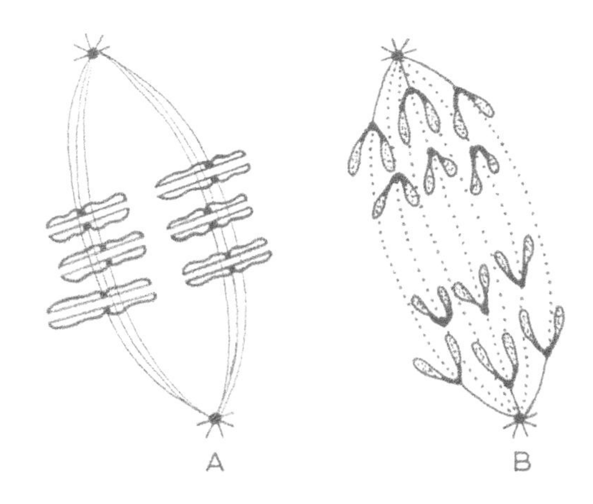

Figure 1.7: A: Division of chromosome into two chromatids. **B:** movement of the chromatids to the poles of the spindle.

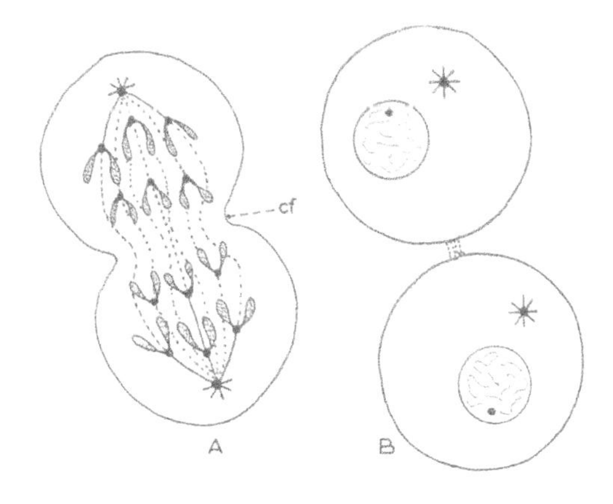

Figure 1.8: Diagram to illustrate the process of division of one cell into two cells. cf = cleavage furrow.

As in mitosis, the meiotic division may be simplified by dividing it into a number of phases. It should be noted that meiosis in oogenesis is slightly different from that in spermatogenesis. These two types of meiosis will be considered separately.

'Oogenic' Meiosis

The oogonia reach the gonad by migrating along the dorsal mesentery of the gastro-intestinal tube from the endoderm of the umbilical vesicle **(Fig. 1.9)**. In the gonad (ovary) the oogonia differentiate into primary oocytes within the ovary and remain there, in interphase, until puberty. Under the influence of pubertal hormones, the primary oocytes receive a signal to undergo maturation and, to reach maturity, the cells will be required to pass through two cycles of meiosis called Meiosis I and Meiosis II. The second meiosis is more in the nature of a mitosis.

The process begins with the interphase cell entering the stage of Prophase I. This stage is a somewhat lengthy process and is traditionally divided into a number of 'substages' – leptotene, zygotene, pachytene, diplotene and diakinesis. In this text, for the sake of brevity we shall attempt to consolidate them into a single stage.

Prophase I

The chromatic material of the nucleus of the oocyte condenses to form chromosomes. These

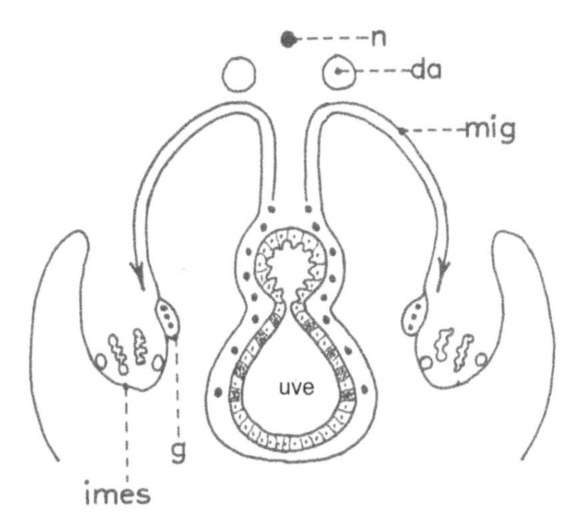

Figure 1.9: Cross section of embryonic abdominal cavity to indicate the pathway of migration of primordial germ cells. da = dorsal aorta; g = gonad; imes = intermediate mesoderm; mig = indicates pathway of migration; n = notochord; uve = umbilical vesicle.

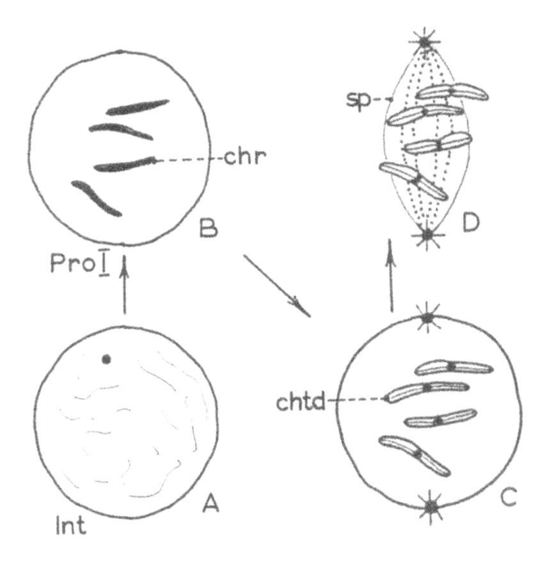

Figure 1.10: A description of Prophase I. chr = chromosomes; chtd = chromatids; Int = nucleus of interphase cell; Pro I = prophase I; sp = spindle.

chromosomes split longitudinally into two sister chromatids attached to one another at a centromere **(Fig. 1.10)**. Thus, each cell nucleus contains 2n chromosomes (the diploid number). The nuclear membrane disappears leaving the chromosomes and their chromatids floating free in the cytoplasm. During this period, the centrosome has split into two centrioles and a spindle forms between the centrioles.

Metaphase I

The free-floating chromatids align themselves in the equatorial plane of the spindle **(Fig. 1.11A)**.

Anaphase I

The homologous (structurally alike) chromatids are drawn by the spindle to the opposite poles of the cell. This takes place without the chromatids splitting off from one another, so that each bundle contains only half (1n) of the original number of chromosomes. This is the **reduction division** and takes place when the primary oocyte divides to form a secondary oocyte **(Fig. 1.11B)**.

Note the difference between this type of cell division and that of mitosis. In mitosis, the chromatids split off the original chromosomes resulting in double the number of chromosomes (4n) and division results in equal numbers (2n) going to each 'daughter-cell'. In the initial meiotic division, the chromatids do not split away, keeping the chromosome number at 2n. Equal division

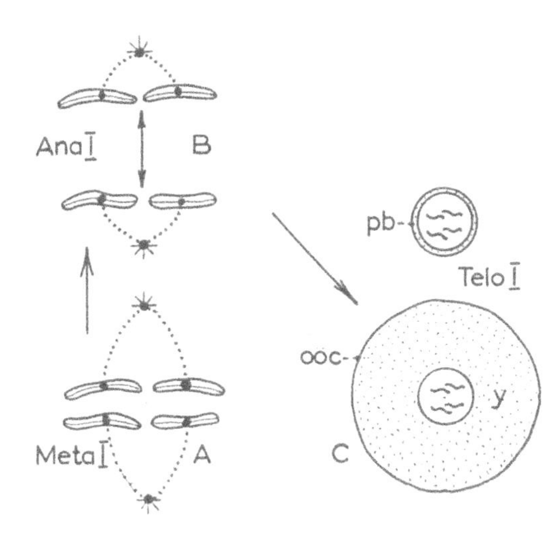

Figure 1.11: Mitotic metaphase (A), anaphase (B) and telophase (C). ooc = oocyte; pb = polar body; y = yolk.

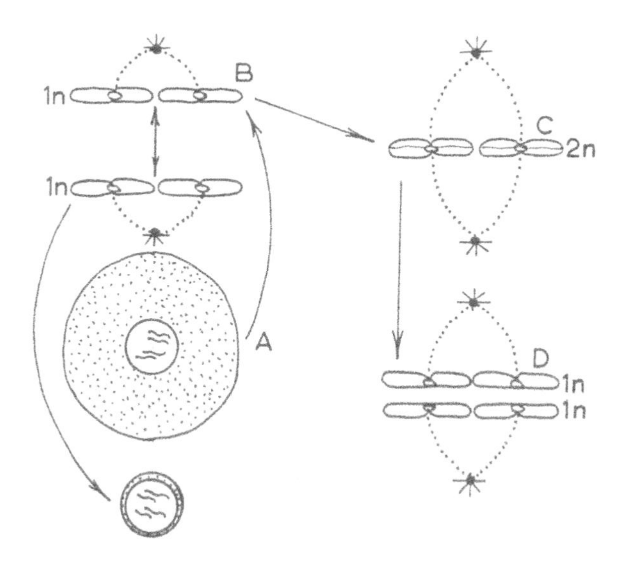

Figure 1.12: Meiosis II. This process is equivalent to mitosis. Note the formation of chromatids which duplicate DNA and which separate so that each chromatid forms 2N.

must, therefore, result in 1n chromosomes going to each 'daughter-cell'. This is the fundamental difference between mitosis and meiosis.

Telophase I

The oocyte divides in a special way. The nucleus and, therefore, the chromosomes divide equally (1n to each 'daughter-cell') but the cytoplasm divides very unequally. The 'daughter-cell' which is to become the true and proper oocyte retains the bulk of the cytoplasm while the other is severely deprived of cytoplasm. The deprived cell is called a **polar body** (the first polar body) (**Fig. 1.11C**). Although the first polar body degenerates fairly soon after its formation, it lasts long enough in most mammals to undergo another division. In the human, however, it disintegrates rapidly and disappears.

It seems reasonable to suppose that the unequal cytoplasmic division is a method for preserving 'yolk' for the subsequent activity of the fertilised oocyte. Likewise, it would seem futile to provide cells (the polar bodies) with valuable yolk when they are to be discarded.

On the completion of Telophase I, the secondary oocyte divides again to form the definitive, fertilisable oocyte. The secondary oocyte thus enters Meiosis II and rapidly goes through the stages of Prophase II, Metaphase II, Anaphase II and Telophase II. In this second series of divisional events, the chromosomes form chromatids *which separate from their parent chromosomes*, producing double the number of chromosomes (**Fig. 1.12**).

Thus, Meiosis II is a process which resembles mitosis, in the sense that the chromatids physically separate from the original chromosomes to form 2n chromosomes. Half of the 2n goes to a polar body (the second polar body) and half goes to the oocyte. Thus, the polar body and the oocyte each contain 1n chromosomes. Again, the cytoplasm is unequally divided, the greater part going to the oocyte.

Note that Meiosis II only occurs if the secondary oocyte produced by Meiosis I is fertilised. Thus, the presence of a spermatozoon within the cytoplasm of the oocyte provides the stimulus for the oocyte to complete Meiosis II and form the second polar body. In the human, the second polar body survives the first few cleavages of the zygote and then disintegrates.

'Spermatogenic' Meiosis

At puberty, under the influence of hormones, the spermatogonia in the seminiferous tubules of the testes begin to undergo a process of maturation which is similar to that of the oogonia in the ovary. The spermatogonia are found at the basal surface of the epithelium of the seminiferous tubules in close association with the base of one or more Sertoli cells (**Fig. 1.13**). The spermatogonia multiply by mitosis, one of the 'daughter-cells' remaining *in situ* for future division and the other passing

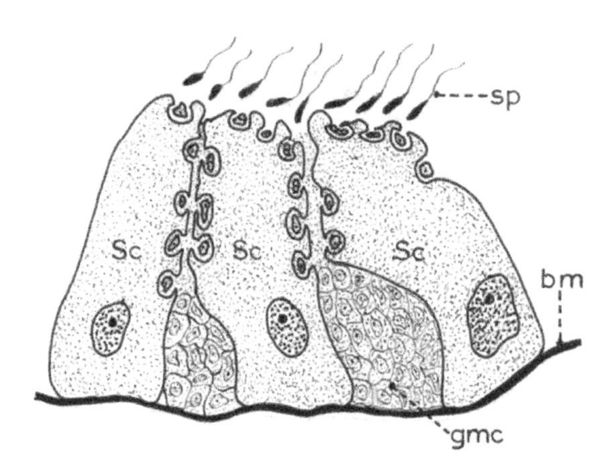

Figure 1.13: Topographical section of part of a seminiferous tubule indicating the passage of developing spermatozoa towards the lumen of the tubule. bm = basement membrane; gmc = germ cells; Sc = Sertoli cell; sp = spermatozoa.

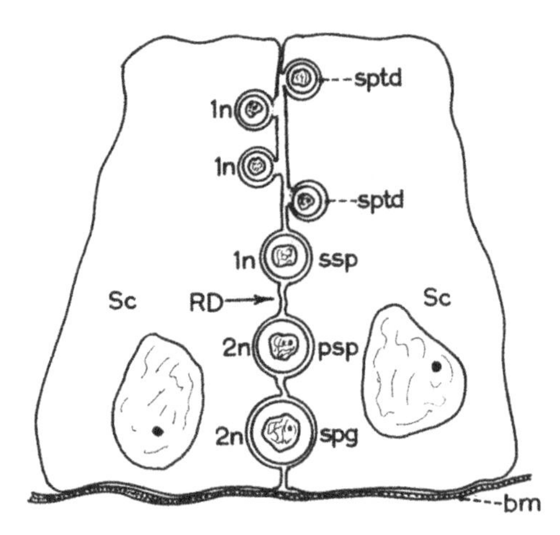

Figure 1.14: Progression of male sex cells towards the lumen during maturation. bm = basement membrane; psp = primary spermatocyte; RD = position of reduction division; Sc = sertoli cell; spg = spermatogonium; ssp = secondary spermatocyte; sptd = spermatid.

between the surfaces of adjacent Sertoli cells, towards the lumen. This rising cell may be called a primary spermatocyte and as it passes between adjacent Sertoli cells, the tight junctions between the Sertoli cell membranes break down and allow the spermatocytes to pass between the cells (**Fig. 1.14**). During this passage, the spematocyte passes through all the stages of Meiosis I (described under oogenic meiosis) and results in the formation of two secondary spermatocytes. The primary spermatocyte contains the diploid (2n) number of chromosomes and having undergone a reduction division, each of the two 'offspring-cells' (secondary spermatocytes) will contain the haploid (1n) number. Each secondary spermatocyte undergoes mitotic division (Meiosis II) to form two spermatids. It should be obvious that four spermatids will result from the original primary spermatocyte. Thus, four cells called spermatids are formed, each containing 1n chromosomes (**Fig. 1.14**).

These spermatids become embedded in the cytoplasm of the Sertoli cells where they undergo a series of morphological changes to become spermatozoa.

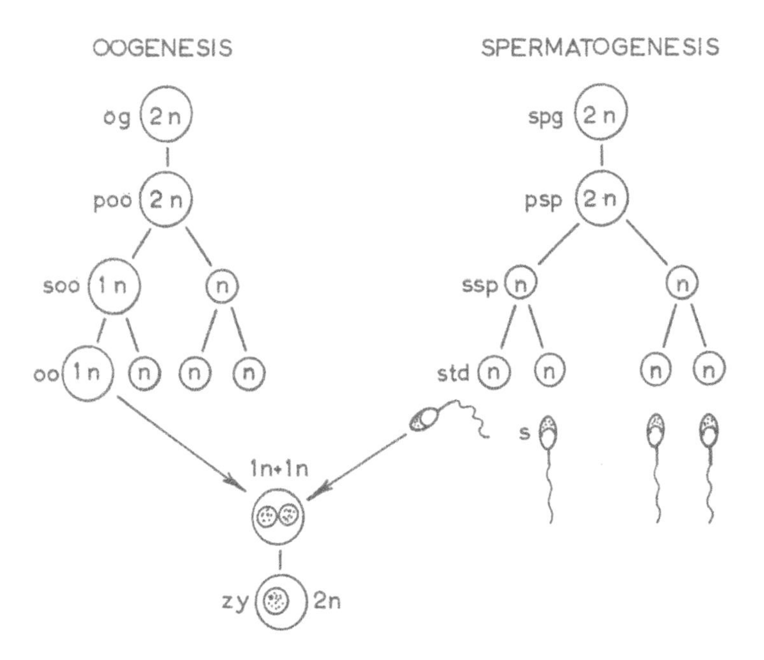

Figure 1.15: Fusion of the ovum and the sperm to form a zygote. og = oogonium; oo = oocyte; poo = primary oocyte; psp = primary spermatocyte; s = spermatozoon; soo = secondary oocyte; spg = spermatogonium; ssp = secondary spermatocyte; std = spermatid; zy = zygote.

Note that at the reduction division, two secondary spermatocytes are formed, whereas in oogenesis, one secondary oocyte and one polar body are formed. At the next division, both secondary spermatocytes give rise to two spermatids, whereas in oogenesis the one secondary oocyte gives rise to a second polar body. In spermatogenesis, there is no 'wastage' of cells, but in oogenesis, the polar bodies are wasted.

The development of the oocyte and spermatozoon is summarised in **Figure 1.15**.

Gametogenesis

This is the term applied to the development of the sex cells or gametes, of which one is derived from the male (spermatozoon) and one is derived from the female (oocyte). They are responsible for the initiation and subsequent growth of the embryo.

The number of chromosomes in the ordinary (somatic) human cell is 46 but in the case of the sex cells, each contains 44 autosomes (ordinary chromosomes) plus two 'sex chromosomes' which are designated in the female as XX and in the male as XY. Thus, the genotype of the female sex cell is 44XX and that of the male 44XY. These chromosomes are responsible for determining the sex of the embryo and it is evident that the male sex cell carries the sex determinant chromosome (Y).

Because conjugation of the sex cells takes place in the formation of a new individual, it is necessary for each to have half (1n) the original number of chromosomes so that when they combine, the original number of 44XX or 44XY (2n) will be reconstituted (**Fig. 1.16**).

Since the gametes are different from the ordinary run of body cells, we should perhaps ask where they originally came from. They arise as specialised cells called **primordial germ cells** in the endoderm of the wall of the umbilical vesicle, adjacent to the exit of the allantoic diverticulum. They move from this region into the mesentery of the gut and around the coelomic cavity until they reach the **urogenital fold** and finally the genital ridge which is destined to be the site of the gonad (testis or ovary) (**Fig. 1.9**).

Let us consider, in more detail, the maturation of the primordial germ cells:

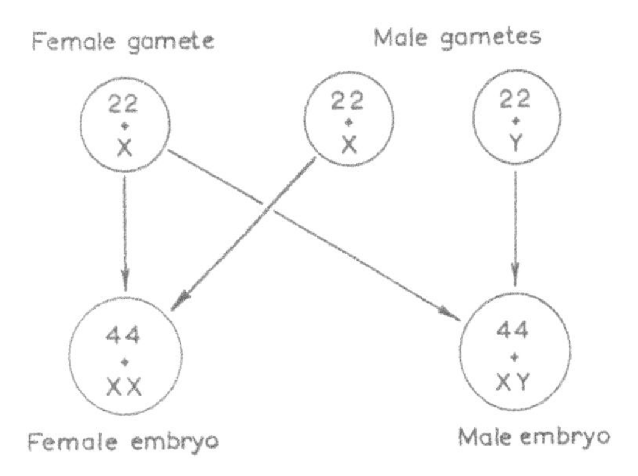

Figure 1.16: The formation of male and female embryos from the relevant gametes. X = female sex chromosome; Y = male sex chromosome.

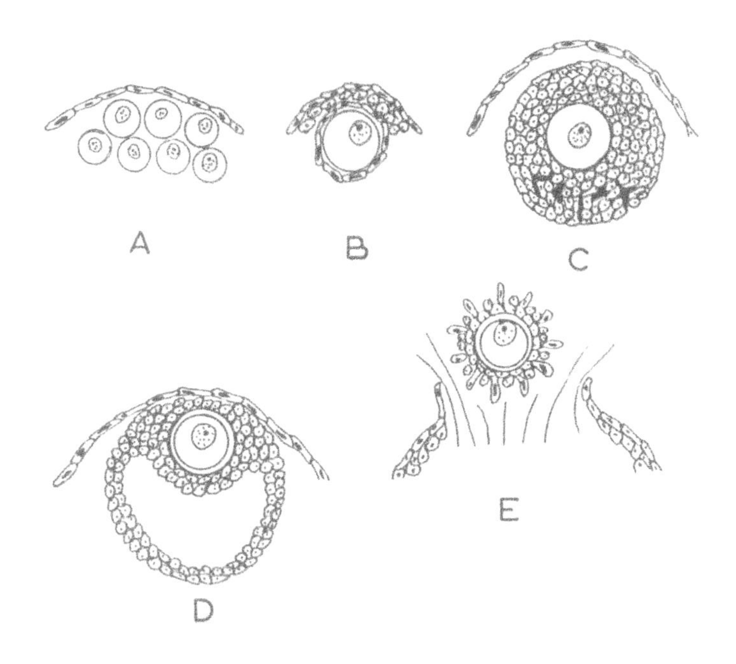

Figure 1.17: A series of changes to show the development of the oocyte and extrusion thereof. **D** depicts the mature follicle.

(a) The ovum develops in the cortical layer of the ovary as one of a group of **oogonia**. These differentiate into **primary oocytes** which become invested with flattened epithelial cells to form primordial follicles which lie dormant in the ovary until puberty. At the time of puberty, the epithelial cells proliferate to form a single layer and then several layers of cuboidal follicular cells around the oocyte. Subsequently, a cavity filled with follicular fluid (liquor folliculi) forms around the oocyte, resulting in the formation of the **mature (Graafian) follicle** (**Fig. 1.17**). While still within the mature follicle the oocyte begins its meiotic division (Meiosis I) with the formation of two cells, one being the definitive oocyte, the other being the first polar body. This is a reduction division. Immediately after the first polar body is extruded, the oocyte enters into the second meiotic division and reaches metaphase. The growing mature follicle will rupture and the oocyte surrounded by its crown of follicular cells, the **corona radiata**, is shed into the peritoneal cavity of the female individual together with the follicular fluid. Here the oocyte is caught up by the **fimbria** of the uterine tube and transported by ciliary action into the uterine tube (**Fig. 1.18**). The human oocyte, immediately after ovulation, is in the stage of metaphase of

Meiosis II. Only if **fertilisation** (entry of sperm into the oocyte) occurs, will the second meiotic division be completed and the second polar body will form. At this time, the oocyte completes this phase of meiosis which is not a reduction division but a mitotic division, resulting in the formation of a second polar body.

Meiosis is a reduction division with the formation of two polar bodies (defunct cells) allowing for the removal of half the number of chromosomes. While the nucleus undergoes equal division, the cytoplasm undergoes unequal division, resulting in the oocyte remaining large while the polar body is small.

Note. The process of cell division is not an immediate process but takes several hours to complete; halving the number of chromosomes produces a haploid number (1n); when the original number of chromosomes is restored by fertilisation, the diploid number (2n) results.

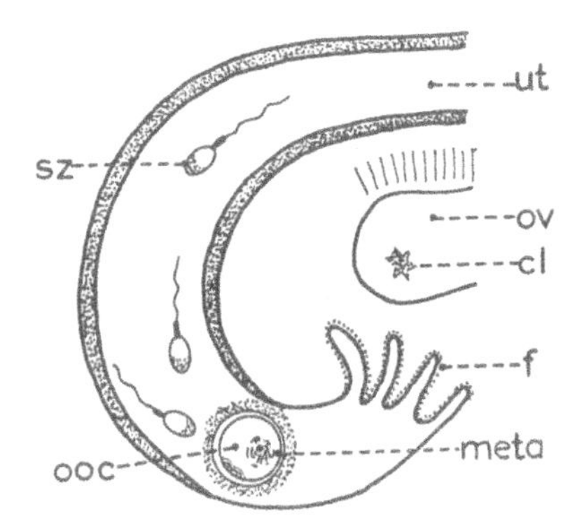

Figure 1.18: The meeting of the oocyte and the spermatozoon in the distal part of the uterine tube. cl = corpus luteum; f = fimbria; ooc = oocyte; ov = ovary; sz = spermatozoon; ut = uterine tube. *Note* the oocyte is at metaphase.

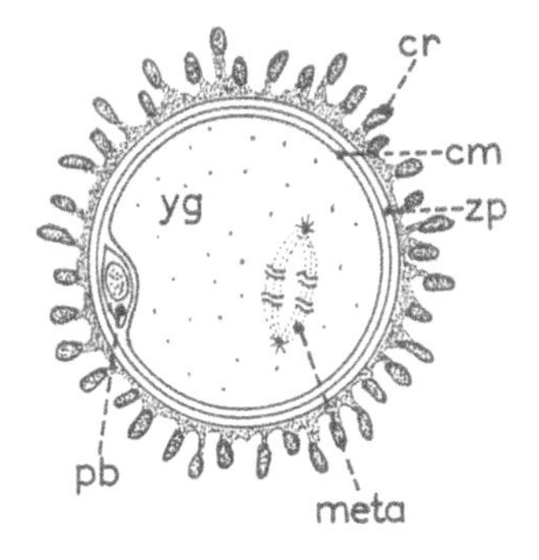

Figure 1.19: Structure of the fertilisable oocyte in metaphase. cm = cell membrane; cr = corona radiata; meta = metaphase spindle; pb = polar body; yg = yolk granules; zp = zona pellucida.

Both oocyte and polar bodies each contain 22+X chromosomes and the oocyte, lying in the outer one-third of the uterine tube, is now ready for fertilisation. The oocyte now consists of a large cell surrounded by a clear zone, the **zona pellucida** which in turn, is surrounded by the corona radiata of follicular cells (**Fig. 1.19**). At this time, the viability of the oocyte is limited; if it is fertilised, it is retained, if not, it is discarded at the next menstrual period. At this stage, the female gamete is about 130 microns in diameter. This is a large cell and its cytoplasm contains a number of yolk granules. These granules are necessary for the survival of such a large cell. When implantation into the uterus occurs, the fertilised cell will also obtain nourishment from the succulent endometrium.

(b) The spermatozoon (sometimes called 'sperm' for short) arises in the **seminiferous tubules** of the testis as a **spermatogonium**. The spermatogonium divides, forming two cells of which one becomes a **primary spermatocyte**. The other cell remains as a 'stem' cell, which will be responsible for further primary spermatocyte formation. The primary spermatocyte undergoes reduction division to become two **secondary spermatocytes**, each of which contains half the number of chromosomes (22+X) or (22+Y). These then undergo further division (not a reduction division) to form **spermatids**, some of which have the X chromosome and others the Y chromosome. The spermatic series of cells proceeds towards the lumen of the seminiferous tubule between the existing **Sertoli cells** (which are important in the formation of the blood-testis barrier). The spermatids undergo differentiation to become spermatozoa (**Fig. 1.14**). This takes place as follows.

The spermatid has a central nucleus and a well-marked Golgi body. Discrete granules form in the Golgi body and these coalesce to form an **acrosomal granule** which becomes covered by a membrane. This membrane spreads over one pole of the nucleus to become the **acrosomal cap** of the sperm, covering about one-half of the nucleus (Fig. 1.20A-D). This cap contains enzymes which are capable of disolving the corona radiata of the oocyte and possibly also the zona pellucida. The part of the nucleus covered by the acrosomal cap becomes the 'leading-end' of the spermatozoon. The centrosome splits into two centrioles, one of which lies at the 'tail-end' of the nucleus and is called the proximal centriole, while the other forms the anulus of the body (Fig. 1.20E,F). An axial bundle of fibres, the **axoneme** forms at the 'tail-end' of the nucleus. The axoneme becomes surrounded by mitochondria and these accumulate to form the middle piece of the sperm. The anulus is pushed away from the proximal centriole until it lies at the distal end of the middle piece (Fig. 1.20F). The fibres of the axoneme continue beyond the end of the middle piece for about 40 microns to form the 'tail-piece' of the sperm which has mobile properties. The fibres of the tail-piece are kept in alignment by a surrounding sheath.

When the spermatozoon is about to leave the space between adjacent Sertoli cells and enter the lumen of the seminiferous tubule, it is still encumbered by a piece of spermatid cytoplasm. This is called the **residual body** (Fig 1.20G) which is bridged to the cytoplasm of the apex of the Sertoli cell. As the spermatozoon enters the lumen of the seminiferous tubule, the residual body disappears into the Sertoli cell cytoplasm.

In structure, the spermatozoon consists of a head, a middle piece and a tail (Fig. 1.20F). The head is partially covered by the acrosome which contains enzymes to enable the sperm to penetrate the corona radiata and the zona pellucida to gain entrance to the cytoplasm of the oocyte. The head contains the nuclear matter of the cell in the form of chromosomes 22+X or 22+Y. The middle piece contains mitochondria to supply energy for movement as well as a centriole, which assists in subsequent division of the nuclear material. The tail, in the form of a flagellum, provides the motive force required to drive the sperm from the vagina to the outer end of the uterine tube. The viability of the spermatozoon is variable. Motile sperm have been found in the cervical mucus several days after deposition but the ability of these to fertilise an oocyte has been questioned.

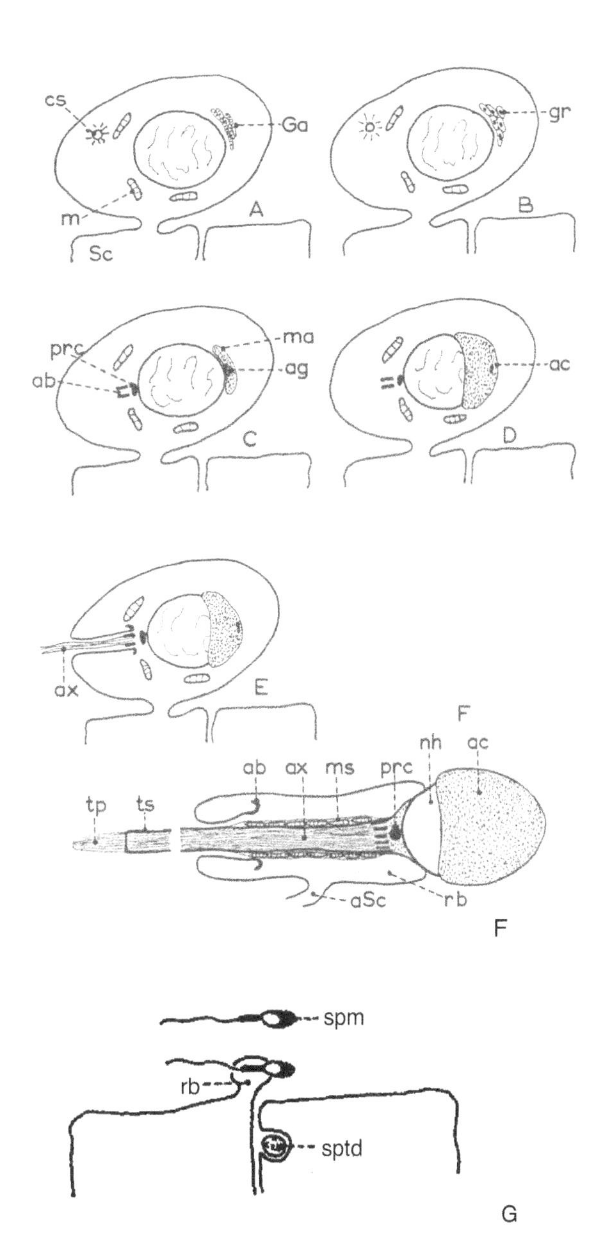

Figure 1.20: Spermiogenesis and anatomy of the spermatozoon. ab = axial bundle; ac = acrosomal cap; ag = acrosomal granule; aSc = attachment to Sertoli cell; ax = axoneme; cs = centriole; Ga = Golgi apparatus; gr = granules; m = mitochondrion; ; ma = acrosomal membrane; ms = mitochondrial sheath; nh = nucleus of head; prc = proximal centriole; rb = residual body; Sc = Sertoli cell; spm = spermatozoon; sptd = spermatid; tp = tail piece; ts = tail sheath.

Fertilisation and Cleavage

Fertilisation is the term applied to the process of fusion of the male and female gametes. After release from the mature follicle, the oocyte is gathered up by the fimbriated end of the uterine tube (Fig.1.18) and conveyed by ciliary action of the epithelial cells of the tube, along the tube. The oocyte (the female gamete) lies in the outer end of the uterine tube and has matured as far as the metaphase stage of Meiosis II. If at this time, spermatozoa are in the vicinity, the necessary ingredients for the formation of a new individual are present.

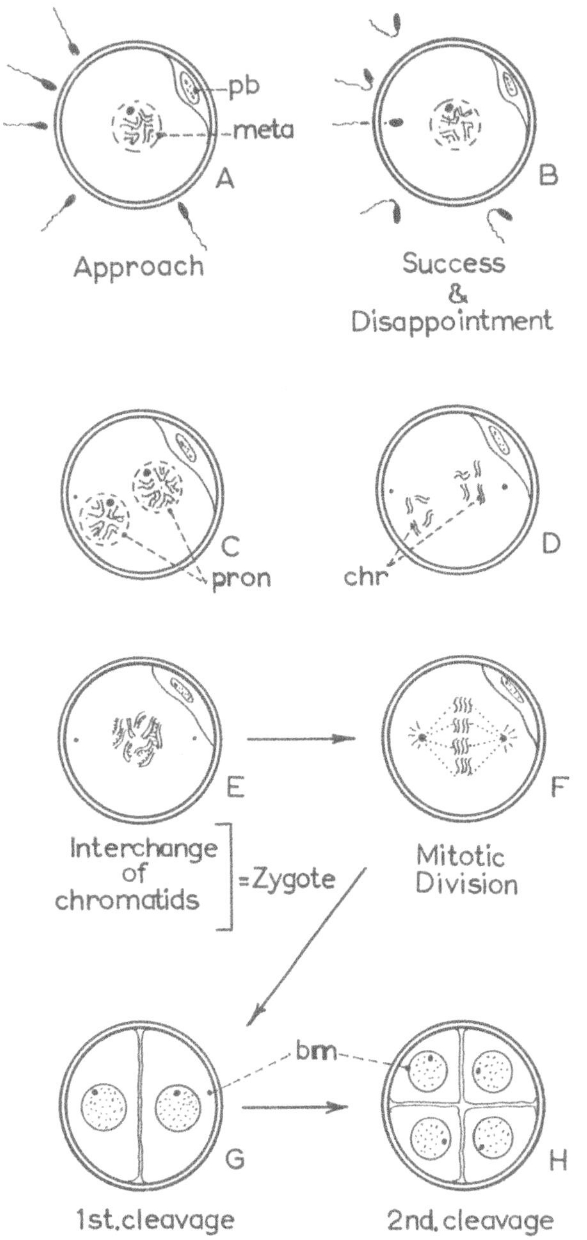

The secretions of the seminal vesicle (gland) and prostate gland are said to have a 'livening' effect upon the spermatozoa and the alkalinity of the vagina and cervical mucus spur the spermatozoa to added activity.

While in the uterus and/or uterine tube, the sperm undergoes a series of events which give it the capacity to enter and fertilise the ovum. These events are known as **capacitation.** When the head of the sperm enters the corona radiata a series of processes known as the **acrosomal reaction** take place. These processes enable the sperm to further penetrate the protective wall of the ovum. The processes are not fully understood but seem to be vested in the acrosomal material. Hyaluronidase (an enzyme) is probably responsible for the dissolution of the cells of the corona radiata and it seems possible that an antigen-antibody reaction may have a part to play in attracting the sperm to the ovum.

By lashing of the tailpieces (flagella) coupled with the muscular activity of the uterus and uterine tube, the spermatozoa pass through the uterine cavity and reach the vicinity of the oocyte in the uterine tube in about 70 minutes (Fig.1.21A). When a sperm has breached the zona pellucida and the membrane of the oocyte, it loses its tail (Fig. 1.21B). The properties of the membrane of the oocyte and zona pellucida are altered at this time to prevent other spermatozoa from entering the oocyte.

The nuclear material of the spermatozoon and oocyte enlarges to form male and female **pronuclei** (Fig. 1.21C), which undergo fusion to form a single nucleus and cell, which is called a **zygote** (Fig. 1.21D, E). While the fusion is taking place, the chromosomes in the nuclei become mixed resulting in mixing of the characteristics of the mother and father as conveyed by the genetic material of the gametes. This also results in a cell with the normal (diploid) number of chromosomes (46 – either 44XX or 44XY). Thereafter, the zygote undergoes ordinary mitotic division to form 2,4,8,16,32, etc. cells (Fig.1.21F-H). This process is called **cleavage** and the resulting cells are called **blastomeres.** The process of cleavage continues until 32 or more cells are formed. The structure is then called a **morula** because it resembles a mulberry (Fig. 1.22). The zona pellucida is still present surrounding the blastomeres, but will soon begin to disintegrate.

Figure 1.21: Fertilisation and cleavage.
bm = blastomeres; chr = chromosomes;
meta = metaphase; pb = polar body; pron = pronuclei.

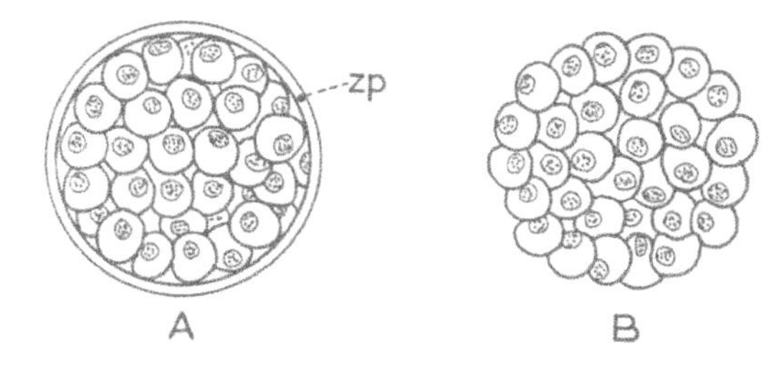

Figure 1.22: The formation of the morula and loss of the zona pellucida (zp).

Important consequences of fertilisation are:

(a) the combination of genetic material from mother and father leads to a desirable variation in the offspring and in the species. However, it also lends itself to the tranmission of undesirable features and diseases such as Tay-Sachs disease, achondroplasia (autosomal dominant), multiple neurofibromatosis (autosomal dominant), Roussay-Levy disease (autosomal dominant);

(b) the combination establishes genotypic sex;

(c) the diploid number (46) of chromosomes is re-established;

(d) cleavage is activated.

CHAPTER TWO

Formation of the Embryo and Implantation

FORMATION OF THE EMBRYO

Although the zygote may be regarded as the initial stage in the formation of the embryo – that is, after conjugation of the male and female pronuclei – the 'working parts' of the embryo are formed later, over a period of approximately eight weeks. The formation of the embryo may be regarded as taking place in three stages:

1. Formation of a two-layered embryonic disc;
2. Conversion of the two-layered disc into a **three**-layered embryonic disc;
3. Conversion of the 'flat' three-layered embryonic disc into a **tubular** embryo by lateral growth. Flexion or 'bending' of the embryo in a cranio-caudal direction also takes place.

While these processes are taking place, the developing embryo is **implanting** itself into the uterine mucosa. It passes through the following three stages:

Stage 1: While the fertilised oocyte is being rolled along the uterine tube, it continues to develop by cleavage. By about the 4th day pf, it has reached a multicellular form called the **morula** stage of development (Fig. 2.1A,B). The morula is a solid mass of cells (**blastomeres**) which is ready (due to nutritional needs) for implantation into the succulent mucous membrane (**endometrium**) of the uterus.

The morula is still surrounded by the zona pellucida. On entering the uterine cavity on day 5 pf, the zona pellucida begins to degenerate, thus allowing fluid to penetrate between the blastomeres and enter the morula (Fig. 2.1C). The accumulation of fluid forms a fluid-filled cavity, the **blastocoel**. The entire structure is now called the **blastocyst**. The blastocoel of the blastocyst is eccentric in position so that internal cells of the blastocyst are displaced to one side forming a mound of cells (Fig. 2.1D). This mound of cells is called the **inner cell mass**. From the internal fluid accumulation, the outer cells of the blastocyst become flattened to form the **trophoblast** (Fig. 2.1D). The inner cell mass will form the embryo while the trophoblast will form the extra-embryonic membranes (including the greater part of the placenta). The blastocyst presents 'two poles' of orientation; the pole at the attachment of the inner cell mass to the trophoblast is called the **embryonic pole** and that at the opposite side of the blastocyst, the **abembryonic pole**.

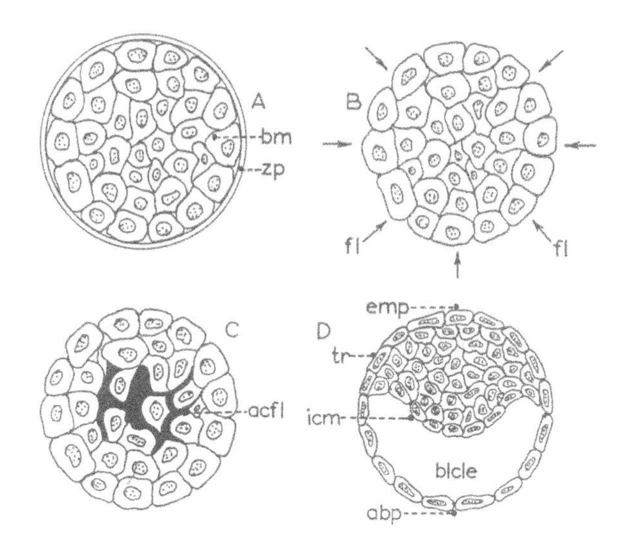

Figure 2.1: Formation of the blastocyst (A-D).
abp = abembryonic pole; acfl = accumulation of fluid; blcle = blastocoel; bm = blastomeres; emp = embryonic pole; fl = inflow of fluid; icm = inner cell mass; tr = trophoblast; zp = zona pellucida.

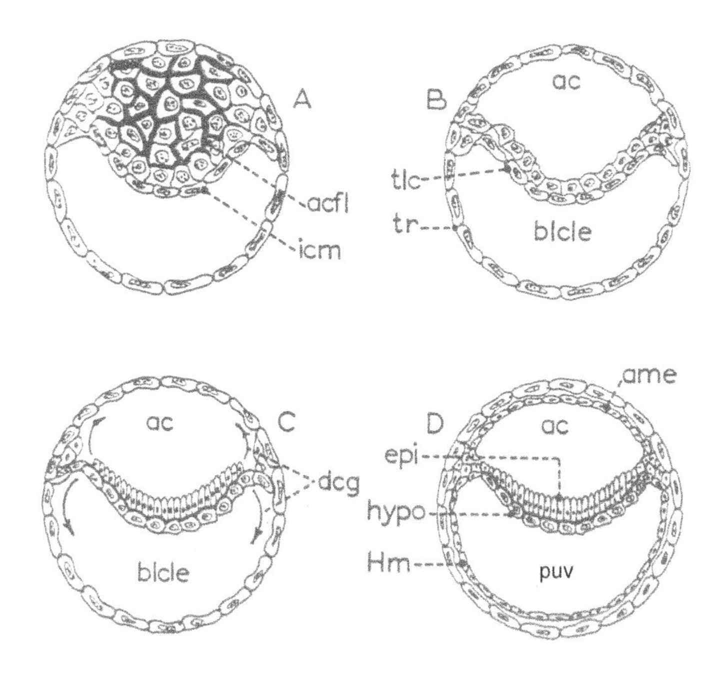

Figure 2.2: Formation of the amniotic cavity and embryonic disc (A-D).
ac = amniotic cavity; acfl = accumulation of fluid; ame = amniotic ectoderm; blcle = blastocoele; dcg = direction of cell growth; epi = epiblast; Hm = Heuser's membrane; hypo = hypoblast; icm = inner cell mass; puv = primitive umbilical vesicle; tlc = thin layer of cells; tr = trophoblast.

Within a short time, fluid begins to accumulate between the cells of the inner cell mass (Fig. 2.2A). This will separate the inner cell mass from the trophoblast resulting in the formation of the **amniotic cavity** which becomes lined by amniotic ectoderm (**Fig. 2.2B**). The inner cell mass now forms two distinct layers, the one layer becomes columnar and is called the **epiblast**. This layer is continuous with the flattened layer of amniotic ectodermal cells on the inner surface of the trophoblast. The layer subjacent to the epiblast is called the **hypoblast**. Thus, a two-layered embryonic disc consisting of epiblast and hypoblast has been formed (**Fig.2.2B**). The hypoblast now produces a layer of flattened cells which form a membrane (**Heuser's membrane**) on the inner surface of the trophoblast. This membrane lines the blastocoel which is now called the **umbilical vesicle**.

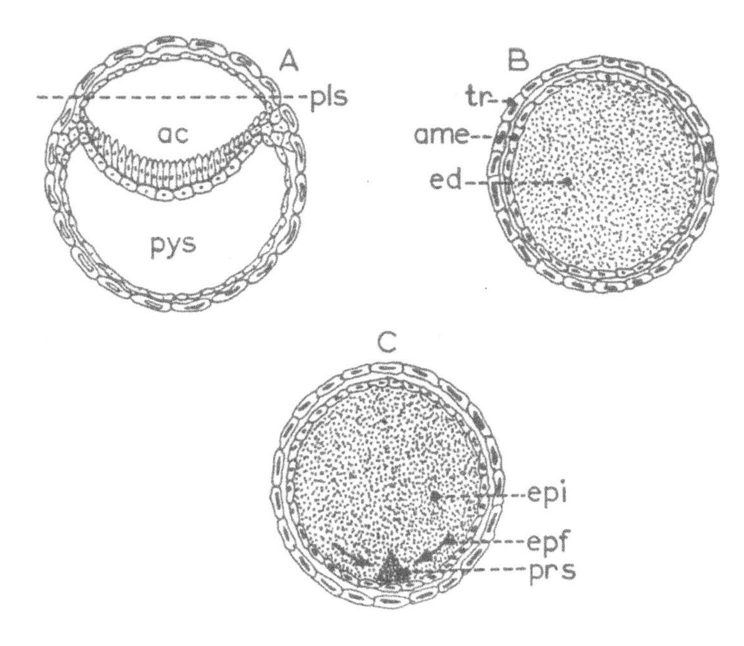

Figure 2.3: A section through the developing embryo (A) and a dorsal view of the embryonic disc (B and C). A indicates the plane of section (pls) for **B** and **C.** ame = amniotic ectoderm; ed = embryonic disc; epf = flow of epiblast; epi = epiblast; prs = primitive streak; tr = trophoblast.

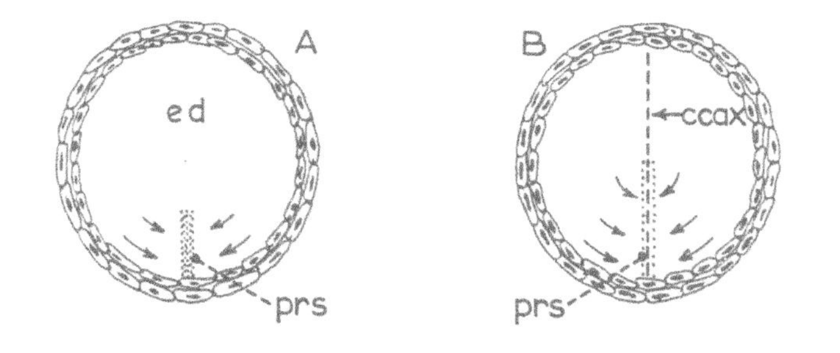

Figure 2.4: Extension of the primitive streak in a cranial direction. ccax = cranio-caudal axis of embryo; ed = embryonic disc; prs = primitive streak.

If a section is made through the inner cell mass to include the amniotic cavity on approximately day 15 pf (**Fig. 2.3A**) and one looks into the cavity from above, a circular mass of epiblast is seen in the lower part of the cavity. This is the embryonic disc seen from the embryonic pole of the embryo (**Fig. 2.3B**). Thus, a circular two-layered embryonic disc is present and as one observes it from its upper surface, the formation of a small ridge at one point on its perimeter is noted (**Fig. 2.3C**). This is the beginning of the formation of a structure called the **primitive streak** and is due to a 'heaping-up' of epiblast cells flowing in from the periphery of the disc, to the region of the streak (**Fig. 2.3C**). As epiblast cells are added, the streak slowly extends from the margin of the disc towards the centre of the disc (**Fig. 2.4A**). The 'end' at the centre of the disc represents the 'cranial' end of the streak while that near the margin is the 'caudal' end of the streak. Thus, a central **'cranio-caudal' axis** for the embryonic disc is formed (**Fig. 2.4B**).

If we now make a longitudinal section through the centre of the disc in the line of the streak, the formation as shown in Figure 2.5A is seen. If a fine needle is inserted between the epiblast and the hypoblast, it becomes evident that the two layers are separable in the centre (**Fig. 2.5B**), but not at the far cranial or far caudal ends of the disc. The adherence at the cranial end of the disc is called the **prechordal plate** and that at the caudal end, the **cloacal plate (Fig. 2.5C)**.

Stage 2: As the primitive streak grows in the cranio-caudal axis of the disc, a groove is formed in its midline – the **primitive groove** (**Fig. 2.6A**). The primitive groove forms a region through which the epiblast cells migrate. The first epiblast cells that enter the groove, migrate directly to the underlying hypoblast where they form the cells of the **endoderm** (**Fig. 2.6B**). These endodermal cells occupy space and thus tend to push the hypoblast cells in a lateral direction.

Further epiblast cells migrating through the primitive groove extend laterally between the epiblast (now **ectoderm**) and endoderm. These cells derived from the epiblast will form the middle layer of the embryo. This middle layer forms the **intra-embryonic mesoderm**. Note that all three germ layers are thus, derived from epiblast.

The 'floor' of the amniotic vesicle consists of a layer of embryonic ectoderm (formerly epiblast) and the 'roof' of the primary umbilical vesicle consists of embryonic endoderm. These

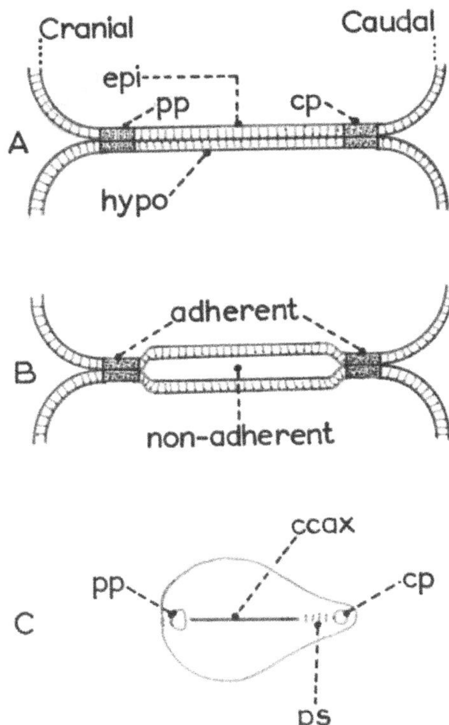

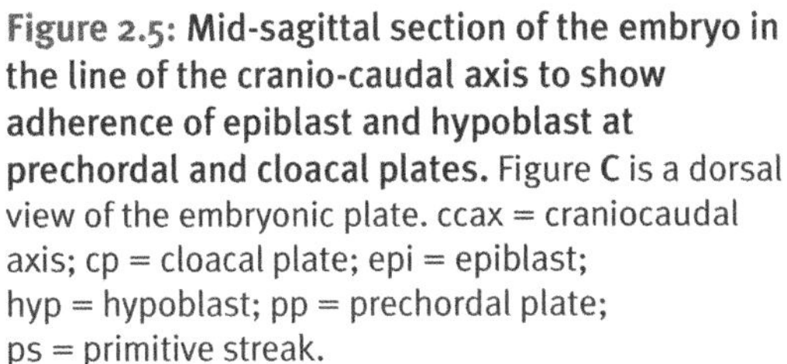

Figure 2.5: Mid-sagittal section of the embryo in the line of the cranio-caudal axis to show adherence of epiblast and hypoblast at prechordal and cloacal plates. Figure **C** is a dorsal view of the embryonic plate. ccax = craniocaudal axis; cp = cloacal plate; epi = epiblast; hyp = hypoblast; pp = prechordal plate; ps = primitive streak.

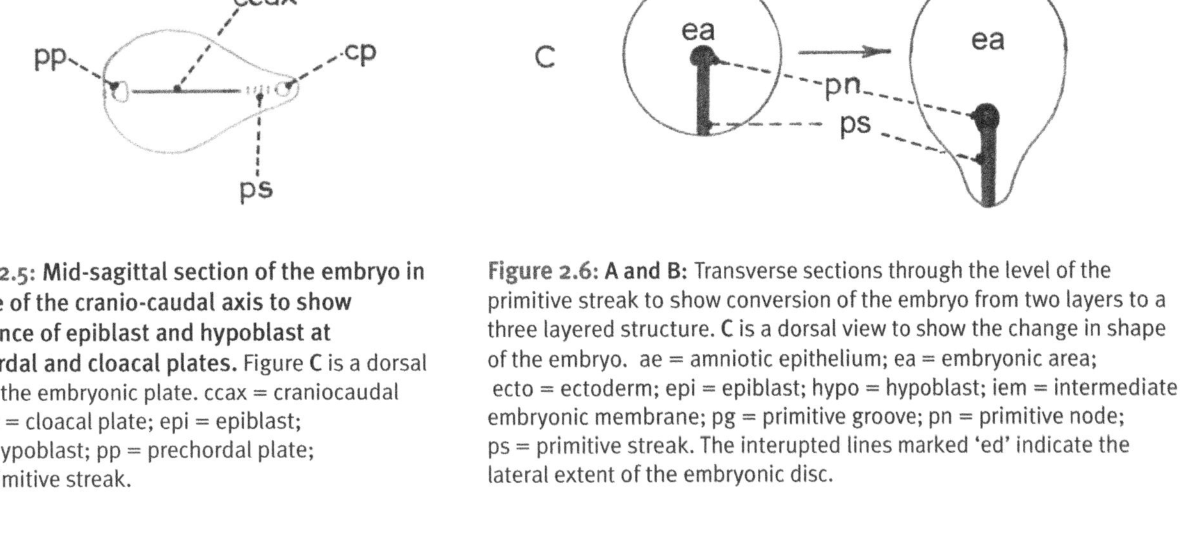

Figure 2.6: A and B: Transverse sections through the level of the primitive streak to show conversion of the embryo from two layers to a three layered structure. **C** is a dorsal view to show the change in shape of the embryo. ae = amniotic epithelium; ea = embryonic area; ecto = ectoderm; epi = epiblast; hypo = hypoblast; iem = intermediate embryonic membrane; pg = primitive groove; pn = primitive node; ps = primitive streak. The interupted lines marked 'ed' indicate the lateral extent of the embryonic disc.

two layers are separated by a layer of intra-embryonic mesoderm which is confined to the embryonic area. The whole structure is surrounded by the trophoblast (Fig.2.7A).

Thus, the embryonic disc is now **trilaminar**, consisting of a dorsal layer of ectoderm, a ventral layer of endoderm separated by a layer of intra-embryonic mesoderm (Fig. 2.6B). The formation of the primitive streak causes the embryonic disc to change its shape from circular to elongated. The caudal end is narrower and thus results in the disc now being in an inverted pear shape (Fig. 2.6C).

The structure of this embryo now consists of a dorsal vesicle (amniotic cavity) and a ventral vesicle (the primary umbilical vesicle) with the trilaminar embryonic disc sandwiched between them. At about the 12th day pf, a layer of **extra-embryonic mesoderm** forms between the primary umbilical vesicle and the trophoblast, as well as between the amniotic cavity and the trophoblast (Fig.2.7A,B). The origin of the extra-embryonic mesoderm is controversial but is most likely to arise from the trophoblast. The extra-embryonic mesoderm develops lacunae which coalesce to form the **extra-embryonic coelom** (Fig. 2.7C). During this coalescence a part of the primary umbilical vesicle is 'cut off' and the reduced umbilical vesicle is now called the **definitive umbilical vesicle** (Fig. 2.7D,E). The endodermally-lined roof of the umbilical vesicle will give rise to the gut tube and its accessory organs. The remnant of the umbilical vesicle (Fig. 2.7E) will gradually be absorbed and disappear.

When the primitive streak reaches the centre of the embryonic disc, the epiblast cells become heaped up to form a knot. This is the **primitive node** (Fig. 2.6C). On the dorsal surface of the node an indentation occurs. This is the **primitive pit**

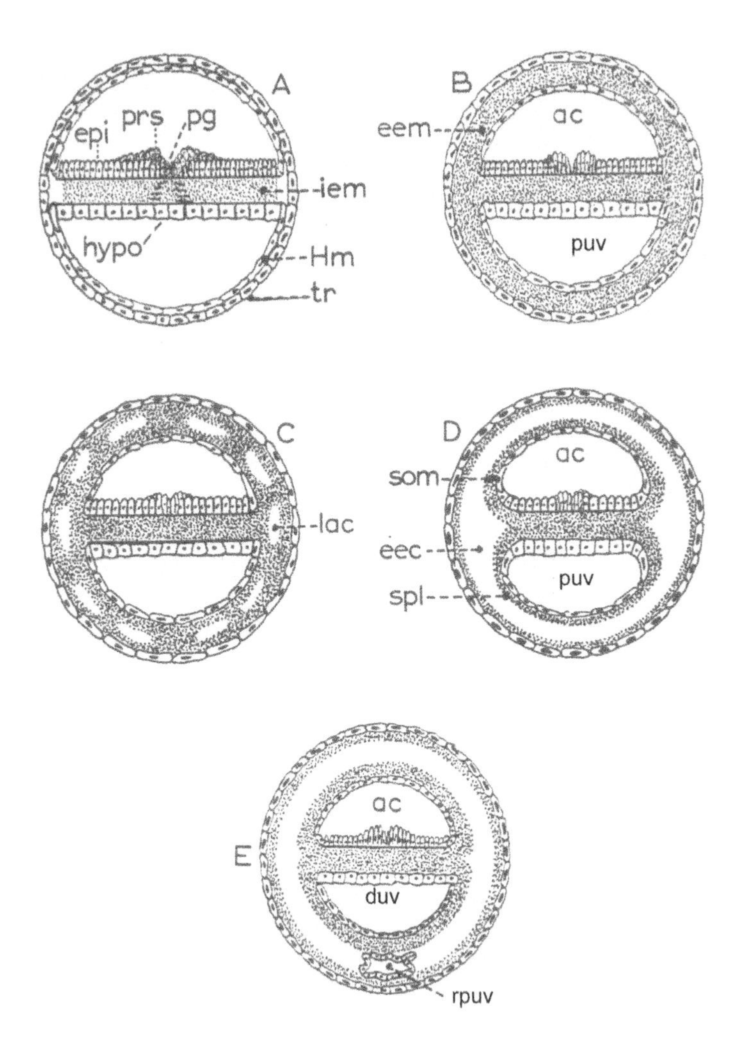

Figure 2.7A-E: Transverse sections to show formation of the extra-embryonic coelom (eec) in the extra-embryonic mesoderm (eem). Figure E indicates the remnant of the primitive umbilical vesicle (rpuv) and the definitive umbilical vesicle (duv). Hm = Heuser's membrane; iem = intra-embryonic mesoderm; lac = lacunae; prs = primitive streak; som = somatic mesoderm; spl = splanchnic mesoderm.

(**Fig. 2.8**). The node produces a solid axial rod of mesodermal cells to form the **notochord** which grows cranially to reach the impenetrable edge of the prechordal plate (**Fig 2.9; 2.10**). The primitive node continues to add cells to the caudal end of the notochord so that the notochord is extended caudally, pushing the primitive node and primitive streak in a caudal direction. This causes elongation of the embryonic disc.

At the same time, the primitive pit burrows into the solid notochordal rod converting it into a tube (**Fig. 2.11**). The tube becomes adherent to the underlying endoderm and this 'adherence' breaks down in one or more places transiently to create a communication between the umbilical vesicle and the hollow tube.

By this time, the neural plate of **neurectoderm** has begun to form in the ectodermal floor of the amniotic cavity and the amniotic cavity communicates with the tubular notochord through the primitive pit. Thus, there is now a communication between the amniotic cavity and the umbilical vesicle via the hollow notochord. This communication is called the **neurenteric canal** (**Fig 2.12A-E**) which persists until the notochord is fully formed.

Note: The thickened neural plate continues to develop by forming two neural folds between which lies a neural groove. The folds gradually approach each other in the midline, starting between somite levels 4-6, and proceeding cranially and caudally. This results in the formation of the **neural tube** which gives rise to the entire central nervous system.

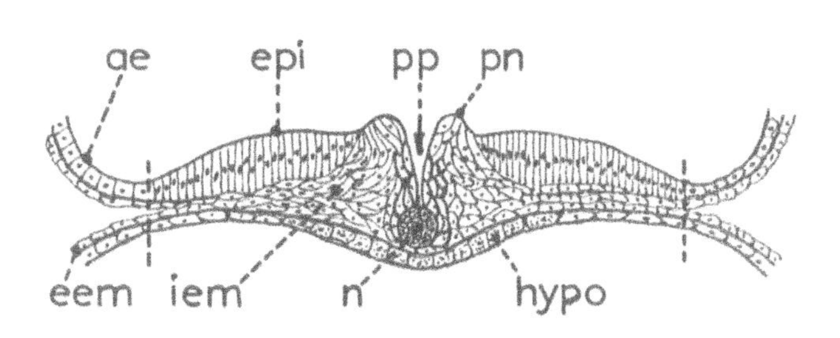

Figure 2.8: Transverse section of the embryonic disc showing the condensation of the mesoderm to form the notochord (n). Vertical interrupted lines indicate the extent of the embryonic disc.
ac = amniotic disc; ae = amniotic ectoderm; eem = extra-embryonic mesoderm; epi = epiblast; hypo = hypoblast; iem = intra-embryonic mesoderm; n = notochord; pp = primitive pit; pn = primitive node.

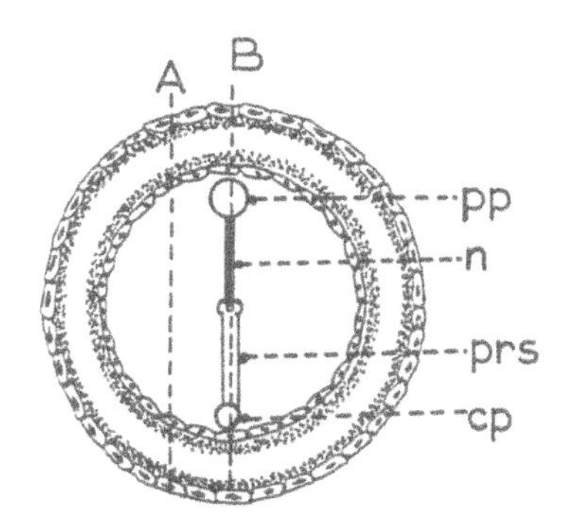

Figure 2.9: Dorsal view of the embryonic disc to show the position of the primitive streak (prs) and the notochord (n). cp = cloacal plate; n = notochord; pp = prechordal plate; prs = primitive streak.

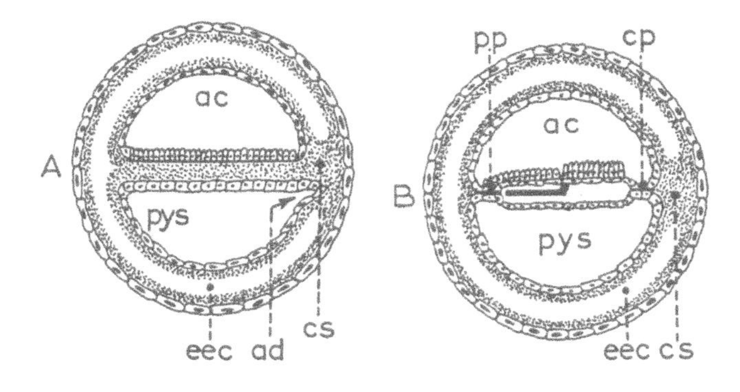

Figure 2.10A: Sagittal section through A in Figure 2.9 to show the formation of the connecting stalk (cs) and allantoic diverticulum (ad). Figure 2.10B is a mid-sagittal section through B of Figure 2.9. This shows the position of the notochord (n) in relation to the prechordal plate (pp) and cloacal plate (cp).

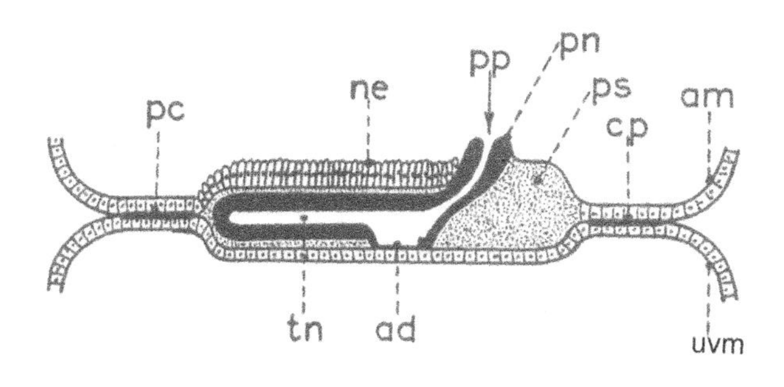

Figure 2.11: Mid-sagittal section depicting the canalisation of the notochord (tn). ad = adherence of the floor of the notochord to the roof of the primitive umbilical vesicle (gut tube); am = amniotic membrane; cp = cloacal plate; ne = neurectoderm; pc = prechordal plate; pn = primitive node; pp = primitive pit; ps = primitive streak; uvm = umbilical vesicle membrane.

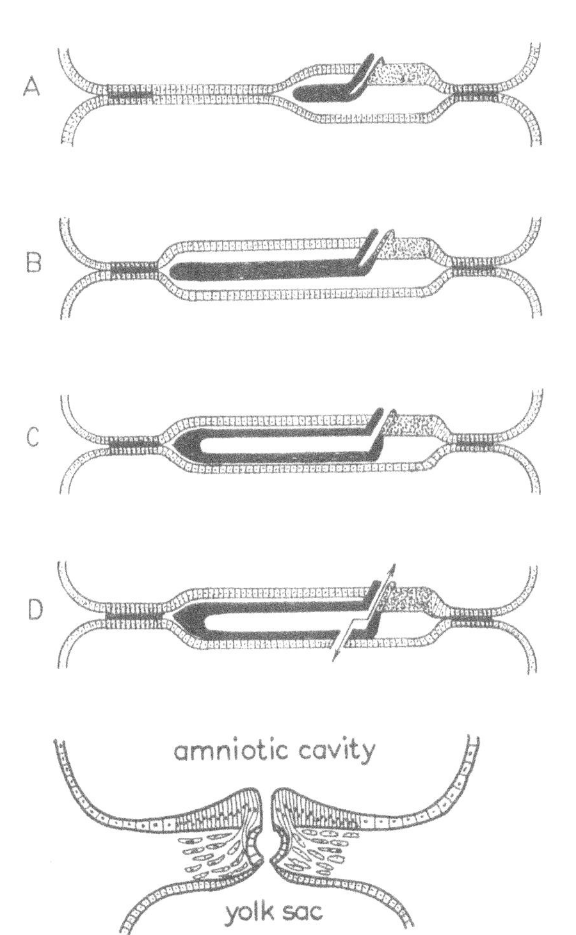

Figure 2.12 A-D: Sagittal sections showing the cranial extension of the notochord, canalisation and final dissolution of the adherent area to form the neurenteric canal. Arrow indicates the neurenteric canal. Figure E is a transverse section to indicate the neurenteric canal.

The intra-embryonic mesoderm which migrates through the primitive groove, passes laterally to the edge of the embryonic disc where it becomes co-extensive with the extra-embryonic mesoderm (**Fig. 2.8**). The intra-embryonic mesoderm will also extend cranially and caudally (**Fig. 2.13**). This mesoderm will pass cranial to the prechordal plate and caudal to the cloacal plate. The intra-embryonic mesoderm fills the region between the ectoderm and the endoderm except at three places:

1. The prechordal plate;
2. The cloacal plate;
3. In the region where the notochord exists.

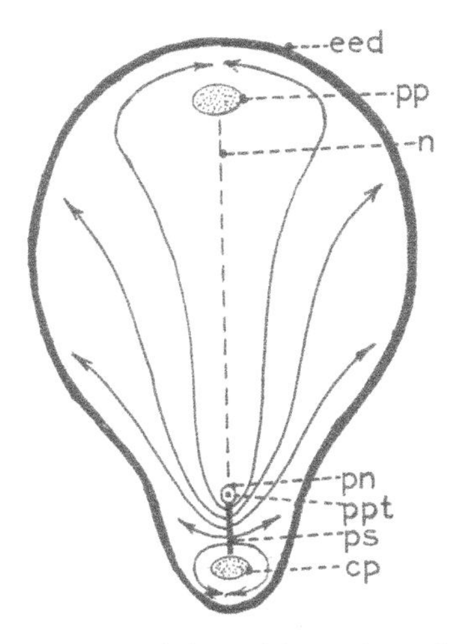

Figure 2.13: Dorsal view of the embryonic disc depicting flow (by arrows) of the intra-embryonic mesoderm from the primitive streak (ps).
cp = cloacal plate; eed = edge of embryonic disc; n = notochord; pn = primitive node; pp = prechordal plate; ppt = primitive pit.

In a sagittal section through the midline of the embryo at this stage (**Fig. 2.14**), the prechordal and cloacal plates and the notochord are indicated.

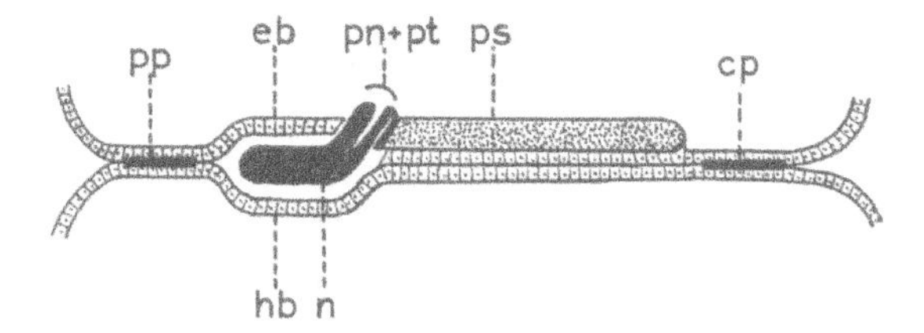

Figure 2.14: Mid-sagittal section depicting the notochord and prechordal and cloacal plates. Note that the ectoderm and endoderm are firmly adherent at the prechordal (pp) and cloacal plates (cp). eb = epiblast; hb = hypoblast; n = notochord; pn + pt = primitive node + primitive pit; ps = primitive streak.

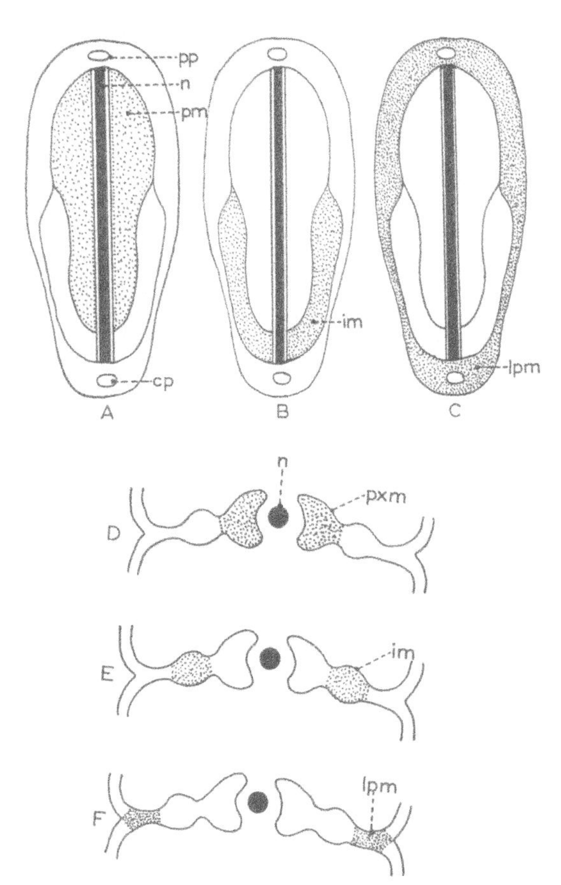

Figure 2.15: A-C: Dorsal view of the embryonic disc depicting formation of the mesodermal regions. pm = paraxial mesoderm; im = intermediate mesoderm; lpm = lateral plate mesoderm. **D-F: Transverse sections of the embryonic disc showing the relative positions of the three mesodermal masses.** cp = cloacal plate; im = intermediate mesoderm; lpm = lateral plate mesoderm; n = notochord; pm + pxm = paraxial mesoderm; pp = prechordal plate.

The ectoderm and endoderm are firmly adherent at the prechordal and cloacal plates. The mesoderm continues to be formed but cannot cross the midline because of the presence of the notochord (**Fig. 2.13**).

The intra-embryonic mesoderm will differentiate into three bands (**Fig. 2.15A-F**) on either side of the notochord. These will be the **paraxial**, **intermediate** and **lateral plate mesoderm.** Later, an **intra-embryonic coelom** will form within the lateral plate mesoderm and become co-extensive with the **extra-embryonic coelom** at the lateral surface of the embryo (**Fig. 2.29**). At the far cranial end of the embryonic disc, the intra-embryonic coelom will form the **pericardial coelom** which has in its floor a condensation of mesoderm known as the **cardiogenic plate**.

Note too, from Figure 2.16, that by this stage an extension of the endoderm has penetrated into the connecting stalk which suspends the embryo in the extra-embryonic coelom (see page 28). This endodermal extension will form the **allantois**, which is rudimentary in man. The connecting stalk attaches the embryo to the developing placenta and will later form part of the umbilical cord.

Stage 3: At the end of 20 days pf, the three-layered, pear-shaped embryonic disc begins to undergo accelerated growth which alters its shape so that it 'bends' both lengthwise and transversely. This is thought to be due to the difference in proliferation of the ectodermal and endodermal cells in the disc, much like the bending of a bimetal strip when heated, which is due to the differential expansion of the dissimilar metals. The central strip of ectoderm will thicken into neurectoderm which will form the central nervous system and these are the cells which show accelerated growth. The so-called 'bending' of the embryo brings about certain changes in the character of the embryo (**Fig. 2.16**):

Changes resulting from lengthwise (craniocaudal) elongation and bending of the embryonic disc are:

(a) The dorsal surface of the embryo bulges into the amniotic cavity.
(b) The prechordal plate comes to lie ventral to the developing nervous system, and the cardiogenic plate and pericardial coelom undergo complete reversal of position from ventral to dorsal.
(c) The cardiogenic plate now lies in the dorsal part of the pericardial coelom.

(d) The mesoderm lying at the extreme cranial end of the disc (Fig. 2.16) is displaced by the bending so that it now lies caudal to the cardiac tubes and the pericardial coelom, in a transverse position. This transverse bar of mesoderm is now called the **septum transversum.**

The new arrangement of positions indents the adjacent endoderm of the roof of the umbilical vesicle so that its cranial end is narrowed and elongated to form the **foregut.** Lying against the floor of the foregut will be the cardiogenic plate and pericardial coelom covered by mesoderm. Dorsal to the foregut is the notochord and the primitive central nervous system. The cavity of the foregut is at this stage separated from the indentation of the ectoderm (**stomodeum**) (Fig. 2.16A-C) by the prechordal plate which has become thinned to form the **oropharyngeal membrane.**

At the tail-end of the embryo, the tail-bud rotates in the opposite direction to the cranial end so that the adjacent part of the umbilical vesicle is narrowed and extended to form the **hindgut.** This rotation is somewhat complicated by the growth of the **allantoic diverticulum** into the connecting stalk (Fig. 2.16C). The cloacal plate now lies in a ventral position and separates the hindgut from a slight ectodermal depression known as the **proctodeum** (Fig. 2.16C).

The intervening part of the umbilical vesicle between the foregut and the hindgut becomes the **midgut** which is joined to each respectively by the **anterior** and **posterior intestinal portals** (Fig 2.17A). The connection between the gut tube and the umbilical vesicle becomes progressively narrowed and forms an **umbilical vesicle stalk** (**vitello-intestinal duct**). Both the umbilical vesicle and the duct normally degenerate completely.

Changes resulting from transverse expansion and folding of the embryonic disc are:

(a) At the centre of the embryo, in the region of the middle of the umbilical vesicle, the transverse folding 'pinches' into the umbilical vesicle, dividing it into a small dorsal part and a large ventral part (Fig.2.17A-F). The small part becomes the midgut and the larger part remains as the umbilical vesicle (Fig. 2.17C). The dorsal part of the umbilical vesicle is thus converted into a tube, blind at either end, but open for a short time, in the ventral midgut region via the umbilical vesicle stalk.

(b). The openings of the intra-embryonic coelom to the extra-embryonic coelom (see page 28; Fig. 2.29B and page 29; Fig. 2.30) are being brought into apposition in a ventral position on the embryonic body (Fig. 2.17D). At the same time, the intra-embryonic coelom is closed off from the extra-embryonic coelom by constriction of the opening of the ventral embryonic body wall.

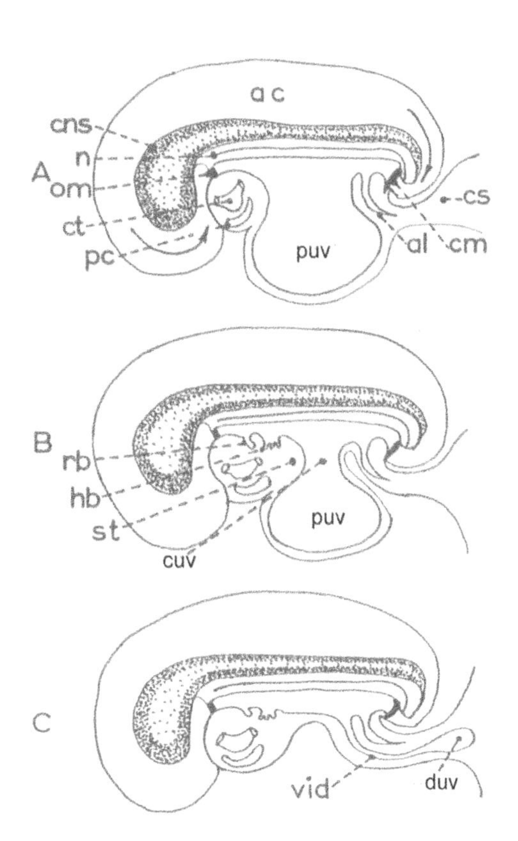

Figure 2.16: Sagittal sections depicting the consequences of cranio-caudal bending of the embryo. ac = amniotic cavity; al = allantoic diverticulum; cm = cloacal membrane; cns = central nervous system; cs = connecting stalk; ct = cardiac tube; cuv = connection of umbilical vesicle; duv = definitive umbilical vesicle; hb = hepatic bud; n = notochord; om = oropharyngeal membrane; pc = pericardial coelom; puv = primitive umbilical vesicle; rb = respiratory bud; st = septum transversum; vid = vitello-intestinal duct.

With the continued longitudinal folding of the tail end of the embryo, the connecting stalk containing the allantoic diverticulum is brought closer to the umbilical vesicle stalk and ultimately the two (the connecting stalk and umbilical vesicle stalk) together will form the definitive **umbilical cord.** The mesoderm which originally was situated at the far caudal end of the cloacal plate is now turned on to the ventral wall of the embryo and will form the anterior abdominal

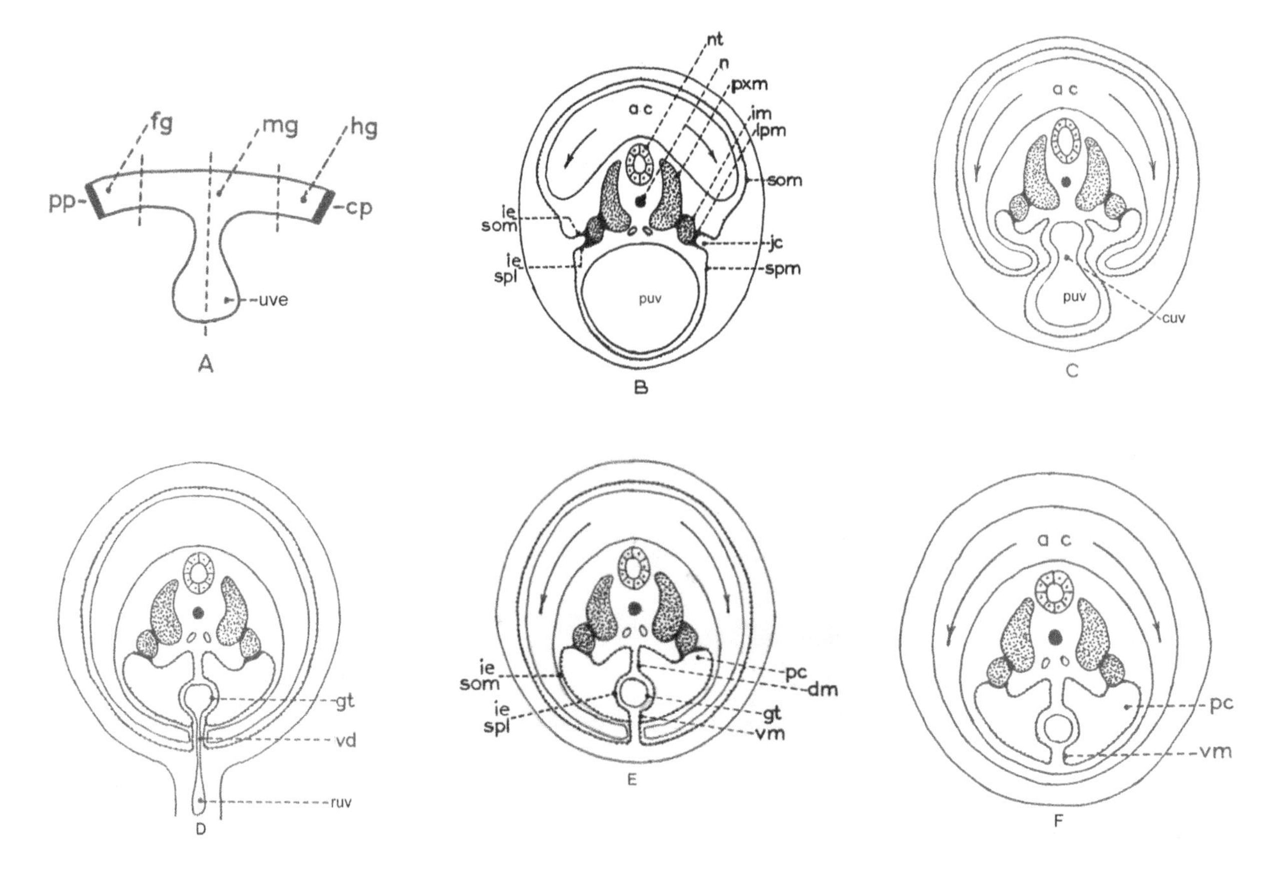

Figure 2.17: A: Lateral view of gut tube and umbilical vesicle. B-F: Transverse sections taken at the level indicated by the interrupted line in Figure A depicting consequences of lateral folding of the embryo. Arrows indicate the direction taken by the amniotic folds. ac = amniotic cavity; cp = cloacal plate; cuv = connection with umbilical vesicle; dm = dorsal mesentery; fg = foregut; gt = gut tube; hg = hindgut; im = intermediate mesoderm; jc = junction between extra- and intra-embryonic coelomata; lpm = lateral plate mesoderm; mg = midgut; n = notochord; nt = neural tube; pc = peritoneal coelom; pp = prechordal plate; pxm = paraxial mesoderm; ; ruv = remnant of umbilical vesicle; som = somatic mesoderm; spm = splanchnic mesoderm; vm = ventral mesentery; uve = umbilical vesicle.

wall below the umbilicus. If this mesoderm does not develop, the anterior abdominal wall will fail to consolidate, exposing the underlying organs.

The Fate of the Mesoderm

The intra-embryonic mesoderm is 'manufactured' from epiblast cells which infiltrate between epiblast and hypoblast via the primitive groove. This mesoderm forms a layer between the ectoderm (formerly epiblast) and the endoderm (formerly hypoblast). The mesoderm soon becomes differentiated into bands which run the length of the embryo and which are continuous with one another on their lateral surfaces. The mesoderm on either side of the notochord is called the **paraxial** (next to the axis) **mesoderm**. This band will later become segmented to form **somites** (**Fig. 2.18**). The next band is the **intermediate mesoderm** (or nephrogenic cord) and the outer band is the **lateral plate mesoderm**. The lateral plate mesoderm, which is in an intra-embryonic position, becomes continuous with the existing extra-embryonic mesoderm at the edges of the embryonic disc (**Fig. 2.18**).

Derivatives of the Paraxial Mesoderm

The presence of the notochord induces the paraxial mesoderm to form segments (somites). These somites develop in a cranio-caudal direction. By the end of the embryonic period, 42-44 pairs of somites will have been formed. In the very early stages of embryonic life, the embryo is so small that the only suitable way of estimating its age is by counting the number of somites under the microscope. The somites can be discerned as bulges seen through the overlying ectoderm. The somites of the paraxial mesoderm differentiate into three parts, the **dermatome** and the **myotome** (which are more dorsolaterally positioned) and the **sclerotome** (which is ventro-medially positioned) (Fig. 2.19). The dermatome gives rise to the dermis of the skin of the whole body, while the epidermis of the skin will be formed from the lateral part of the ectodermal layer (the surface ectoderm). The myotomes will provide the skeletal muscles of the trunk, and possibly the muscles of the limbs. The sclerotomes form the segments of the vertebral column (vertebrae and anulus fibrosis of the intervertebral discs).

Derivatives of the Intermediate Mesoderm

In the early embryo the intermediate mesodermal band is covered on its dorsal surface with ectoderm and on its ventral surface with endoderm (Fig. 2.20). This band of mesoderm will contain cells which form the greater part of the urogenital system. This mesoderm gives rise to:

1. In both sexes: the pronephros (rudimentary in the human), mesonephros, metanephros, metanephric duct, trigone of bladder, ureter, renal calyces and all the tissues of the gonads except the 'sex cells' which migrate to the genital ridge from the region of the umbilical vesicle.
2. In the male: the epididymis, ductus deferens, seminal gland (vesicle), ejaculatory duct and prostatic utricle.
3. In the female: the paramesonephric duct, uterine tube, uterus and part of the vagina.

Derivatives of the Lateral Plate Mesoderm

This part of the intra-embryonic mesoderm is continuous with the lateral edge of the intermediate band medially and with the extra-embryonic mesoderm laterally (Fig. 2.20). The extra-embryonic mesoderm splits to form a **somatic** layer surrounding the amniotic cavity and a **splanchnic** layer surrounding the umbilical vesicle. The intra-embryonic lateral plate mesoderm will also develop clefts which divide it into a somatic mesodermal layer and a splanchnic mesodermal layer. The cleft between these, forms the **intra-embryonic coelom**.

The lateral plate mesoderm thus gives rise to:

1. Intra-embryonic somatic mesoderm: forms the connective tissue, cartilage, bone, muscle and vascular (including lymphatic) elements

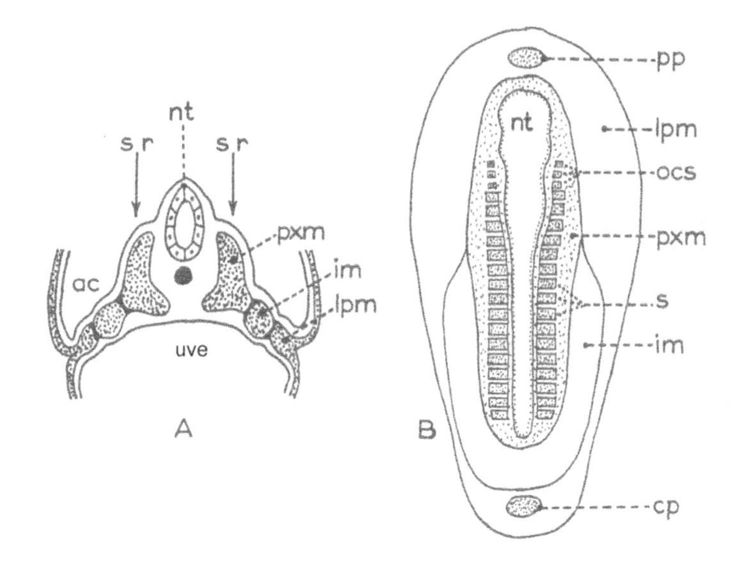

Figure 2.18: Formation of the somites. A: Transverse section of embryo. **B:** Dorsal view of embryo with ectoderm removed. ac = amniotic cavity; im = intermediate mesoderm; lpm = lateral plate mesoderm; nt = position of neural tube; ocs = occipital somites; pxm = paraxial mesoderm; s = somites; sr = somitic ridge; uve = umbilical vesicle.

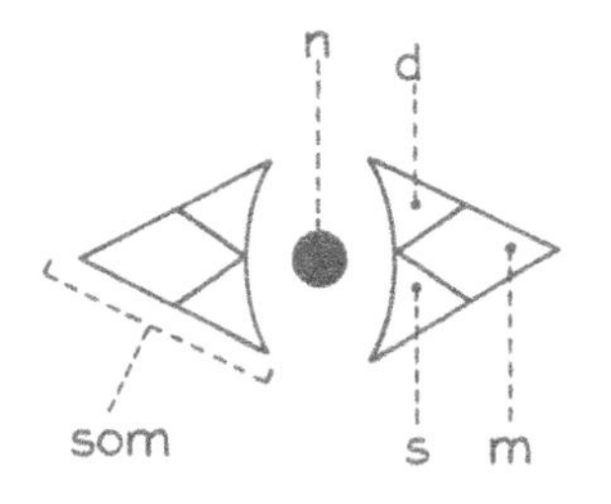

Figure 2.19: Transverse section of the notochord and paraxial mesoderm to indicate the differentiation of the somites. d = dermatome; m = myotome; n = notochord; s = sclerotome; som = somite.

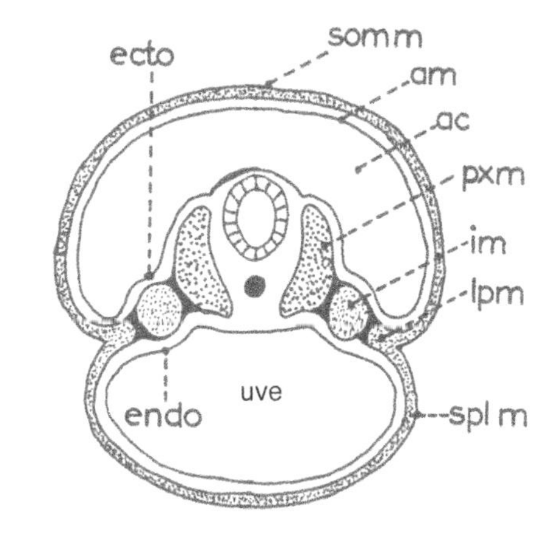

Figure 2.20: A transverse section showing the relationship between the ectoderm (ecto), components of the intra-embryonic mesoderm (pxm, im, lpm) and the endoderm (endo). ac = amniotic cavity; am = amniotic membrane; im = intermediate mesoderm; lpm = lateral plate mesoderm; pxm = paraxial membrane; somm = extra=embryonic somatic mesoderm; splm = extra-embryonic splanchnic mesoderm; uve = umbilical vesicle.

of the body wall, as well as the core of the limb buds.

2. Intra-embryonic splanchnic mesoderm: covers the entire enteric (gut) tube as well as its extensions such as ducts (extrahepatic biliary and pancreatic) and organs (liver and pancreas). The pericardium, lungs and bronchi and peritoneum are also surrounded by this mesoderm.

Mesenchyme is a term frequently used in embryology and consists of stellate cells suspended in a fluid matrix. It is mostly, but not exclusively derived from embryonic mesoderm and forms, or is partly responsible for the formation of a large number of body tissues and organs. Mesenchyme may also form from ectoderm (e.g. ectomesenchyme, see neural crest cell development) and possibly from endoderm.

Implantation, Placentation and Extra-embryonic Membranes

The menstrual cycle is the series of events which occur in the endometrium (uterine mucosa) to prepare it for the possible reception of a fertilised

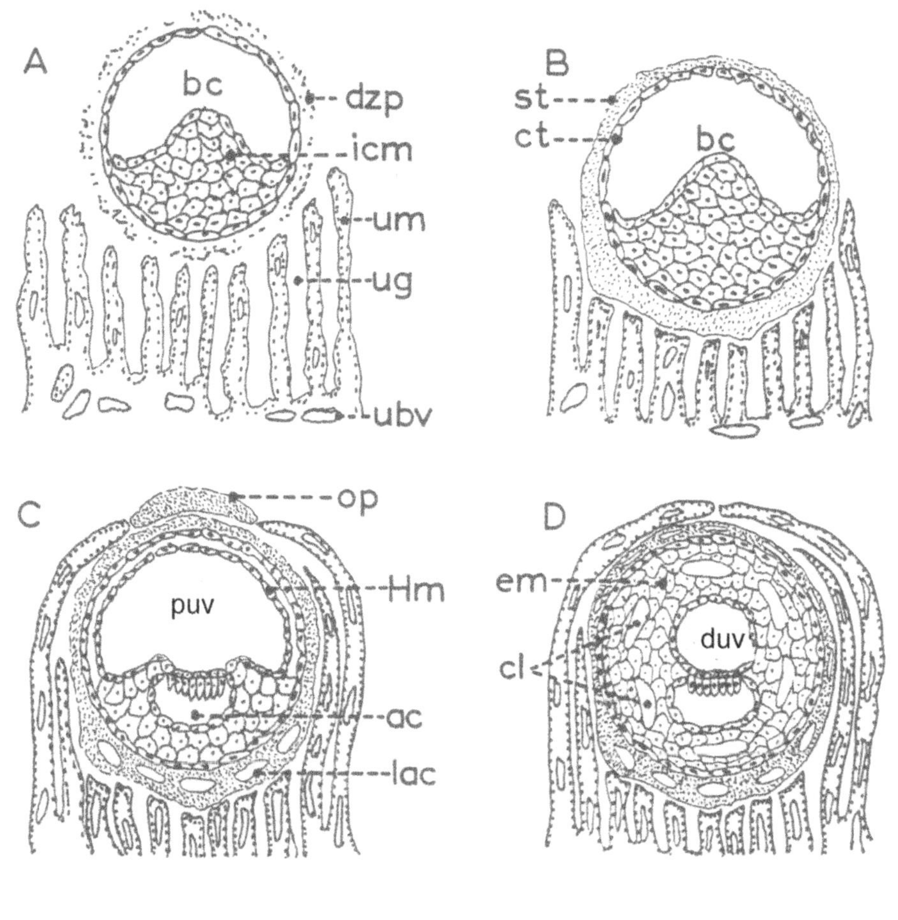

Figure 2.21: A-E: Transverse sections depicting implantation into the endometrium. ac = amniotic cavity; bc = blastocoel; cl = lacunae in extra-embryonic mesoderm; ct = cytotrophoblast; dys = definitive umbilical vesicle; dzp = degenerating zona pellucida; ecs = connecting stalk; eec = extra-embryonic coelom; eem = extra-embryonic mesoderm; Hm = Heuser's membrane; icm = inner cell mass; lac = lacuna; mbv = maternal blood vessel; m-evc = maternal to embryo venous channels; op = operculum; puv = primary umbilical vesicle; st = syntrophblast; ubv = uterine blood vessel; ug = uterine gland; um = uterine mucosa.

oocyte. After menstruation (desquamation of the previously prepared endometrium), the uterine mucosa begins to regenerate by proliferation of its cells and glandular elements. This is called the **proliferative phase** of regeneration and lasts about 14 days from the first day of the immediately preceding menstrual period.

At about this time, a mature follicle in the ovary ruptures, releasing an oocyte and a quantity of liquor folliculi into the peritoneal cavity of the prospective mother. The oocyte is caught up by the fimbria of the uterine tube while the hormones in the liquor folliculi are rapidly absorbed and stimulate the endometrial glands to produce mucous so that the interior of the uterus becomes wet and succulent. This is called the **secretory phase** of the menstrual cycle and is the time during which a fertilised oocyte might reach the endometrium. If the oocyte is *not* fertilised, the remains are discarded at the next menstrual period.

If an oocyte has been fertilised, it passes along the uterine tube until it reaches the endometrium of the uterine cavity. As it passes along the tube it continues to develop and, on reaching the endometrium, it has usually attained the blastocyst stage (**Fig. 2.21A**). Its survival now depends upon its obtaining nutriments by diffusion from the wet mucous surface of the uterus. This, and later changes, are called the **histiotrophic phase** of embryonic development. Later, when the placenta forms, the embryo enters the **placental** (or **haemotrophic**) **phase** (nourishment from the blood vessels of the placenta) and for this purpose the blastocyst becomes embedded in the endometrium.

The blastocyst reaches the uterine endometrium on about the 5th to 6th day after fertilisation when the zona pellucida is in the process of disintegrating. This disappearance allows the trophoblast (outer flattened layer of cells of the blastocyst) to make contact with the endometrium. This contact stimulates the trophoblastic cells to proliferate rapidly. An inner layer of cells called the **cytotrophoblast** which is the proliferative layer, and an outer **syncytiotrophoblast** (**syntrophoblast** for short) develops from the trophoblastic layer. The syntrophoblast cells do not have time to separate properly and a syncytium (multinucleated cytoplasmic mass) is formed on the outer surface of the trophoblast. The trophoblastic proliferation occurs particularly where the blastocyst is in contact with the endometrium and this is usually at the embryonic pole. Implantation begins at about the 6th to 7th day after fertilisation. The part of the blastocyst projecting into the uterine cavity remains relatively thin (**Fig. 2.21B**).

The syntrophoblast contains a proteolytic enzyme which causes destruction of the endometrial cells so that the blastocyst sinks deeper and deeper into the uterine mucosa. The blastocyst is surrounded by the products of enzymatic digestion and these also are related to the histiotrophic phase of survival. The enzymatic action also damages the walls of the blood vessels in the stroma of the uterus resulting in the release of blood. The final deficiency in the endometrium is sealed off by a blood or fibrin clot, overlying the blastocyst. This cover is called the **operculum** (**Fig. 2.21C**). By about 10 to 12 days after fertilisation, the blastocyst is completely encased in the endometrium and thus, implantation is complete (**Fig.2.21D, E**).

The usual place for implantation is in the endometrium of the fundus of the uterus on its posterior surface (**Fig. 2.22A**). Although this is the most common place for implantation, the blastocyst may implant *anywhere* in the endometrium. Other common places are shown in Figure 2.23. Probably, the most dangerous place for implantation within the uterus is in the region of the internal os (**Fig. 2.23**). Implantation here produces the condition of **placenta praevia** and constitutes an obstruction to the birth process. The placenta at this site therefore, undergoes premature separation causing the fetus to be deprived of oxygen and often resulting in its death. The early separation may also result in

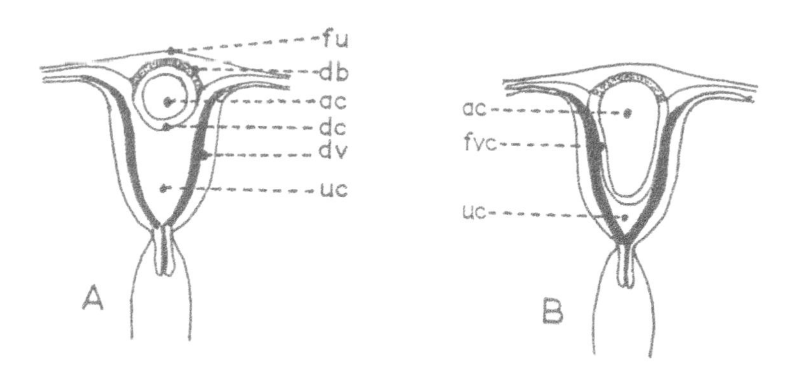

Figure 2.22: Normal position of implantation of the embryo.
ac = amniotic cavity; db = decidua basalis; dc = decidua capsularis; dv = decidua vera; fu = fundus of uterus; fvc = fusion of deciduae vera and capsularis; uc = uterine cavity.

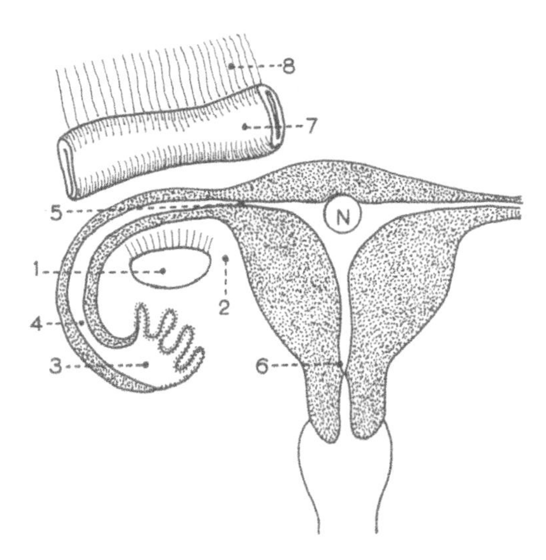

Figure 2.23: Abnormal sites of implantation of the embryo (ectopic implantation). N = normal; 1 = ovarian; 2 = broad ligament; 3, 4 and 5 = within uterine tube; 6 = internal os; 7 = surface of intestine; 8 = mesentery.

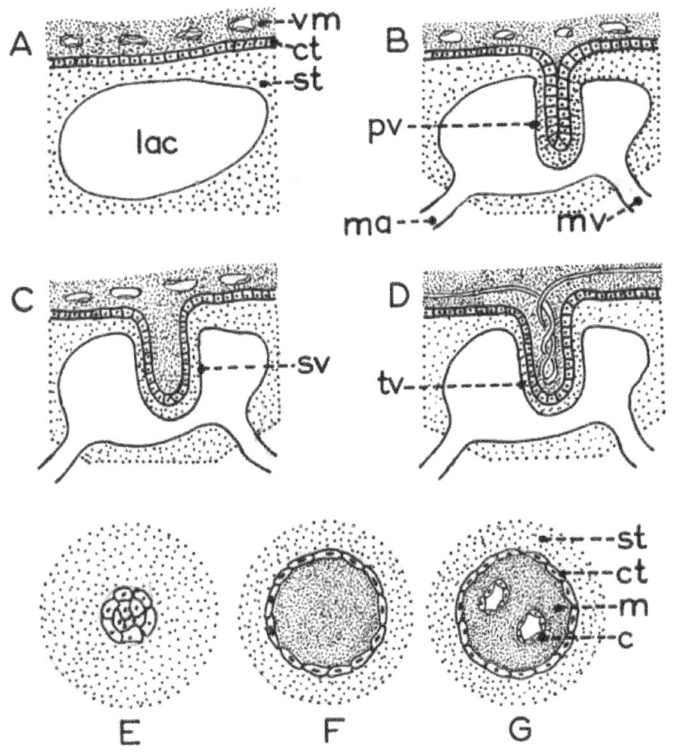

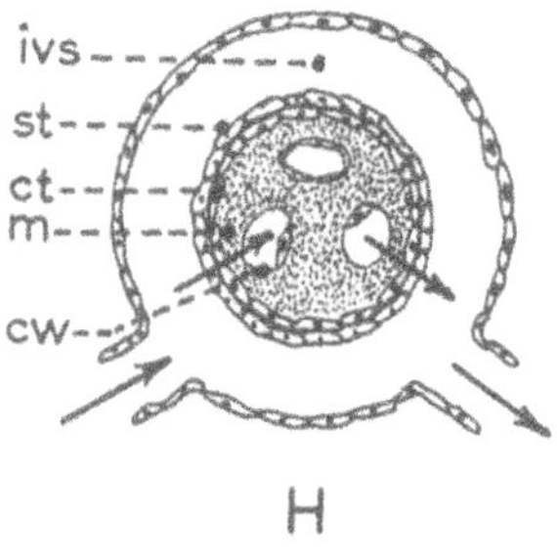

Figure 2.24: Formation of the placental villi. A-D in lateral view and **E-H** in transverse section. c = fetal capillary; ct = cytotrophoblast; cw = capillary wall; ivs = intervillous sinus; lac = lacuna; m = mesenchyme; ma = maternal artery; mv = maternal vein; pv = primary villus; st = syntrophoblast; sv = secondary villus; tv = tertiary villus; vm = maternal vessels.

the mother suffering a torrential haemorrhage from the placental site, which may result in her death and in the death of the fetus. Sometimes implantation occurs outside the uterus and this is known as **ectopic implantation** (Fig. 2.23).

Placenta Formation

By the time the blastocyst is completely embedded in the endometrium, an important series of events has taken place:

(a) After extrusion of the oocyte from the mature follicle, the **granulosa cells** of the lining of the cavity of the mature follicle proliferate to fill the cavity and form the **corpus luteum**. The main activity of the corpus luteum is to form progesterone which is said to be responsible for the continuation and maintenance of the pregnancy.

(b) Under the influence of hormones (oestrogen and progesterone), the glandular and vascular elements of the endometrium undergo proliferation and become demarcated into deep and superficial layers. This vascular and glandular endometrium is now called the **decidua**.

Note. The term *decidua* indicates that after the pregnancy terminates, the lining membrane of the uterus 'falls off' (desquamates).

The parts of the decidua are named by their relationship to the implanted blastocyst:

- **Decidua basalis**: at the base of the embryonic pole of the blastocyst;
- **Decidua capsularis**: that covering the blastocyst;
- **Decidua vera**: that covering the remainder of the uterine cavity. This will fuse with the decidua capsularis as the embryo and amniotic cavity enlarge (Fig. 2.22B).

(c) The syntrophoblast and cytotrophoblast cover the entire blastocyst. At the embryonic pole of the blastocyst, the cytotrophoblast sends extensions into the substance of the syntrophoblast to form **primary villi** (Fig. 2.24B,E). This process of development is more pronounced at the embryonic pole than at the abembryonic pole. Simultaneously, the syntrophoblast develops a series of clefts in its substance called lacunae (Fig. 2. 24A). The maternal (uterine) blood vessels (arteries and veins) break into the lacunae so that maternal blood circulates through the lacunar network (Fig. 2.24B). The lacunae become confluent to form large channels resembling sinuses (intervillous sinuses) which surround the primary villi (Fig. 2.24B). Soon the centres of the primary villi are invaded by extra-embryonic mesoderm and these are termed **secondary villi** (Fig. 2.24C,F). As the embryo develops its own circulation, embryonic blood vessels appear in the extra-embryonic mesoderm of the secondary villi thus converting them into **tertiary villi** (Fig. 2.24D,G). The trophoblast plus extra-embryonic mesoderm together with the embryonic blood vessels constitute a membrane called the **chorion**.

It follows, therefore, that in the region of the decidua basalis a double circulation is established – one from the maternal side and the other from the embryonic side (Fig. 2.24H). The decidua basalis together with its blood vessels and chorion, initiate the formation of a structure which is called the **placenta**. This continues to enlarge as the embryo and fetus enlarge. As the embryo and fetus continue to increase in size, the umbilical cord (which will form from the connecting stalk and the umbilical vesicle stalk), connects the embryo/fetus to the placenta.

In early development the tissue barrier between the maternal and embryonic circulations consists of syntrophoblast, cytotrophoblast, extra-embryonic mesoderm and the endothelium of the embryonic capillary. As the fetus continues to grow, thinning of the barrier will result, to facilitate increased exchange of nutrients and waste products between fetus and mother. The new barrier will now consist of a thin syntrophoblastic layer and the endothelium of the fetal capillary. Physiologically, these layers are called the **placental barrier**.

Substances of low molecular weight such as sodium, potassium, chloride and iodide pass through the barrier in response to the diffusion gradient. Larger and more complex molecules pass through by an active metabolic process. Drugs, alcohol and the products of smoke inhalation as well as bacteria, viruses, spirochaetes and malaria parasites are known to pass through the barrier in particular circumstances.

The cells of the cytotrophoblast and syntrophoblast are known to synthesise various hormones. Since these are rapidly growing cells, they are capable of forming very malignant tumours such as chorionic carcinoma and chorioepithelioma.

The Fate of the Extra-embryonic Membranes

The extra-embryonic membranes consist of the amniotic membrane, the umbilical vesicle membrane and the chorion.

The amniotic membrane encloses the amniotic sac which is filled with amniotic fluid and which is attached to the edge of the embryonic disc (Fig. 2.25). It was formerly the ectodermal vesicle. Because of this, the dorsal part of the embryo lies originally within the amniotic sac but with the bending and growth of the embryo, the whole embryo and the umbilical cord becomes covered with amniotic membrane (Fig.2.26).

The syntrophoblast, cytotrophoblast, underlying extra-embryonic mesoderm and capillaries form the **chorionic membrane**. The extra-embryonic coelom lying within the extra-embryonic mesoderm forms the **chorionic cavity** (Fig. 2.26A). As the embryo/fetus grows and the amniotic cavity expands circumferentially outwards, the chorionic cavity is slowly reduced in size until amnion and chorion fuse, so that the chorionic cavity (formerly extra-embryonic coelom) is obliterated (Fig. 2.26D).

Since the blastocyst has completely embedded itself in the endometrium, now called the

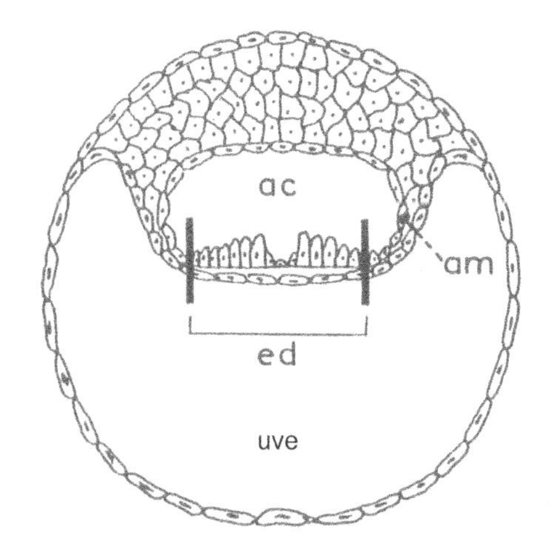

Figure 2.25: Transverse section of the early embryo to show the position of the membranes of the amniotic cavity and the umbilical vesicle (uve). am = amniotic mesoderm; ed = embryonic disk.

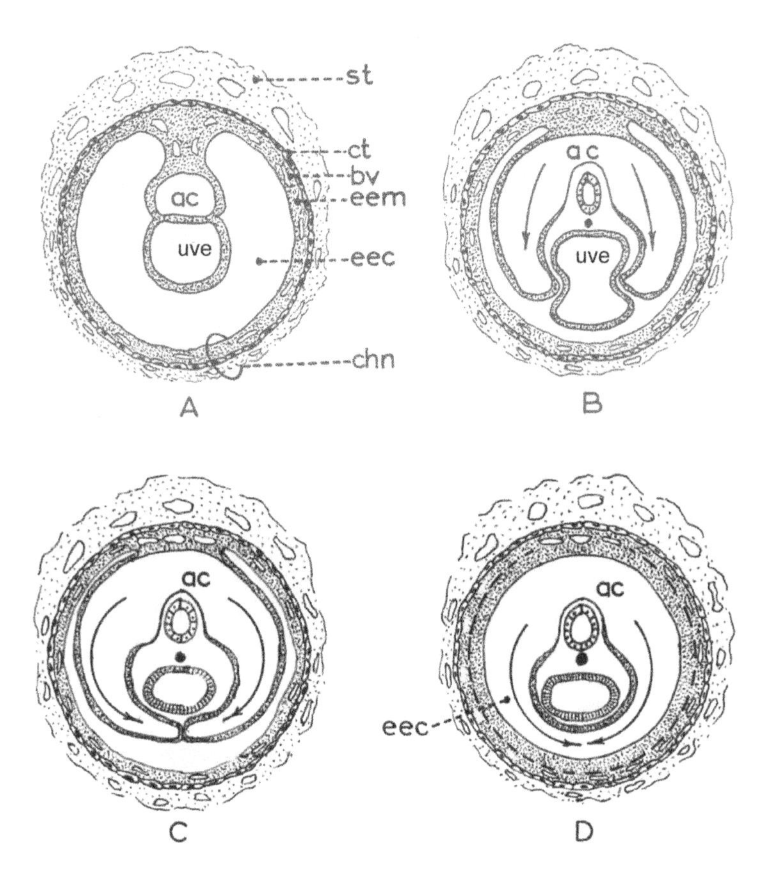

Figure 2.26: A: Sagittal section depicting the development of the chorion (chn), umbilical vesicle (uve) and amnion (ac). **B,C,D:** Transverse sections showing the expansion of the amniotic cavity. bv = blood vessels; chn = chorion; ct = cytotrophoblast; eec = exta-embryonic coelom; eem = extra-embryonic mesoderm; st = syntrophoblast.

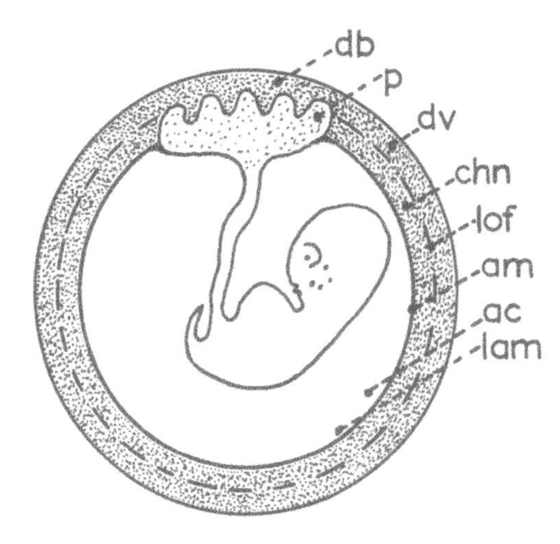

Figure 2.27: Obliteration of the uterine cavity. ac = amniotic cavity; am = amniotic membrane; chn = chorion; db =decidua basalis; dv = decidua vera; lam = layer of amniotic membrane; lof = line of fusion; p = placenta.

decidua, it is covered by the **decidua capsularis**, which, covering the developing embryo, protrudes into the cavity of the uterus. The residual cavity of the uterus is lined by **decidua vera.** As the fetus grows and the amniotic cavity expands, the decidua capsularis and the decidua vera make contact and by the end of the 4th month, they fuse so that the cavity of the uterus is obliterated (**Fig. 2.27**). Gradually, the combined layers (decidua capsularis and decidua vera) undergo atrophy from the pressure of the amniotic fluid in the expanding sac.

After parturition, the uterus is reduced in size by muscular contraction. This results in separation of the placenta and decidua vera from the muscular uterine wall. The placenta and membranes are then extruded as the 'afterbirth'. Examination of the placenta and membranes following the birth is a routine procedure by the attending doctor, to obviate the possibility that a part of each may possibly have been retained within the uterus. Retention of part of the placenta or membranes (amnion, chorion, decidua) may result in persistent post-partum haemorrhage and/or infection of the inner 'raw' surface of the uterus. Both these conditions are dangerous to the well-being of the mother.

Examination of the placenta should show that it is normally discoid in shape, about 15 to 20cm in diameter and about 5cm in thickness. The umbilical cord usually emerges from its inner surface at the centre, and this surface is covered with amniotic membrane which extends on to the umbilical cord (**Fig. 2.28**). Occasionally, the umbilical cord emerges from the edge of the placenta and this is known as a 'battledore' placenta. The uterine surface of a normal fresh placenta is dull red in colour and is divided by sulci into a number of quadrilateral areas known as **cotyledons.** Close examination reveals that this surface is covered with a thin layer of greyish tissue which is the adherent **decidua basalis.** On touching the surface with the finger, there are hard gritty areas in the cotyledons; these are areas of calcification.

The amniotic and chorionic membranes are adherent to one another, the inner (amniotic) surface being smooth and glistening while the outer (uterine) surface is rough and somewhat ragged, being attached to the inner surface of the uterus. These membranes have a rent in them through which the fetus was born. During parturition the head of the fetus descends into the lower part of the uterus pushing before it a pocket of amniotic fluid. As the cervix of the uterus dilates, the membrane covering the pocket loses its lower support and may rupture with the sudden release of amniotic fluid to the exterior. This in layman's terms is known as the 'breaking of the waters'.

FORMATION AND FATE OF THE INTRA-EMBRYONIC AND EXTRA-EMBRYONIC COELOMATA

The word 'coelom' is used for a cavity and is applied especially in embryology. In the embryo there are two coelomata:

- The extra-embryonic coelom which is external to the embryonic area;
- The intra-embryonic coelom which is a cavity formed within the lateral plate mesoderm of the embryonic area.

Formation of the Extra-embryonic Coelom

In the early embryo a new membrane from cells of the hypoblast grows around the inside of the blastocoele, converting it into a new cavity called the **primary umbilical vesicle**. In the 'space' between this membrane and the trophoblast, cells will proliferate and infiltrate to form a layer between the primary umbilical vesicle and the existing trophoblast. Similarly, the infiltration will proceed to the embryonic pole, where the cells will form a layer between the amniotic cavity and the trophoblast. This layer is the **extra-embryonic mesoderm**. Spaces form in this extra-embryonic mesoderm and these coalesce to form a single cavity – the **extra-embryonic coelom** (Fig. 2.26A).

Cavitation of the extra-embryonic mesoderm does not occur at the **connecting stalk** which remains intact to suspend the developing embryo in the extra-embryonic coelom.

During the formation of the extra-embryonic coelom, a part of the umbilical vesicle is 'cut-off' from the primary umbilical vesicle, converting the primary umbilical vesicle into a **definitive umbilical vesicle**. The part that is 'cut-off' disintegrates and disappears.

The extra-embryonic coelom splits the extra-embryonic mesoderm into two parts:

- The **somatic layer**, which overlies the amniotic cavity as well as forming a lining for the trophoblast (Fig. 2.17);

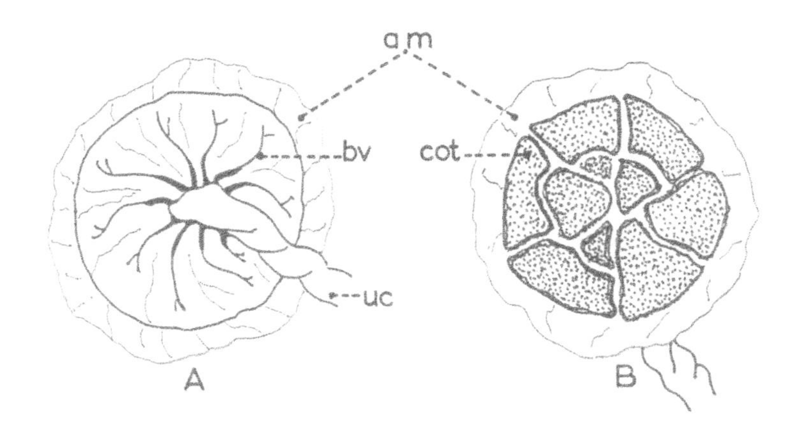

Figure 2.28: Placenta. am = amniotic membrane; bv = blood vessel; cot = cotyledon; uc = umbilical cord.

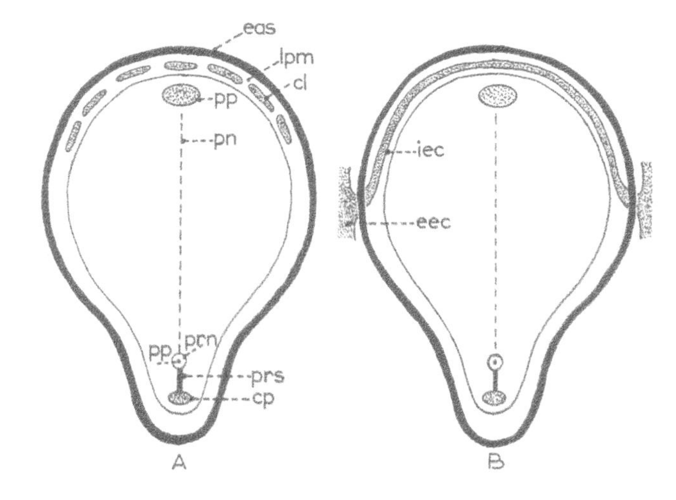

Figure 2.29: A and B: Dorsal views of the embryonic plate to show the formation of the intra-embryonic coelom. cl = clefts occurring in lpm; eas = edge of embryonic plate; eec = extra-embryonic coelom; iec = intra-embryonic coelom; lpm = lateral plate mesoderm.

- The **splanchnic layer**, which overlies the umbilical vesicle (Fig. 2.17).

Note. The term **somatopleure** denotes the combination of somatic mesoderm plus (amniotic) ectoderm, while the term **splanchnopleure** denotes the combination of splanchnic mesoderm plus (umbilical vesicle) endoderm.

Formation of the Intra-embryonic Coelom

The intra-embryonic coelom develops in the intra-embryonic mesoderm which is derived from primitive streak epiblast. The intra-embryonic mesoderm has paraxial, intermediate and lateral plate components (Fig. 2.20). The lateral plate mesoderm forms the peripheral band

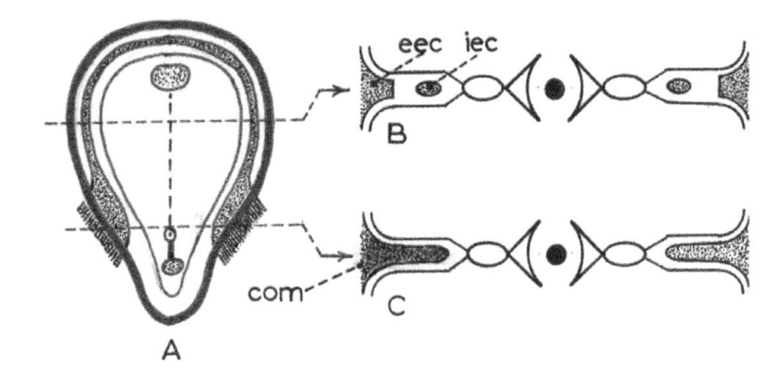

Figure 2.30: A: Dorsal view of the embryonic plate indicating by interrupted lines the levels at which the transverse sections in B and C are taken. C: The level at which communication between the intra- (iec) and extra- (eec) embryonic coelomata will occur. com = communication between the intra-embryonic and extra-embryonic coelomata.

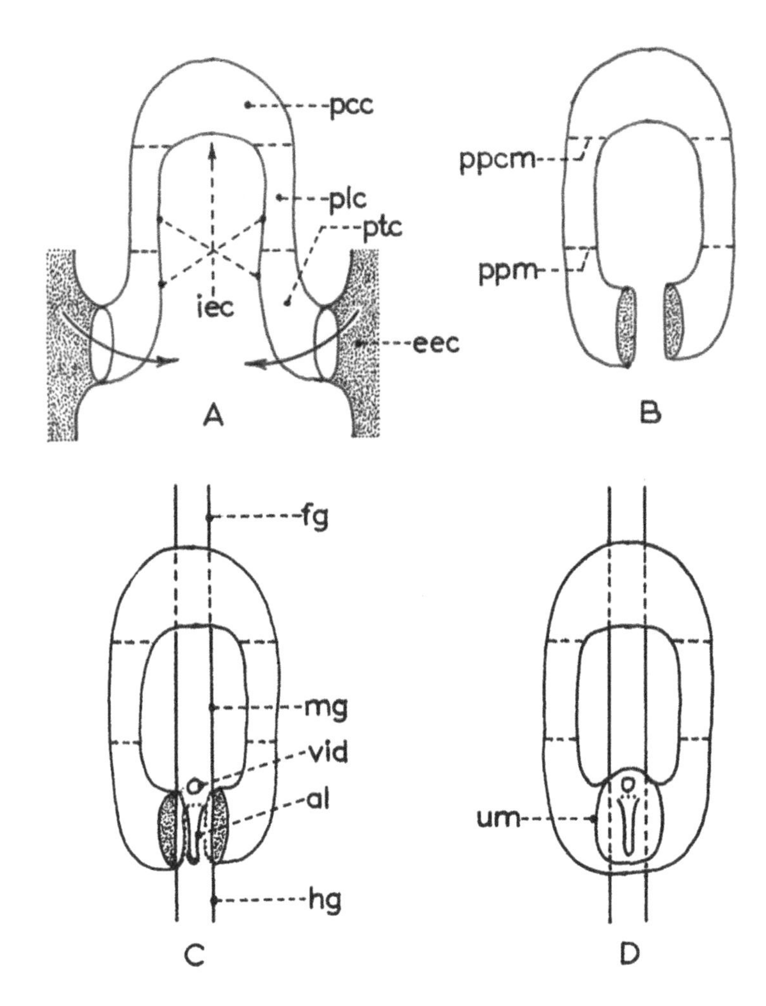

Figure 2.31: Ventral views of the separation of the intra-embryonic coelom (iec) from the extra-embryonic coelom (eec). The arrows indicate the direction of folding. al = allantois; ccv = common cardinal vein; fg = foregut; hg = hindgut; mg = midgut; pcc = pericardial coelom; plc = pleural coelom; ptc = peritoneal coelom; ppcm = pleuropericardial membrane; ppm = pleuroperitoneal membrane; vid = vitello-intestinal duct; um = umbilicus.

of mesoderm (**Fig. 2.20**). Beginning at the cranial end, the lateral plate mesoderm develops a series of spaces and these combine to form a 'tube' (the **intra-embryonic coelom**) (**Fig. 2.29A,B**).

The tube thus formed has the shape of an inverted U, with the bend of the U surrounding the cranial part of the embryonic plate (**Fig. 2.29B**) and the limbs of the U passing along the sides of the flat, discoid embryo. Half-way along the sides of the embryonic plate, the lateral plate mesoderm will degenerate so that the limbs of the intra-embryonic coelom open to the extra-embryonic coelom at these places. Thus, at this stage of embryonic development, the intra-embryonic and extra-embryonic coelomata communicate with each other on either side of the embryonic plate (**Fig. 2.30**).

The Fate of the Intra-embryonic Coelom

The intra-embryonic coelom is at first, a single cavity which will be divided into four separate parts later. These parts are the **pericardial** cavity, the **pleural** cavities (of which there are two) and the **peritoneal** cavity (the four Ps).

During the phase of development which follows, the embryo begins to undergo longitudinal and transverse growth which will ultimately result in the complete separation of the extra-embryonic from the intra-embryonic coelom. This will result in a 'circular-type' of intra-embryonic coelom within the now 'cylindrical' embryo. The pericardial cavity will form from the cranial region of the coelom (the curve of the inverted U), while the peritoneal cavity will form from the more caudal part of the coelom. The intervening part of the coelom, joining the pericardial and peritoneal cavities, will form the pleural tubes (cavities) on either side.

The following transverse sections and ventral views of the coelomata attempts to explain the events that lead to the sealing off of the intra-embryonic coelom from the extra-embryonic coelom (**Figs 2.17B-F; Figs 2.31A-D; Fig.2.32**).

(a) In the early embryo, the opening between the intra-embryonic and the extra-embryonic coelomata is in a lateral position of the embryonic disc (**Fig. 2.17B** and **2.31A**).
(b) With transverse folding of the embryo, the two lateral openings are brought into a ventral position (**Fig. 2.17C** and **2.31B**).

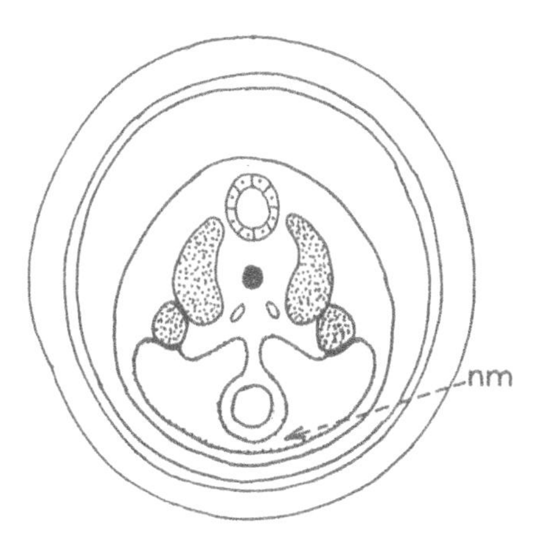

Figure 2.32: A transverse section showing the disappearance of the ventral mesentery.
nm = no mesentery.

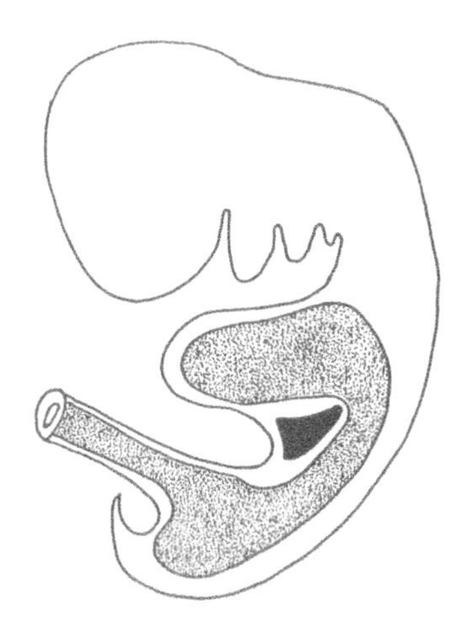

Figure 2.33: The position of the coelom in the embryo.

(c) The layers of the coelom and amnion constrict around the connecting stalk (which contains the allantois and umbilical vessels), the degenerating umbilical vesicle stalk and accompanying vitelline vessels (Fig. 2.17F and 2.31C,D) which together form the umbilical cord. Above and below the umbilical opening, the body wall closes.

Note that the layer of somatic mesoderm fuses with the splanchnic mesoderm on either side of the ventral midline as the intra-embryonic coelom closes off from the extra-embryonic coelom (Fig. 2.17D,E). The two layers of splanchnic mesoderm which are now apposed ventral to the gut tube, result in the formation of a temporary **ventral mesentery**. Note that a mesentery will only persist if a specific structure is present within it. In the case of the ventral mesentery, no such content exists below the level of the vitello-intestinal duct and the mesentery therefore, will disappear (Fig. 2.32). From Figure 2.17D note that dorsal to the gut tube, the two layers of splanchnic mesoderm together form the **dorsal mesentery** (Fig 2.17D,E). This persists because it contains the blood vessels and nerves of the gut.

Figure 2.33 illustrates the 'appearance' of the intra-embryonic coelom following transverse and longitudinal folding of the embryo. Figure 2.34 illustrates the closure of the body wall above and below the region of the umbilical cord.

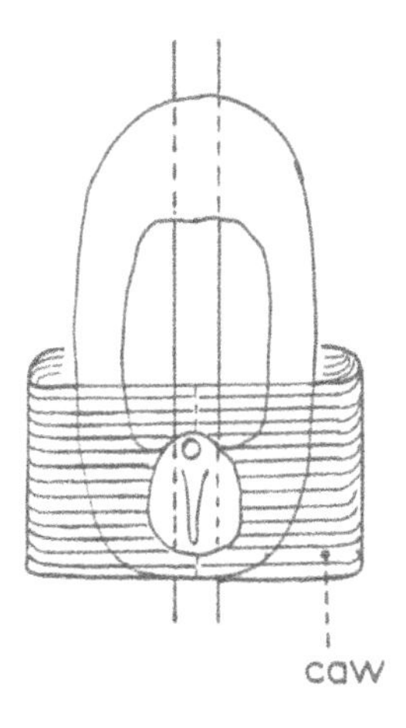

Figure 2.34: Closure of the anterior abdominal wall (caw).

The separation between the pericardial and the pleural coelomata is brought about by lateral pressure from the common cardinal veins (ducts of Cuvier) accompanied by the phrenic nerves (Fig. 2.31B). The separation forms the **pleuro-pericardial membranes.** The separation between the pleural coelomata and the peritoneal coelom is brought about by the formation of the **pleuro-peritoneal membranes** which are raised by the dorsally placed urogenital ridges (Fig. 2.31B). The pleuro-peritoneal openings and the pleuro-pericardial openings are thus closed off. The two pleural cavities are bilateral and narrower than the pericardial and peritoneal cavities (Fig. 2.35).

The heart, lung buds and gut tube grow into the developing cavities and are covered by the splanchnic mesoderm and epithelial covering of the coelom to provide membranes necessary for their proper function. These membranes give rise to the pericardium and epicardium, the visceral and parietal pleural membranes and the various mesenteries in the peritoneal cavity.

The intra-embryonic splanchnic mesoderm may be called the 'organ layer' of the embryo since all organs are either formed or housed within it. In the adult, this layer is the extra-pleural and extra peritoneal layer of the body. In the thorax, it lies between the inner surfaces of the ribs and intercostal muscles on the one hand and the pleural membrane on the other. This layer is called the **endothoracic fascia**.

In the abdomen, the intra-embryonic splanchnic mesoderm lies between the abdominal musculature and the peritoneum and is called the **transversalis fascia**. The blood vessels which supply and drain all the organs run in the retrocoelomic mesoderm and utilise the mesenteries, in some cases, to reach the organs.

The lining of the coelomic cavities is a simple squamous epithelium (mesothelium) which secretes fluid. It may be said that the pericardial, pleural and peritoneal cavities contain nothing but a small amount of fluid to lubricate the surfaces. There is, however, one exception to this. In the female, the mature follicle of the ovary bulges into the peritoneal cavity. When the follicle ruptures, the oocyte is ejected into the peritoneal cavity where it is caught up by the fimbriated end of the uterine tube. Thus, on a monthly basis, the female peritoneal cavity temporarily contains a minute living structure – the oocyte.

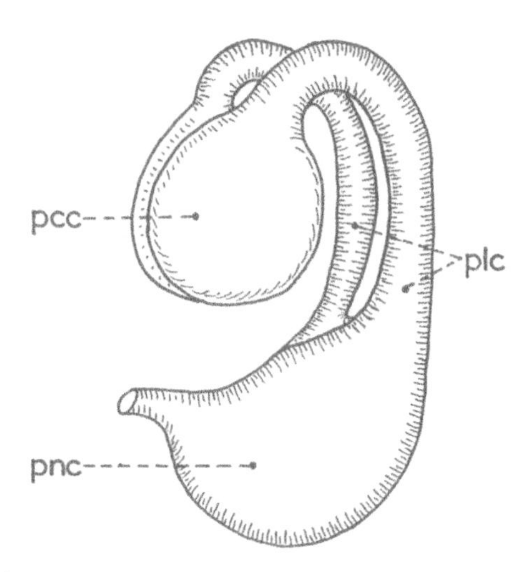

Figure 2.35: A three-dimensional view of the isolated intra-embryonic coelom showing the pericardial cavity (pcc), the pleural cavities (plc) and the peritoneal cavity (pnc).

CHAPTER THREE

Organogenesis

THE CARDIOVASCULAR SYSTEM

This system consists of the **heart**, the **arteries** and the **veins** of the body, as well as the **lymphatic channels**.

The Earliest Blood Vessels

These vessels begin in the extra-embryonic mesoderm by the formation of **blood islands**. The earliest islands are found on the surface of the umbilical vesicle and the connecting stalk in the so-called **area vasculosa** (Fig. 3.1A). The islands contain cells called **haemangioblasts** which are capable of differentiating into **angioblasts** (to form blood vessels) and **haemocytoblasts** (to form blood cells) (Fig. 3.1B,C). The angioblasts form a network of vessels in the mesoderm over the umbilical vesicle and in the connecting stalk. The network of vessels join into two main vessels, the **vitelline veins** which pass along the **vitello-intestinal duct** (umbilical vesicle stalk) to open into the caudal end of the cardiac tube (future heart). Similarly, venous channels form in the connecting stalk and also enter the caudal end of the cardiac tube (Fig. 3.2). These are the right and left **umbilical veins** from the developing placenta. Thus, the caudal end of the cardiac tube initially receives four veins.

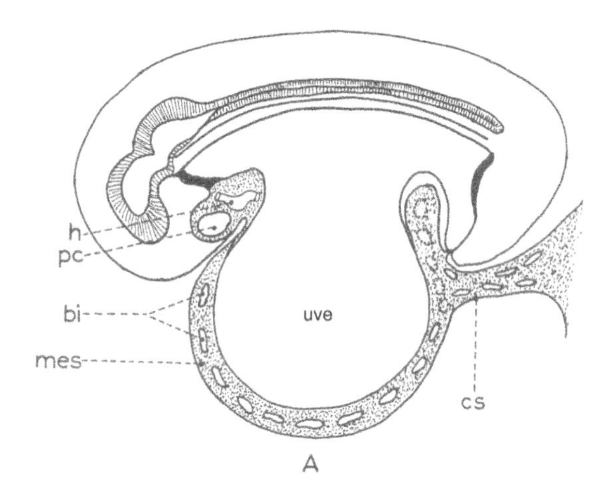

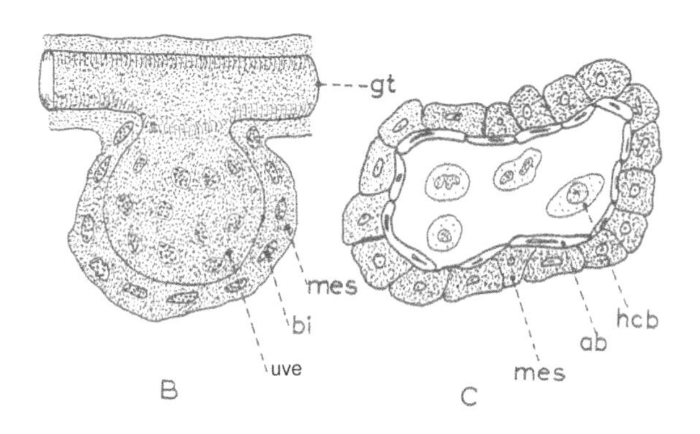

Figure 3.1: A: Longitudinal section indicating the reversal of the position of the developing heart (h). **B:** The blood islands cover the entire surface of the umbilical vesicle. **C:** Blood islands (haemangioblasts) form blood vessels lined by angioblasts (ab) and containing haemocytoblasts (hcb). bi = blood islands; cs = connecting stalk; mes = mesoderm; gt = gut tube; pc = pericadial coelom; mes = mesoderm; uve = umbilical vesicle.

Early Heart Development

In the cranial part of the intra-embryonic mesoderm, ventral to the future pericardial coelom, a condensation occurs in the splanchnic mesoderm (Fig. 3.3A). This is the **cardiogenic plate** from which the heart tube will develop. As the embryo grows and begins its folding process, the cardiogenic plate undergoes a reversal of position to come to lie on the dorsal aspect of the **pericardial coelom** (Fig. 3.3B). During this time the cardiogenic plate undergoes bilateral canalisation to form two **cardiac tubes**. The cranial parts of these tubes will form the **ventral aortae** on either side. Connections will later join the ventral aortae (aortic sac, Fig. 3.4 A,B) to the existing **dorsal aortae**. These connections will form the arteries of the pharyngeal arches. Caudally the tubes will join with the vitelline and umbilical veins (Fig. 3.2). The cardiac tubes fuse to form a single heart tube which gradually sinks into the underlying coelom which is now called the **pericardial coelom** (future pericardial sac)

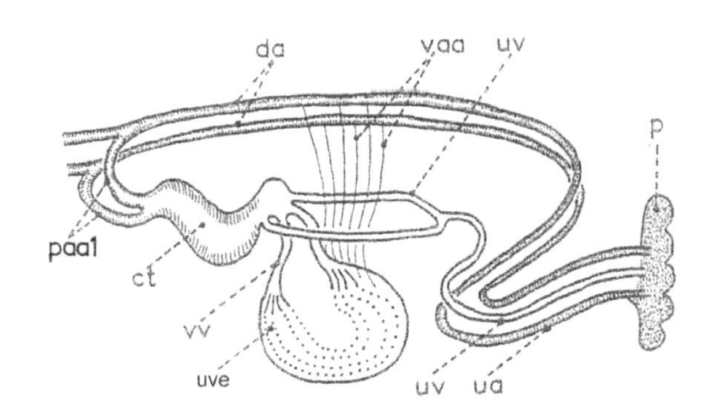

Figure 3.2: Longitudinal diagram illustrating the vasculature of an early embryo. ct =cardiac tube; da = paired dorsal aortae; p = placenta; paa1 = first pharyngeal arch artery 1; ua = umbilical arteries; uv = umbilical veins; vaa = vitelline arteries; vv = vitelline veins; uve = umbilical vesicle.

(Fig. 3.4C). A mesentery, the **mesocardium,** is thus formed on the dorsal surface of the heart tube. Soon after the formation of the mesocardium, an aperture develops in the mesocardium (Fig. 3.5). This aperture is the **transverse pericardial sinus**. The function of such an aperture is not clear, but it probably creates a degree of flexibility in the dorsal aspect of the bending and twisting heart tube.

At this stage, the heart consists of an endothelial tube surrounded by a visceral layer of the pericardium (Fig. 3.6), which is called the **epicardium**. From the time when the cardiac tubes begin to fuse, their walls undergo fibrillary movement, which is the forerunner to cardiac contraction. (It is noteworthy that contraction of the cardiac elements takes place at a very early stage in development; as development proceeds and the myocardium matures, the contractions become more pronounced and rhythmical.) When the heart begins to beat, it draws in blood from the vitelline and umbilical veins.

Later, the narrow space between the epicardium and the cardiac endothelium becomes filled with a jelly-like material (so-called **cardiac jelly**), which is invaded by cells from the deep layer of the epicardium. These are **myoblasts** which will ultimately form cardiac muscle cells. The combined layer of epicardium and invaded jelly is called the **myoepicardial mantle** (Fig. 3.6). The epicardial layer will also give rise to blood islands from which a vascular network will form, and from which the blood supply (coronary vessels) to the myocardium will result.

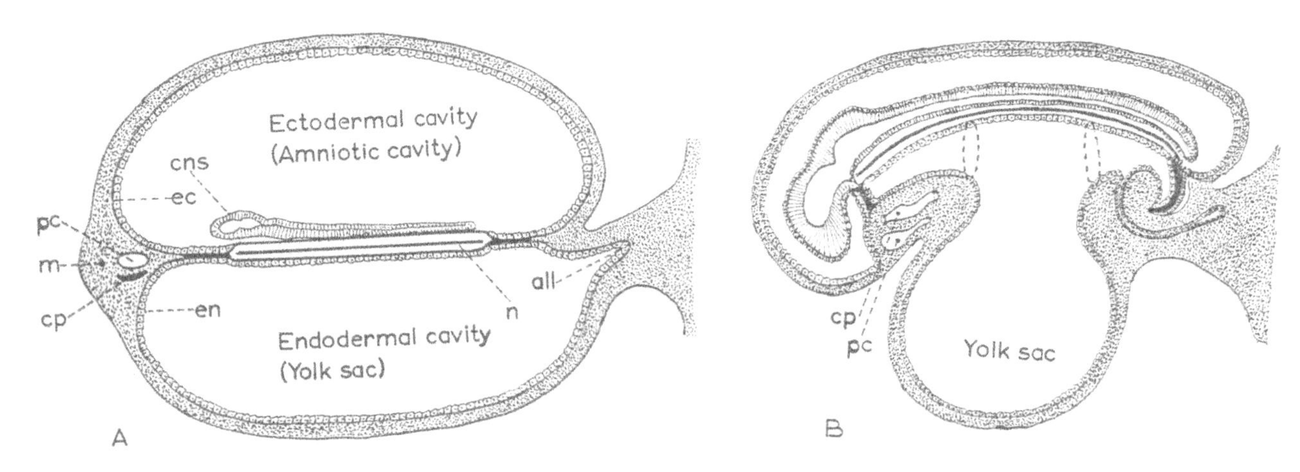

Figure 3.3: Longitudinal section showing the original positions of the cardiogenic plate (cp) and pericardial coelom (pc). all = allantois; cns = central nervous system; ec = ectoderm; en = endoderm; m = mesoderm; n = notochord.

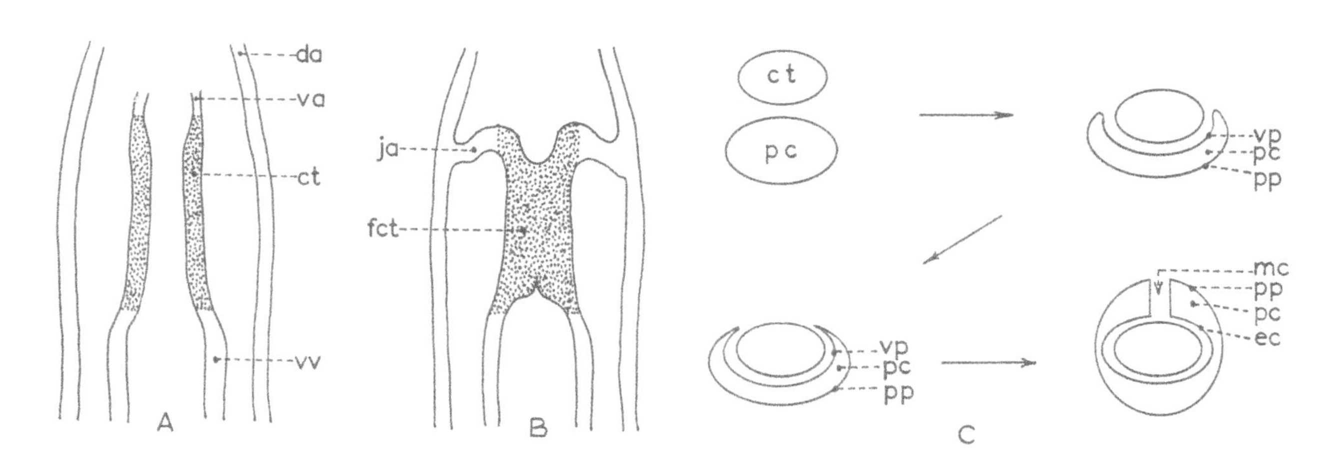

Figure 3.4: A and B illustrate the fusion of the two cardiac tubes (ct). C: Serial diagrams illustrating the cardiac tube 'sinking' into the pericardial coelom (pc). da = dorsal aortae; ec = epicardium, fct = fused cardiac tubes; ja = junctional arteries; mc =dorsal mesentery (mesocardium); pp = parietal pericardium; va = ventral aortae; vp = visceral pericardium; vv = vitelline veins.

Now that the heart tube has acquired a 'lubricating sleeve' (the pericardial coelom) (Fig. 3.5), its cranial and caudal ends become anchored to the pericardium and at this stage, the tube begins to undergo **external** and **internal** changes.

External Changes

By this time, the placenta is fairly well established and the heart must develop rapidly to provide a suitable blood circulation through it, since the embryo is beginning to 'outgrow' its histiotrophic phase of nourishment. The heart tube now develops a series of constrictions which divide it into anatomical parts (Fig. 3.7). From the cranial end to the caudal end, these are:

- the **truncus arteriosus**, where the outflow tract of the heart will develop;
- the **bulbus cordis**, between the truncus and the single ventricle;
- the **ventricle**, which is the largest part of the tube and which shows the greatest degree of expansion during the tortuous growth of the heart; this is continuous with the bulbus cordis via a bulboventricular canal;
- the **atrium**, which is continuous with the ventricle via an atrioventricular canal;
- the **sinus venosus**, the caudal end of the tube, receives blood from the four entering veins and transmits it to the atrium via a sino-atrial canal.

Note. When blood forms it will enter at the caudal end (inflow tract) of the primitive heart tube and leave at the cranial end (outflow tract).

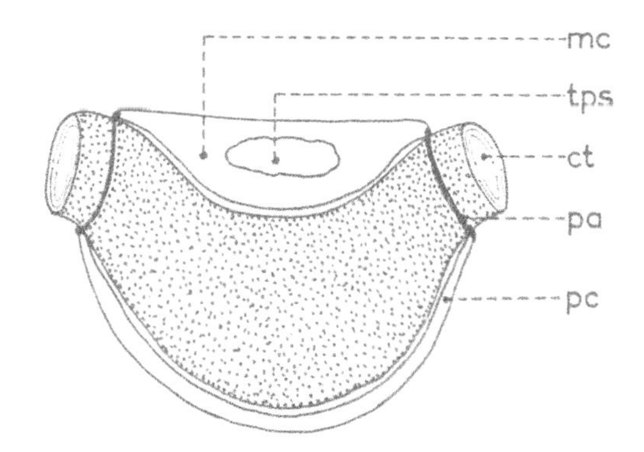

Figure 3.5: A lateral view of the developing heart (ct) within the pericardial cavity (pc). ct = cardiac tube; mc = mesocardium; pa = pericardial attachments; pc = pericardial cavity; tps = transverse pericardial sinus.

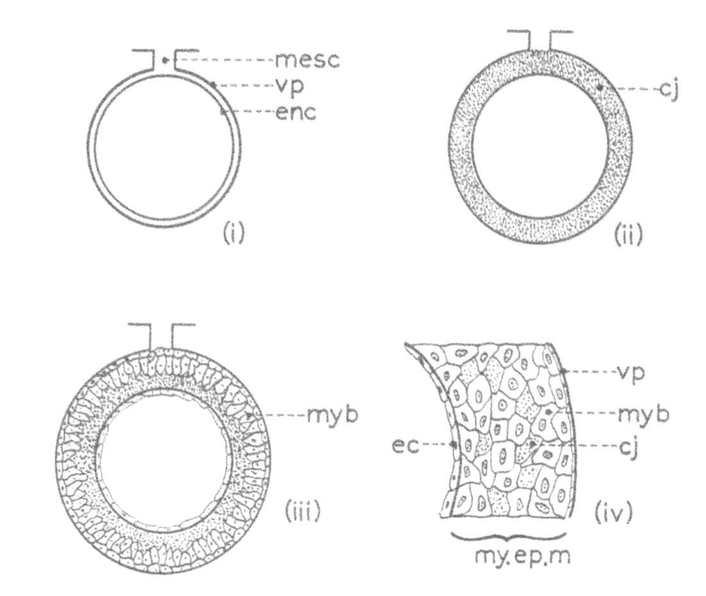

Figure 3.6: Formation of the myoepicardial mantle (my.ep.m). cj = cardiac jelly; ec = endocardium; enc = endocardium; mesc = mesocardium; myb = myoblasts; vp = visceral pericardium (epicardium).

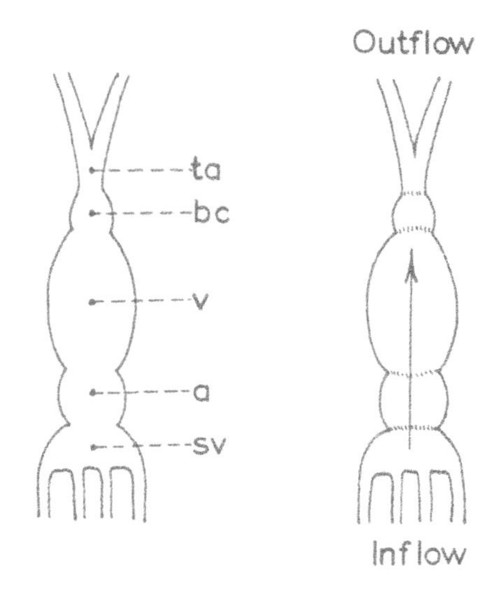

Figure 3.7: Anatomical divisions of the heart from the cranial to the caudal region. a = atrium; bc = bulbus cordis; sv = sinus venosus; ta = truncus arteriosus; v = ventricle.

The growth of the heart results in it 'bending' and 'twisting' within the pericardial coelom, possibly due to its ends being tethered by the vessels of the inflow and outflow tracts. The mesocardium and foregut lie posterior to the heart and prevent it from bending dorsally. Thus, in order to accommodate the increase in length of the heart tube, it bends ventrally into the pericardial coelom, creating a space dorsal to it. As the ventricle enlarges it absorbs the lower part of the bulbus cordis to form the **bulboventricular loop** (Fig. 3.8A-C). The ventricle not only enlarges and lengthens but it twists, usually to the left, so that the atrium and sinus venosus come to lie dorsal to the bulbus cordis and lower part of the truncus arteriosus (Fig. 3.8D,E). This procedure seems to be facilitated by the development of a large aperture in the mesocardium (Fig. 3.5). This aperture is the **transverse pericardial sinus.** The presence of the aperture allows flexibility of the bending heart so that the venous inflow end comes to lie directly dorsal to the arterial outflow end (Fig. 3.8E-G). As a result of the twisting of the tube and, when viewing the heart from its ventral surface, the atrium peeps out on the right side (Fig. 3.8E,F), and when viewed dorsally, the sino-atrial opening is also on the right side. However, the atrioventricular opening is still on the left (Fig. 3.8E). Dorsal to the heart, the lungs are developing and their venous return in the form of a single vein enters the atrium slightly to the left side (Fig. 3.9). The formation of four pulmonary veins on the dorsal aspect of the heart results in the formation of the **oblique pericardial sinus** (Fig. 3.10).

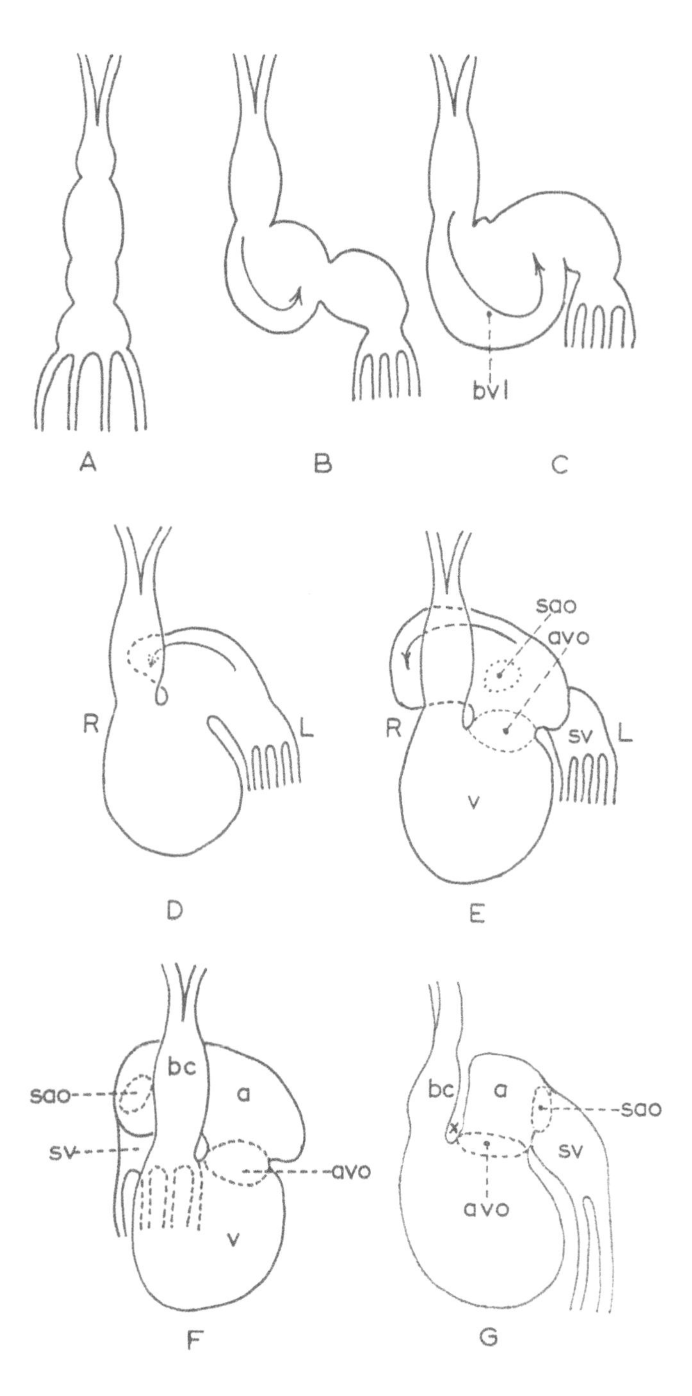

Figure 3.8: A series of diagrams illustrating the bending and twisting of the segmented cardiac tube. A-F: Ventral views. **G:** Lateral view. a = atrium; avo = atrioventricular opening; bc = bulbus cordis; bvl = bulboventricular loop; sao = sino-atrial opening; sv = sinus venosus; v = ventricle; x = position of transverse pericardial sinus.

POSSIBLE ABNORMALITIES

- **Dextro-rotation,** in which case the heart bends and twists to the right. This may be associated with other abnormalities such as **situs inversus** (reversal in position of all the organs in the trunk).
- **Ectopia cordis,** in which excessive ventral bending of the bulboventrcular loop results in the heart protruding through the anterior thoracic wall.

Internal Changes

It is important to realise that in the adult, the heart is basically a 'pump'. The internal changes in the developing heart are designed to convert a single heart tube (pump) into two heart tubes (pumps) so that the one may subserve the pulmonary circuit and the other the systemic circuit. These changes may be listed as follows:

- separation of the single **atrioventricular canal** into two by the growth of **endocardial prominences** (endocardial cushions);
- septation of the atrium;
- septation of the ventricle;
- septation of the truncus arteriosus;
- formation of valves to ensure unidirectional flow in the pumps.

All the above changes occur simultaneously.

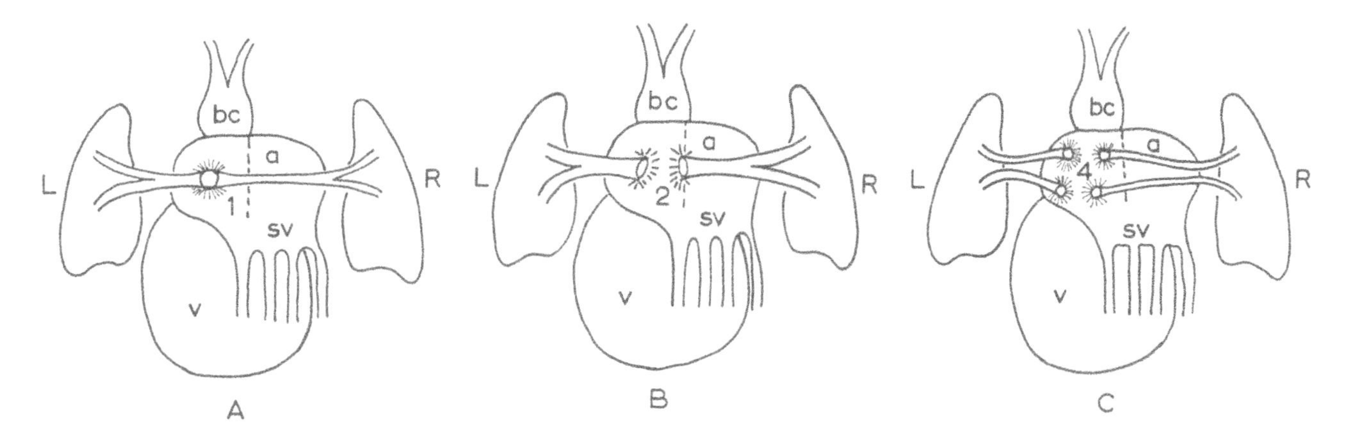

Figure 3.9: The dorsal aspect of the developing heart showing the absorption of the pulmonary veins into the left atrium, resulting in four openings of the veins. a = atrium; bc = bulbus cordis; sv = sinus venosus; v = ventricle.

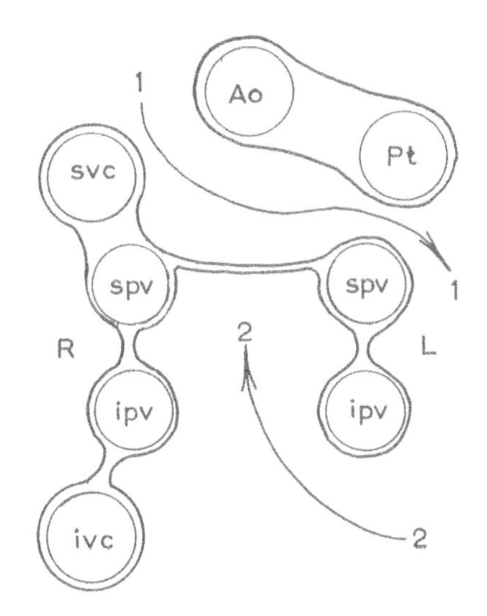

Figure 3.10: Transverse pericardial sinus (1) and oblique pericardial sinus (2) in the adult.
Ao = aorta; ivc = inferior vena cava; ivp = inferior pulmonary veins; Pt = pulmonary trunk; spv = superior pulmonary veins; svc = superior vena cava.

Formation of Bilateral Atrioventricular Canals

While the heart is bending and twisting, the common atrioventricular canal develops two prominences (cushions), ventral and dorsal **(Fig. 3.11A)**. The ventral and dorsal prominences join to form a single bar, the **intermediate bar**, which separates the original canal into two smaller **right** and **left atrioventricular** canals **(Fig. 3.11B)**. The atrioventricular valves will form partly by proliferation of subendocardial tissue and partly from the existing endocardial prominences. (see page 39; Fig. 3.12)

Septation of the Atrium

When the embryo is about 5mm in length, septation of the atrium begins with a first septum (**septum primum**) growing from the roof of the

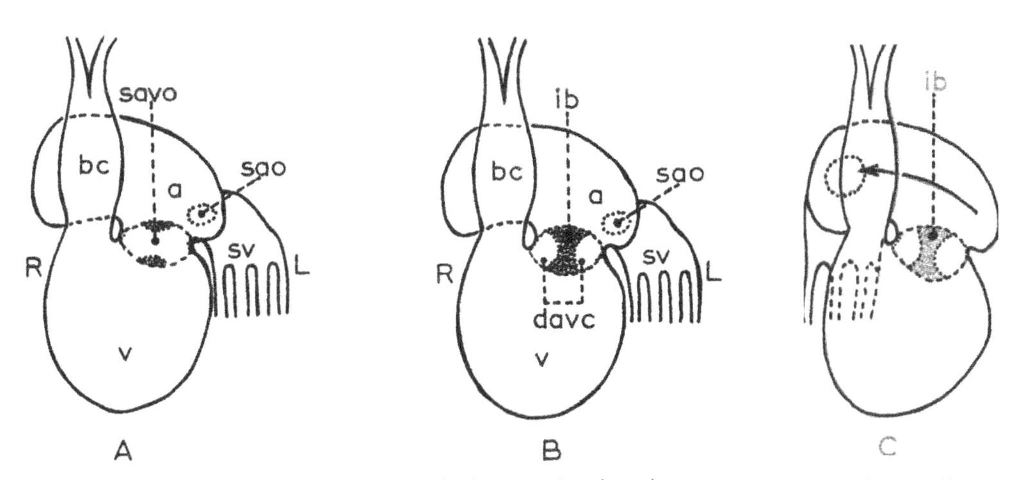

Figure 3.11: Division of the single atrioventricular opening (savo). a = atrium; bc = bulbus cordis; davc = double atrioventricular canals; ib = intermediate bar; L = left hand side; R = right hand side; sao = sino-atrial orifice; sv = sinus venosus; v = ventricle.

atrium towards the intermediate bar. The septum has a crescentic lower edge and as it advances towards the intermediate bar, the ever-decreasing space between it and the bar is known as the **foramen primum** (Fig. 3.13A,D). As the septum primum reaches and fuses with the intermediate bar, the upper part becomes fenestrated (window formation) so that the septum becomes deficient in its upper part (Fig. 3.13B,E). This deficiency is called the **foramen secundum**. Since in the embryo and fetus the lungs are not functional, it is important that an aperture be maintained between the right and left sides of the atrium so that oxygenated placental blood reaches the left side of the heart.

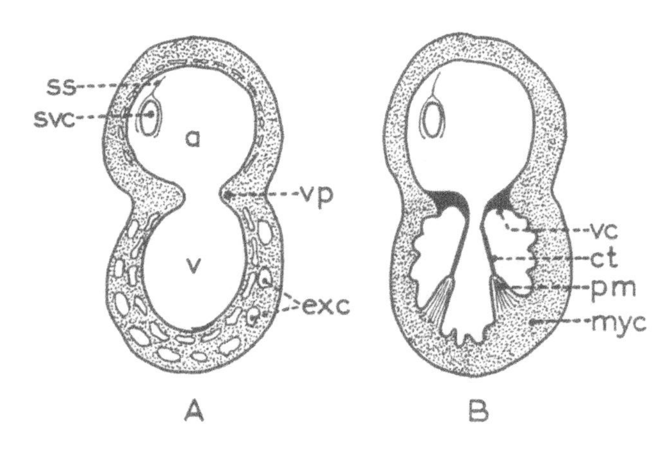

Figure 3.12: Formation of the atrioventricular valves. a = atrium; ct = chordae tendineae; exc = ventricular excavations; myc = myocardium; pm = papillary muscles; ss = septum spurium; svc = superior vena cava; v = ventricle; vc = valve cusps; vp = ventricular prominences.

Simultaneously with the fenestration of septum primum, a second septum (**septum secundum**) appearing on the right side of the first septum, grows from the roof of the atrium towards the intermediate bar. Septum secundum also has a crescentic lower edge which is directed somewhat dorsally (Fig. 3.13C,F). When this septum reaches and fuses with the intermediate bar there is a crescentic deficiency at its dorsal edge and, coupled with the bulging dorsal wall of the right atrium, the opening is oval in shape and is called the **foramen ovale** (Fig. 3.13F). The double septum arrangement produces an incomplete separation of the original embryonic atrium into two smaller **right** and **left atria**. This arrangement allows blood from the right side to reach the left side by passing under the crescentic edge of the septum secundum (foramen ovale) and through the aperture in the septum primum (foramen secundum) (Fig. 3.13G,H). This reduces the vascular resistance produced by the non-functioning lungs. During embryonic and fetal life, the septum primum acts as a valvular flap, preventing reflux of blood from the left atrium into the right atrium via the formamen ovale.

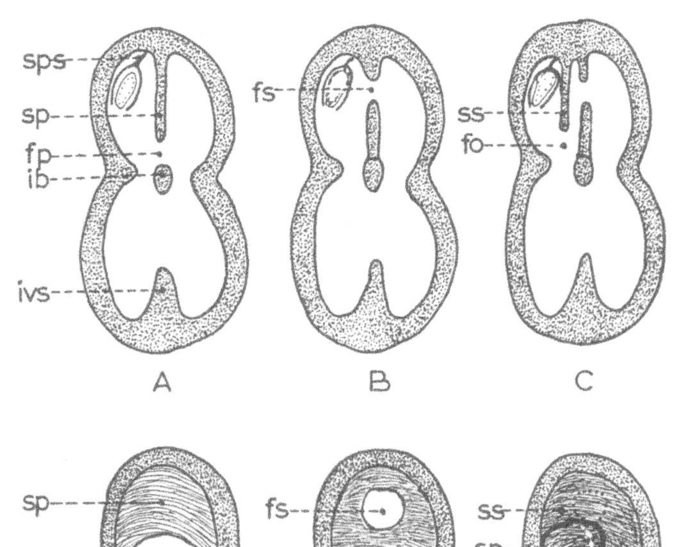

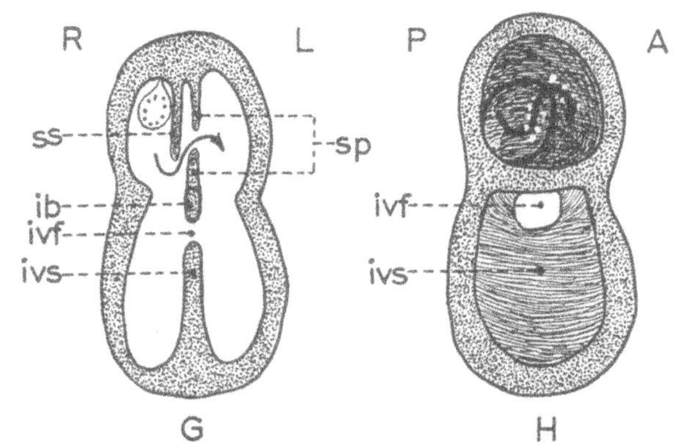

Figure 3.13: A series of coronal (A-C,G) and sagittal (D-F, H) sections of the heart to illustrate septation of the atrium. D–H and F are views from the interior of the right atrium. fo = foramen ovale; fp = foramen primum; fs = foramen secundum; ib = intermediate bar; ivf = interventricular foramen; ivs = interventricular septum; sp = septum primum; sps = septum spurium; ss = septum secundum.

At birth, when the lungs become functional and the pressure in the left atrium rises because of the venous inflow from the lungs as a result of aeration, the septum primum is forced against septum secundum, obliterating the 'right-to-left shunt'. The two septa normally fuse to form a definitive **interatrial septum**, thus effectively preventing the admixture of deoxygenated (from the body) and oxygenated blood (from the lungs). The change from an aquatic to a terrestrial environment which occurs at birth necessitates the closure of the aperture between the right and left atria. This requires the presence of two overlapping septa in the fetus.

In the adult heart, the position of the foramen ovale is indicated by the **fossa ovalis**, an indentation seen on the right side of the interatrial septum. The edge of the foramen ovale is demarcated by the **anulus ovalis** surrounding the fossa (Fig. 3.14). Not infrequently, it is possible to pass a probe, obliquely upwards, under the anulus, without creating an artificial opening, from the right atrium to the left atrium. This minute opening indicates that complete fusion of the septa has not taken place but it produces no physiological changes in the function of the heart because the opening is so small.

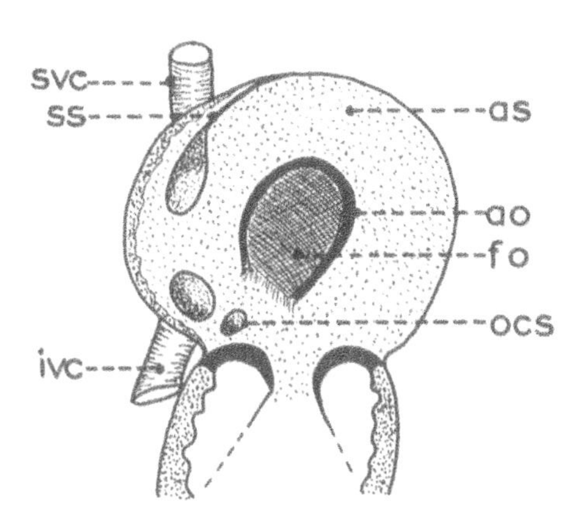

Figure 3.14: Interior of the right atrium of the adult. ao = anulis ovalis; as = atrial septum; fo = fossa ovalis; ivc = inferior vena cava; ocs = opening of coronary sinus; ss = septum spurium; svc = superior vena cava.

POSSIBLE COMMON ABNORMALITIES

Complete failure of septation of the atria results in a **cor triloculare biventriculare,** which is a single atrium with two ventricles.

If one or both of the apertures in septum primum and septum secundum is too large and the one 'overlaps' the other, an opening persists in the interatrial septum. This is called a **patent interatrial septum** which, depending upon its size, may cause serious physiological disturbances in the cardiac haemodynamics. If the opening is large enough, it produces the clinical condition known as 'blue baby' **(patent foramen ovale).** Most cases are amenable to surgical correction.

Fate of the Sinus Venosus

While the septation of the atrium is taking place, the sinus venosus is undergoing changes, one of which is that it is being 'absorbed' into the right atrium. At first, the sinus venosus is simply a venous inlet into the heart. The blood is collected from the paired vitelline and umbilical veins (Fig. 3.15A);

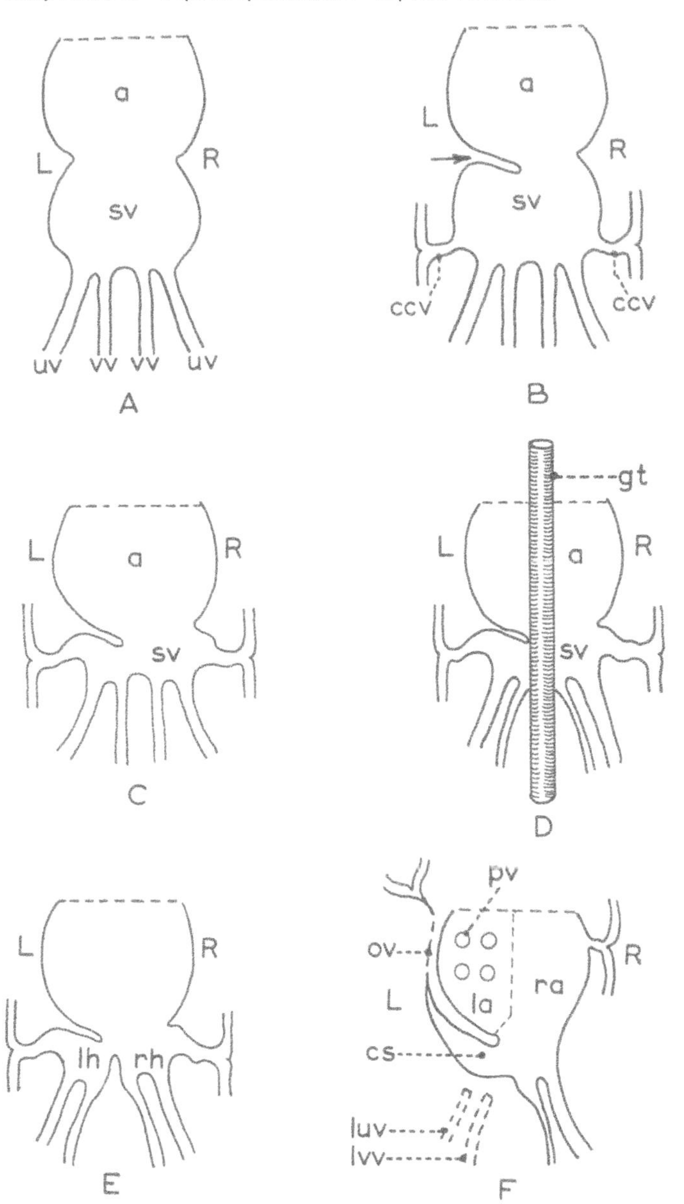

Figure 3.15: A-F: Development of the sinus venosus (sv), as viewed dorsally. a = atrium; ccv = common cardinal veins; cs = coronary sinus; gt = gut tube; L = left hand side; la = left atrium; lh = left horn; lvv = left vitelline vein; luv = left umbilical vein; ov = oblique vein of left atrium; pv = pulmonary veins; R = right hand side; ra = right atrium; rh = right horn.

later, these are augmented by the paired common cardinal veins which enter in the upper part of the sinus venosus (Fig. 3.15B,C).

As the dorsally placed gut tube comes into existence, the vitelline veins are displaced laterally and the sinus venosus splits partially to form right and left 'horns' (Fig. 3.15D,E).

The so-called 'absorption' of the sinus venosus into the right atrium begins with the formation of a deep groove at the left sino-atrial junction (Fig. 3.15B). This groove separates the left horn of the sinus from the atrium so that the blood entering the left horn is progressively diverted to the right horn. Less blood is being returned to the left horn of the sinus venosus, resulting in a reduction in size of the left horn. This leaves a residual right horn. As the left horn becomes smaller, the veins entering it become more and more attenuated until the left common cardinal vein forms the **oblique vein of the left atrium** and the remaining part of the left horn forms the **coronary sinus** (Fig. 3.15F).

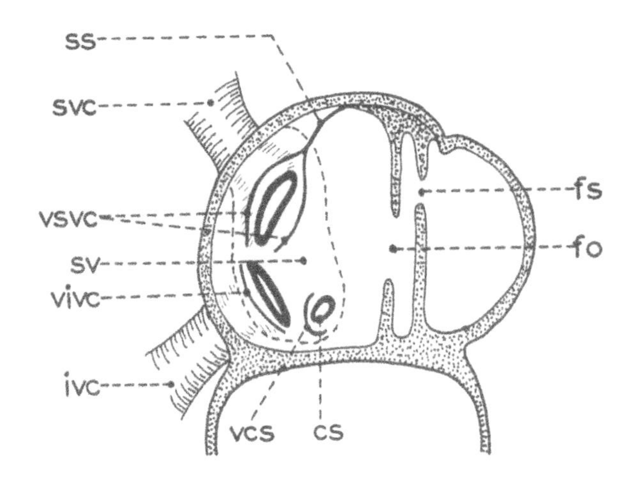

Figure 3.16: Right atrial wall to show the sinus venarum (sv) within interrupted lines.
cs = opening of coronary sinus; fo = foramen ovale; fs = foramen secundum; ivc = inferior vena cava; ss = septum spurium; svc = superior vena cava; vcs = valve of coronary sinus; vivc = valve of inferior vena cava; vsvc = valve of superior vena cava.

As the body of the sinus venosus is absorbed into the right atrium, the entrance of the right common cardinal vein, being at first close to the sino-atrial junction, is drawn into the atrium. The right common cardinal vein thus first appears in the lower right wall of the atrium and is surrounded by two 'valve-like' flaps of endocardium. Gradually, the opening rises up in the wall of the atrium to form the opening of the superior vena cava (Fig. 3.16). The two endocardial 'flaps' join at their cranial ends to form a ridge called the **septum spurium** which extends to the cranial part of the atrium, there to thicken and form the **crista terminalis.** The crista terminalis forms the line of junction between the smooth part (sinus venosus part) and the rough part (definitive atrial part) of the anatomical atrium (Fig. 3.16). The smooth part of the definitive right atrium which is derived from the sinus venosus is called the **sinus venarum**. The primitive right atrium will give rise to the **right auricular appendage.**

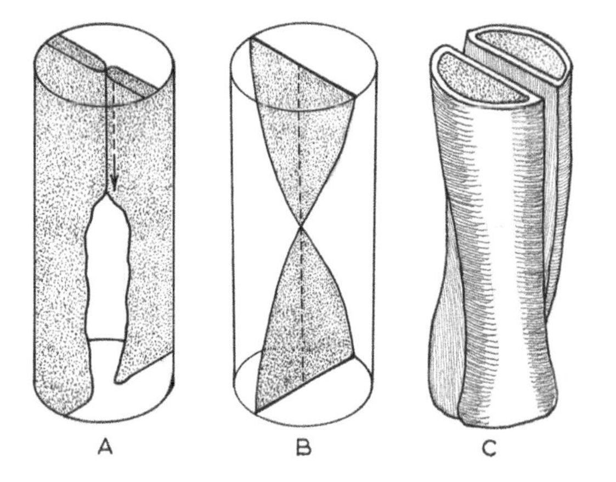

Figure 3.17: Spiral septation of the truncus arteriosus. A: Truncal ridges fusing in a cranio-caudal direction (arrow). **B:** Completed spiral septum (aortico-pulmonary septum). **C:** Division of truncus arteriosus by the spiral septum, producing the pulmonary trunk (ventrally) and the ascending aorta (dorsally).

Expansion of the Left Atrium

Whereas the right atrium has both rough and smooth internal surfaces, the left atrium is smooth throughout with the exception of the left auricular appendage. The left atrium is designed to receive blood from the lungs. Early in development a single pulmonary vein is absorbed into the left atrium thus creating a smooth surface. The single pulmonary vein is formed from four tributaries. With the progressive absorption of this vein into the wall of the left atrium, two openings (right and left) are formed. Further absorption results in the formation of four openings in the posterior wall of the left atrium. These are the definitive **pulmonary veins** (Fig. 3.9).

The left primitive atrium will give rise to the **left auricular appendage.**

Formation of the 'Spiral' Septum of the Truncus Arteriosus

While the septation of the atrium and ventricle is taking place, the truncus arteriosus is simultaneously undergoing its septation. Following partial ventricular septation, it is necessary to provide separate outflow tracts for the two future ventricles. Beginning at the cranial end of the truncus arteriosus, two projections develop **(Fig. 3.17A)**. The projections form a spiral septum **(Fig. 3.17B)**, the **aorticopulmonary septum**, which divides the interior of the truncus arteriosus into separate channels which spiral about each other **(Fig. 3.17C)**. While the spiral is separating the lumen of the truncus arteriosus internally into two separate channels, grooves on the outer surface of the truncus mirror this separation **(Fig. 3.17)**. One channel (pulmonary trunk) will connect to the right ventricle while the other (ascending aorta) connects to the left ventricle **(Fig. 3.18)**. Thus, both ventricles now have individual outflow tracts. When the divided truncus is viewed ventrally at its junction with the ventricles (its root), the channel leaving the left (posterior) ventricle passes posterior to that arising from the right (anterior) ventricle **(Fig. 3.18)**. Thus, the ascending aorta and the pulmonary trunk are spiralled around each other in the adult **(Fig. 3.19)**.

The outflow tracts of the heart are heavily populated by neural crest cells which are responsible for the production of elastic fibres and smooth muscle in the ascending aorta and pulmonary trunk. These fibres provide the function of elastic recoil for these two vessels.

With the completion of the truncal septation, the pars membranacea septi of the interventricular septum may now be completed.

POSSIBLE COMMON ABNORMALITIES

- A failure of the truncal septum to form will result in a **truncus communis** (common trunk).
- Unequal separation, leads to unequal size of the channels of the ascending aorta and pulmonary trunk (eg. pulmonary **stenosis** and enlarged ascending aorta or aortic stenosus and enlarged pulmonary trunk).
- Inverted direction of the spiral leads to the major arteries (ascending aorta and pulmonary trunk) being reversed with respect to their origin from the ventricles. This is called **transposition of the great vessels** of the heart. (Incompatible with life.)
- **Fallot's Tetralogy** is a combination of defects arising primarily from abnormal spiral septation, and consists of additional defects such as pulmonary stenosis, an overiding aorta, a ventricular septal defect and right ventricular hypertrophy. (***Note.*** An overiding aorta is one which receives a contribution of blood from the right ventricle. Normally the ascending aorta would receive blood only from the left ventricle. Overiding occurs because the root of the ascending aorta is wider than the root of the pulmonary trunk and partially 'overlaps' the right side of the heart.)

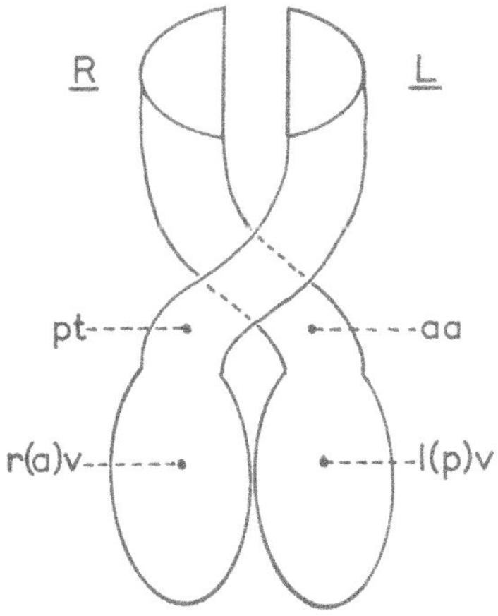

R
L
pt
aa
r(a)v
l(p)v

Figure 3.18: Consequences of the spiral of the septum of the truncus arteriosus. The pulmonary trunk (pt) forms the outflow tract from the right (anterior) ventricle (r(a)v) and the ascending aorta (aa) forms the outflow tract of the left (posterior) ventricle (l(p)v).

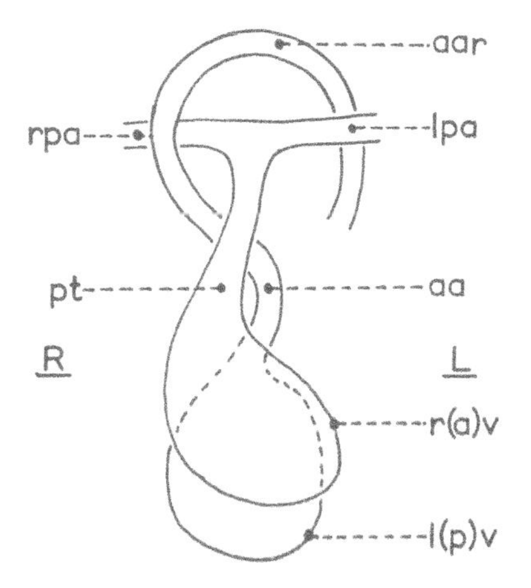

Figure 3.19: The outflow tracts in the adult. aa = ascending aorta; aar = aortic arch; lpa = left pulmonary artery; l(p)v = left ventricle; pt = pulmonary trunk; r(a)v = right ventricle; rpa = right pulmonary artery.

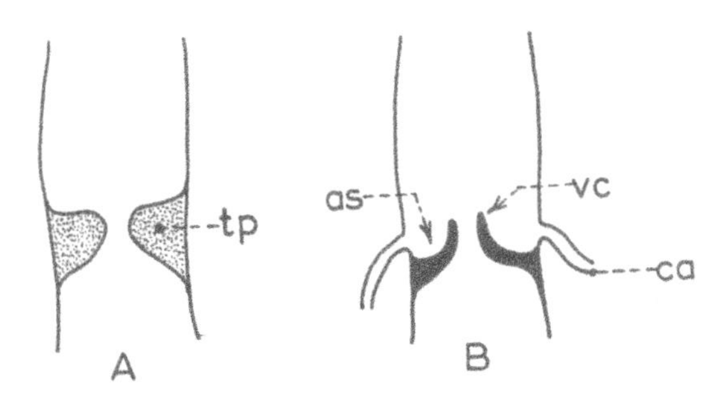

Figure 3.20: A: The position of the truncal prominences (tp). **B:** The valve cusps (vc), aortic sinuses (as) and coronary arteries (ca) emerging from the sinuses.

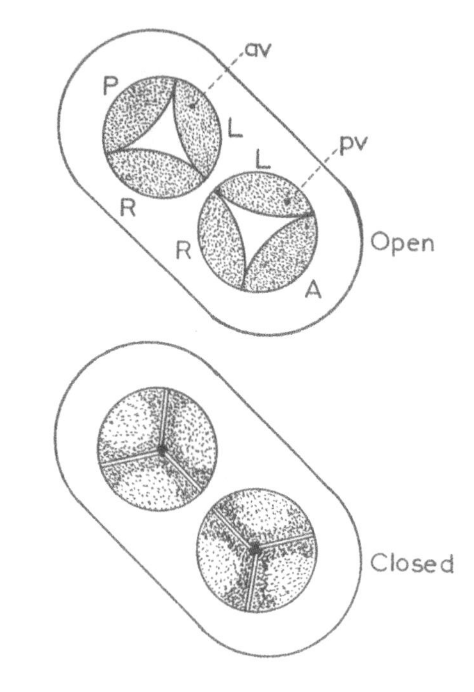

Figure 3.21: Relative positions of aortic valve cusps (av) and pulmonary valve cusps (pv) in the open and closed position in the adult.
A= anterior, P = posterior, R = right, L = left.

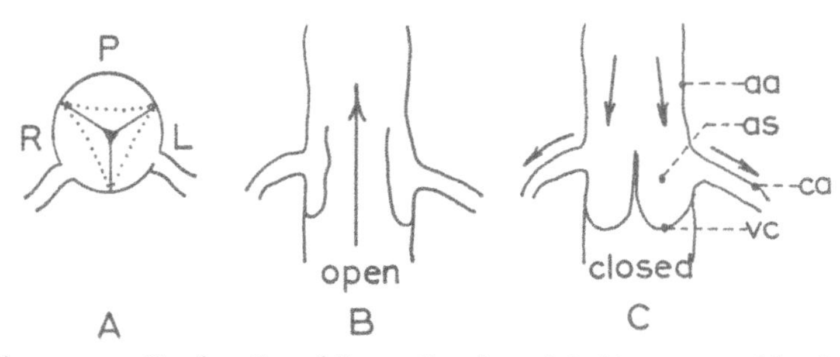

Figure 3.22: The function of the aortic valve related to coronary blood flow. A: Plan-view of cusps and coronary arteries. P = posterior. **B:** Elevation-view of open cusps with outgoing flow. **C:** Reversal of flow closes valve cusps (vc) so that blood flows into coronary arteries (ca) from the ascending aorta (aa) and right and left aortic sinuses (as).

Formation of the Semilunar Valves

The single outflow tract (truncus arteriosus) has now been divided into two (ascending aorta and pulmonary trunk) by the growth of the spiral aorticopulmonary septum (**Fig. 3.17**). Swellings of subendothelial tissue occur on each side of the fused ridges and a similar third swelling occurs in the periphery of each tract (**Fig 3.20**). These swellings are 'hollowed-out' on their superior surfaces to form the three semilunar cusps in each tube (**Fig. 3.20**). Thus three valve cusps are formed in each tract (**Fig. 3.21**).

When closing, the crescentic edges of the three cusps move towards the centre of the channel (**Figs. 3.21, 3.22**), thus closing the channel completely.

Septation of the Ventricle

Septation of the ventricle takes place by means of a single septum consisting of two parts (pars muscularis septi and pars membranacea septi).

While the septation of the atrium is taking place, a septum grows in a cranial direction from the base of the ventricle towards the intermediate bar. This is the muscular part (**pars muscularis septi**) of the **interventricular septum** (**Fig. 3.23A**). The growth of the pars muscularis stops short of the intermediate bar leaving a gap between its upper end and the bar - this is the **interventricular foramen**. Such a gap is necessary because the left ventricle 'to be', does not yet possess an outflow tract of its own and any blood entering this chamber must necessarily escape through the interventricular foramen (**Fig. 3.24**). The remaining part of the interventricular foramen will be closed by the formation of the **pars membranacea septi.** The pars membranacea septi has a somewhat complicated development and is formed by a downgrowth of the intermediate bar with contributions from the anterior and posterior truncal ridges, which are dividing the common outflow tract into ascending aorta and pulmonary trunk (**Fig. 3.23B**). Some embryologists maintain that there is a fourth contribution to the pars membranacea which arises from a further upgrowth of the pars muscularis (**Fig. 3.23C**). Neural crest cells migrate into the truncal ridges and are, therefore, involved in the formation of the outflow tract of the heart and also of the interventricular septum. The pars muscularis septi and the pars membranacea septi together form the **definitive interventricular septum**. The septum creates right and left ventricles. In the

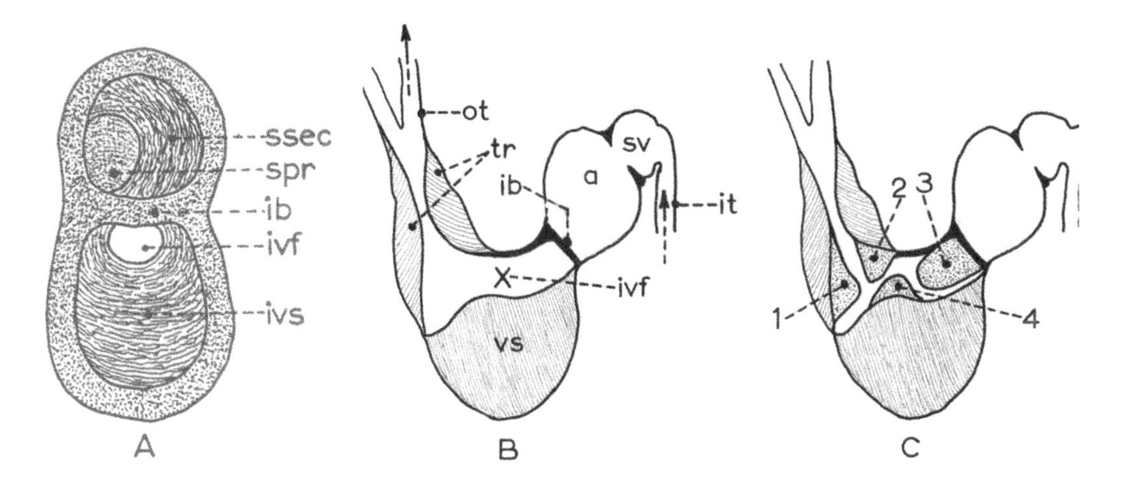

Figure 3.23: **A:** The two septa in the atrium and the interventricular foramen and pars muscularis septi of the developing interventricular septum. **B** and **C:** The closure of the interventricular foramen, after separation of the truncus arteriosus by the spiral septum (sps). a = atrium; ib = intermediate bar; it = inflow tract; ivf = interventricular foramen; ivs = interventricular septum (pars muscularis septi); ot = outflow tract; spr = septum primum; ssec = septum secundum; sv = sinus venosus; tr = truncal ridges; vs = ventricular septum; X = ivf (interventricular foramen); 1 + 2 = extensions of truncal ridges; 3 = extension of intermediate bar; 4 = possible further extension of pars muscularis septi. The closure of the interventricular foramen by the intermediate bar (3) with assistance from the truncal ridges (1 and 2) and the pars muscularis septi (4) is shown.

adult, the wall of the left ventricle is much more robust than that of the right ventricle. The reason for this is that the driving force required to pump blood through the systemic circuit is greater than that required for the pulmonary circuit.

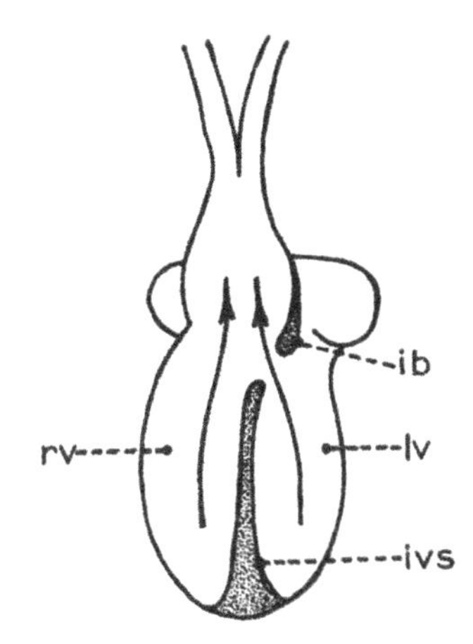

Figure 3.24: Before truncal separation, the left (dorsal/posterior) ventricle (lv) must empty through a patent interventricular foramen, between the intermediate bar (ib) and the interventricular septum (pars muscularis septi) (ivs). rv = right ventricle.

POSSIBLE COMMON ABNORMALITIES

- Complete failure of the formation of the septum of the ventricle results in a **cor triloculare biatriatum,** which is a single ventricle with two atria.
- A **persistent interventricular septal defect** may occur either in the pars muscularis septi (rare) or in the pars membranacea septi (more common, because of its complex development). The latter is compatible with life.

Formation of the Atrioventricular Valves

During its formation, the myocardium develops intramuscular excavations on its deep surface which result in the formation of the **trabeculae carneae** and **papillary muscles**. These excavations extend upwards along the ventricular walls to reach the overhanging dorsal and ventral prominences of the atrioventricular canals, hollowing them out to form valve flaps (Fig. 3.12A,B). In the right atrioventricular canal, three valvular cusps are formed (**tricuspid valve**) and two are formed in the left atrioventricular canal (**bicuspid** or **mitral** valve). At the present time there seems to be no satisfactory explanation for the asymmetry in the number of these valve cusps.

Strands of myocardial tissue remain between the valve flaps and the myocardial wall, resulting in connections between the two. These connections are converted to **chordae tendineae** which attach the edges of the valve cusps to the papillary muscles of the myocardium (Fig. 3.12). These tendons are necessary to prevent the valve flaps from turning 'inside out' when the ventricular pressure exceeds that in the atria.

Formation of the Coronary Arteries and Veins

The coronary arterial supply of the developing heart is formed from a network of vessels arising from blood islands in the epicardium soon after the heart has 'dipped' into the pericardial coelom. These blood vessels penetrate the wall of the heart tube at an early stage, while it is undergoing its initial contractions.

There is some dispute about the origin of the main coronary arteries; one suggestion is that preferential channels spring from the ascending aorta and make connection with the existing network, while another idea is that preferential channels form in the epicardial vascular network and then penetrate the ascending aorta.

Whatever the case, the coronary arteries normally arise anatomically from the ascending aorta above the semilunar valve cusps. Since the heart works constantly, the cardiac muscle requires a steady supply of blood. This is provided by the vast **arteriolar** anastomotic network arising from the coronary arteries. However, there is little effective anastomosis between the larger vessels (coronary arteries) which have been dubbed as 'end-arteries'. This situation has important implications in disease of the coronary arteries.

The fact that the main coronary arteries arise from the ascending aorta in the sinuses created by the curvature of the aortic valve cusps has important physiological implications. It should be obvious that when a muscle contracts (systole) (and this includes the myocardium), the contained blood vessels are compressed, preventing blood flow in them during the contraction period. Clearly, blood flow should take place when the vessels are open and this is during muscular relaxation (diastole). The exit of the coronary arteries from the sinuses of the ascending aorta provides this situation. During relaxation of the myocardium, the blood from the aorta tends to flow back into the heart due to the elastic recoil of the aortic and pulmonary trunks, but this is prevented by closure of the semilunar cusps. Because the coronary arteries arise above the cusps, they are subjected to the full force of the aortic blood pressure and because the myocardial vessels are not compressed, they carry the blood to all parts of the heart.

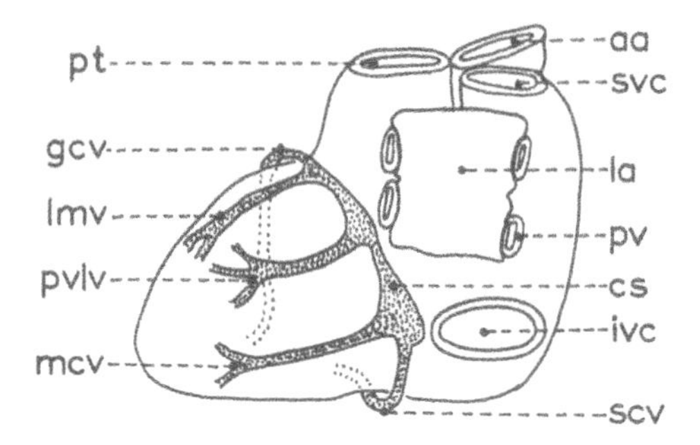

Figure 3.25: Posterior view of the adult heart showing the coronary sinus (cs) and its tributaries. aa = ascending aorta; gcv = great cardiac vein; ivc = inferior vena cava; la = left atrium; lmv = left marginal vein; ; mcv = middle cardiac vein; pt = pulmonary trunk; pv = pulmonary vein; pvlv = posterior vein of the left ventricle; scv = small cardiac vein; svc = superior vena cava.

POSSIBLE ABNORMALITY

In a small proportion of people, one of the coronary arteries may arise from the pulmonary trunk. In this case, the heart muscle is supplied by a proportion of deoxygenated blood. This condition is not totally incompatible with life and is surgically remediable. When both coronary arteries arise from the pulmonary trunk, the condition is not compatible with life.

The venous return from the myocardium is via the cardiac veins which are also derivatives of blood islands of the epicardium. The larger of these veins empty into the coronary sinus (**Fig. 3.25**), which in turn, opens into the right atrium, while the smaller veins (Thebesian) open directly into the heart chambers

Development of the Systemic Arteries

As stated previously (see page 34), the vascular system begins as blood islands in the intra-embryonic mesoderm of the umbilical vesicle and connecting stalk. These give rise to a vascular network in which preferential channels develop. The main arterial channels are the paired dorsal aortae which lie on either side of the embryo between the notochord and the gut tube (**Fig. 3.26A,B**). These run the length of the embryo (**Fig. 3.2**). Between the levels of T4 and L4 (of the adult), the dorsal aortae fuse to form a single channel. The dorsal aortae cranial and caudal to the fusion remain paired. The paired dorsal aortae connect to the ventral **aortic sac** via a series of **pharyngeal**

arch arteries. Caudally, the dorsal aortae give rise to the umbilical arteries (**Fig. 3.2**). Laterally, the dorsal aorta/aortae give rise to a series of arteries known as the intersegmental arteries.

The major embryonic arteries may be broadly classified as:

- pharyngeal arch arteries, between the aortic sac and dorsal aortae;
- trunk arteries between the dorsal aortae and the viscera;
- limb arteries.

Fate of the Pharyngeal Arch Arteries

Before dealing with these in detail, it is necessary to consider the formation of the **aortic sac.** This sac is formed by the dilated distal region of the truncus arteriosus (**Fig. 3.26A**) which lies outside the cranial extremity of the pericardial coelom. With the lengthening of the embryonic neck, the aortic sac becomes drawn out into two 'horns'.

The pharyngeal arch arteries develop in the ectomesenchyme (of neural crest origin) of each pharyngeal arch. The arteries circumnavigate the pharynx (within the pharyngeal arches) to link the aortic sac with the paired dorsal aortae (**Fig. 3.26B**). While six pharyngeal arch arteries arise in the embryo, not all persist throughout embryonic life. The serial appearance of the pharyngeal arch arteries proceeds in a cranio-caudal direction. Remodelling of the arteries into their adult forms begins cranially, so that even as the fourth arch artery is beginning to appear, the first and second arch arteries are being modified into their adult form. This is further evidence for the law of cranio-caudal specialisation in the embryo.

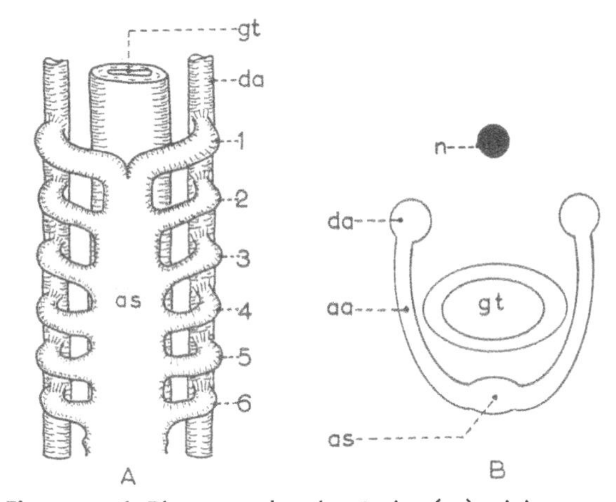

Figure 3.26: Pharyngeal arch arteries (aa) arising from the aortic sac (as) and entering the dorsal aorta (da). A: Ventral view. **B:** Transverse section. gt = foregut tube (pharynx); n = notochord.

The following information should be used in conjunction with Figure 3.27:

(a) First pharyngeal arch artery on both sides: Largely disappears except for a small contribution to the **maxillary artery;**

(b) Second pharyngeal arch artery on both sides: Largely disappears leaving the **stapedial artery;**

(c) Third pharyngeal arch artery on both sides: Persists as the **internal carotid artery** and utilises the pre-existing dorsal aorta on each side to extend into the head. The **common carotid artery** on each side is formed from that part of the bilateral horns of the aortic sac which lie in relation to the third arch artery (**Fig. 3.27B**). The **external carotid artery** forms as a new outgrowth on either side from the root of the internal carotid artery (**Fig. 3.27C**).

(d) Fourth pharyngeal arch artery on the *left side:* Becomes the **distal part** of the **aortic arch;** the proximal part of the adult aortic arch is formed from the left horn of the aortic sac (**Fig. 3.27B**). On the *right side,* the fourth pharyngeal arch artery forms the right **proximal part** of the **subclavian artery;** the distal part of that subclavian artery being formed by the **7th intersegmental artery** (**Fig. 3.27C,D**). On the *left side,* the subclavian artery is formed purely from the **left 7th intersegmental artery** which becomes continuous with the aortic arch.

Due to the change in the direction of the flow of blood in the third and fourth pharyngeal arch arteries, the dorsal aorta on both sides between the third and fourth arch arteries degenerates (**Fig. 3.27B**). The **brachiocephalic artery** is formed from the lower part of the right horn of the aortic sac (**Fig. 3.27D**).

By this stage the truncus arteriosus has been separated into two and the aortic arch is directly continuous with the ascending aorta. The aortic arch leads into the dorsal aorta on the left side, which now becomes the **descending aorta.**

(e) Fifth pharyngeal arch artery on both sides: Disappears completely;

(f) Sixth pharyngeal arch artery: The aorticopulmonary septum has divided the truncus arteriosus into two channels (the ascending aorta and the pulmonary trunk) and the sixth arterial arches on both sides become continuous with the pulmonary trunk from which further branches are diverted into the lungs as the **pulmonary arteries.**

The more dorsal part of the left sixth pharyngeal arch artery remains connected to the dorsal aorta.

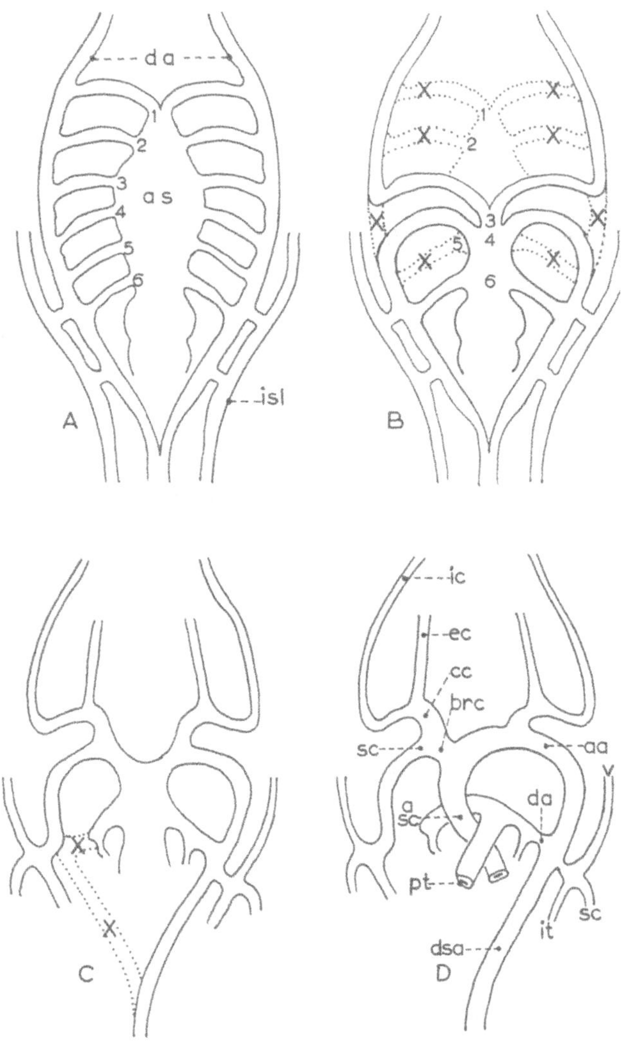

Figure 3.27: Development of the pharyngeal arch arteries.
A: Indicates the six (1-6) arterial arches between the aortic sac (as) and the dorsal aortae (da) and secondarily, to the outer longitudinal system, the intersegmental line (isl). **B:** Indicates in 'x's' and interrupted lines those channels that disappear. **C:** Part of the right 6th arch and the entire right dorsal aorta caudal to the 6th arch, disappear. **D:** Adult arrangement. The ascending aorta (asc) and pulmonary trunk (pt) are derived from the truncus arteriosus and are continuous with the 4th and 6th arches respectively. aa = aortic arch; brc = brachiocephalic artery; cc = common carotid arteries; da = ductus arteriosus between left pulmonary artery and descending aorta; dsa = descending aorta; ec = external carotid; ic = internal carotid arteries; it = internal thoracic artery; sc = subclavian artery; v = vertebral artery.

This connection is called the **ductus arteriosus** and is responsible for diverting blood from the lungs to the dorsal aorta in the embryo and fetus. The reason for the presence of this 'short-circuit' is that the non-functional lungs present a considerable resistance to the flow of blood through them. At birth, with the expansion of the lungs, the reduction of pulmonary arterial pressure makes the presence of the ductus arteriosus no longer necessary and it begins to close. In the adult a fibrous structure remains as the remnant of this duct and is known as the **ligamentum arteriosum.**

On the left side, the single dorsal aorta persists between the ductus arteriosus and the confluence of the dorsal aortae to form the **descending aorta (Fig. 3.27).** On the right side, the single dorsal aorta between the 6th pharyngeal arch artery and the fused part of the dorsal aorta degenerates due to flow of the blood to the lungs **(Fig 3.27; X-X).**

Note. In the adult, the positions of the recurrent laryngeal nerves which carry fibres of the cranial part of the accessory nerves, are related to the structural changes that take place in the pharyngeal arch arteries. Thus, on the right side, the recurrent laryngeal nerve hooks around the right subclavian artery, while on the left side the recurrent laryngeal nerve hooks around the aortic arch, distal to the ligamentum arteriosum **(Fig. 3.28).**

COMMON ABNORMALITIES

- The two subclavian and common carotid arteries may arise independently from the aortic arch.
- The two common carotid arteries may arise from a common trunk **(bicarotid).**
- A left brachiocephalic artery may arise from the aortic arch.
- The right subclavian and right and left common carotid arteries may arise by a common trunk.
- The aortic arch may be right sided **(dextro-aorta)** or double. This results from the persistance of the right sigle dorsal aorta below the level of the fourth pharyngeal arch artery.
- Additional arteries may arise from the aortic arch, these being the left vertebral and less commonly, the thyroidea ima artery.
- Stenosis (narrowing) of the thoracic aorta may occur in relation to closure of the ductus arteriosus. There are two types, known as pre- or postductal **coarctation of the aorta.**

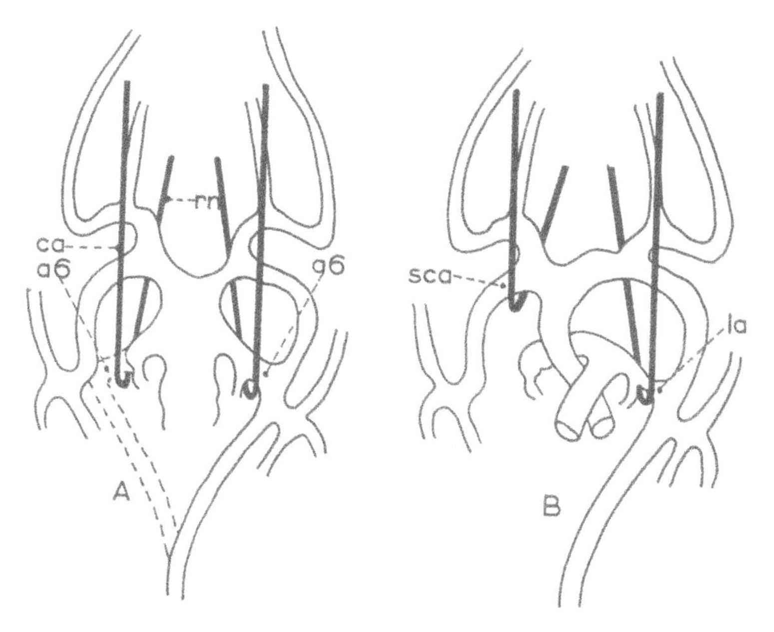

Figure 3.28: Early (A) and adult (B) positions of the recurrent laryngeal nerves. a6 = sixth pharyngeal arch artery; ca = cranial accessory nerve (recurrent laryngeal nerve); la = ligamentum arteriosum; rn = recurrent nerve; sca = subclavian artery.

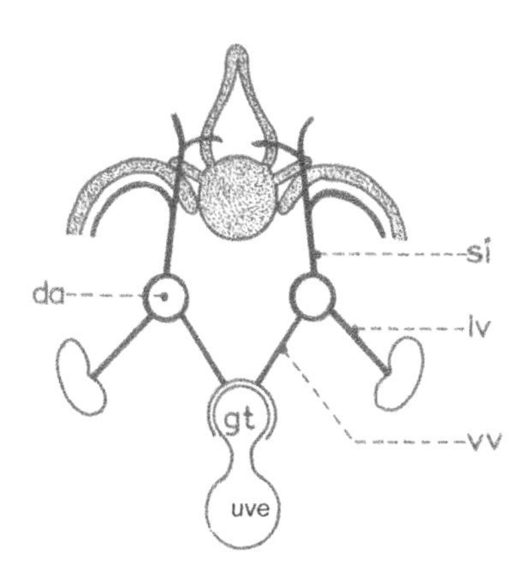

Figure 3.29: Branches of the dorsal aorta. gt = gut tube; si = somatic intersegmental; lv = lateral visceral; vv = ventral visceral; uve = umbilical vesicle.

Fate of the Branches of the Dorsal Aorta

At first the dorsal aortae are paired throughout the length of the embryo (Fig. 3.29), and give bilateral branches to the various structures of the embryo. Later, the aortae fuse to form a single channel from about the adult level of T4 to the level of L4 and the fused part then gives preferential channels to the embryonic organs. These branches (Fig. 3.29) may be classified as follows:

- **Somatic (intersegmental)**: Intercostal and lumbar arteries;
- **Ventral visceral**: Coeliac, superior mesenteric, inferior mesenteric, bronchial, oesophageal;
- **Lateral visceral**: Inferior phrenic, renal, suprarenal, gonadal.

Special features of some of these branches: The superior mesenteric artery (fused vitelline arteries) is the branch around which the midgut loop rotates. The median sacral artery is a branch from the lower junction of the dorsal aortae and extends to the coccyx. The umbilical arteries are the continuation of the two dorsal aortae to the placenta and as they pass the superior part of the urinary bladder, they supply it via the superior vesical arteries. When the umbilical cord is severed, the parts of the umbilical arteries beyond the bladder become obliterated forming the two **obliterated umbilical arteries.** In the adult, the obliterated umbilical arteries raise the medial umbilical peritoneal folds which contain the **medial umbilical ligaments** (fibrotic umbilical arteries).

Development of Limb Arteries

In general, the definitive arterial supply of the limbs is preceeded by the formation of a capillary plexus in which an axial artery is formed. Branches from this artery form the final supply to the limb.

Upper Limb

When the upper limb emerges from the embryonic trunk, it is supplied by a plexus of capillaries. As the limb enlarges, a preferential channel develops in the capillary network and ultimately gives rise to the **brachial, axillary** and **subclavian** arteries (Fig. 3.30A,B). The subclavian artery reaches the brachiocephalic artery on the right side and the aortic arch on the left side. In the region where the elbow joint is to develop, the axial artery gives off the **anterior interosseous artery**, which travels along the developing forearm to reach the hand where it forms part of the **deep palmar arch.** Within the forearm, the axial artery gives a **posterior interosseous artery** and a

median artery, the latter running with the median nerve to reach the hand. The **radial** and **ulnar arteries** are the final branches to form from the brachial part of the axial artery and these arise in the region which is to become the elbow (Fig. 3.30C). The radial and ulnar arteries enlarge to form the main arteries of the forearm and the hand, leaving the anterior, posterior and median vessels as relatively small channels. The ulnar artery supplies to a large extent, the structures in the palm of the hand via the **superficial palmar arterial arch** while the radial artery supplies the posterior aspect of the forearm and hand, as well as taking part in the formation of the **deep palmar arterial arch** (Fig. 3.30C).

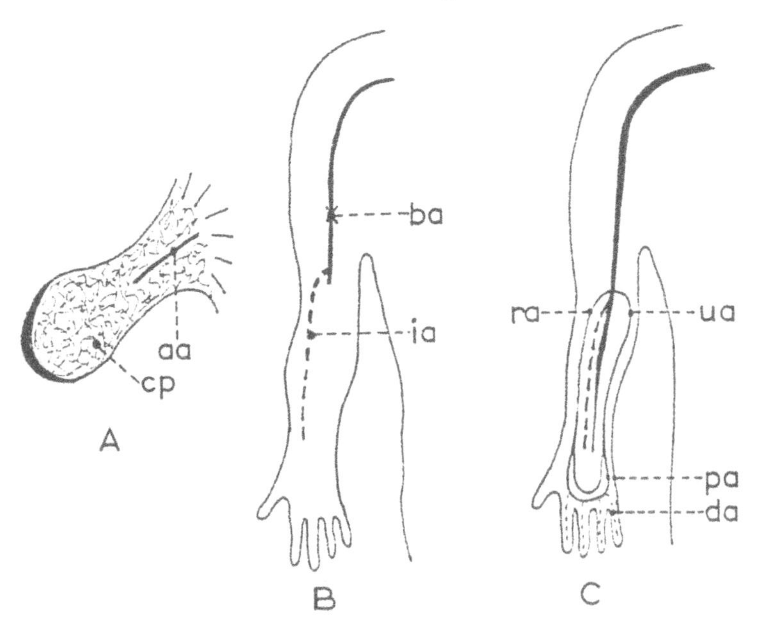

Figure 3.30: Arterial development in the upper and lower limbs.
A: Limb bud showing a capillary plexus with early formation of the axial artery (aa). **B:** Developed upper limb with brachial artery (ba) and interosseous artery (ia). **C:** Brachial artery dividing into the radial artery (ra) and the ulnar artery (ua) which form the palmar arches (pa) and digital arteries (da).

Lower Limb

Recall that the dorsal aortae fuse to form a single vessel at the adult level of vertebra L4, the median sacral artery being the caudal persistence of the fusion. The unfused parts of the dorsal aortae give rise to the **umbilical arteries.** The **common iliac arteries**, the **internal iliac arteries** and **external iliac arteries** arise from the intersegmental arterial system. The axial artery of the lower limb arises from the internal iliac artery and courses deeply along the posterior aspect of the thigh to the popliteal fossa and thence into the leg and foot where it forms a capillary plexus (Fig. 3.31). This axial artery is the **sciatic artery** (arteria ischiadica).

Since flexion of the hip joint would place a dorsal artery at a risk of obstruction, it becomes necessary to replace it with an anterior channel which would maintain its lumen during flexion of the hip. The replacement artery is the **femoral artery** which lies on the ventral surface of the upper part of the limb bud (thigh) and initially is a preferential channel in a capillary network between the external iliac artery and the sciatic artery (Fig. 3.31B). This artery forms the **profunda femoris** artery. In the popliteal fossa, the femoral artery gives two branches, the **posterior tibial** and the **fibular arteries** which course along the developing limb to reach the foot.

In the popliteal fossa, the axial artery joins the fibular artery which, by now, has given off the **anterior tibial branch**. The three arteries of the leg have now been established (anterior tibial, posterior tibial, fibular) and the axial artery, having completed its task, partially disappears (Fig. 3.31C). The upper part of the sciatic artery is retained as the **inferior gluteal artery** and its lower part retains an anastomosis with the perforating branches of the **profunda femoris** artery. The inferior gluteal artery forms part of the **cruciate anastomosis** of the buttock.

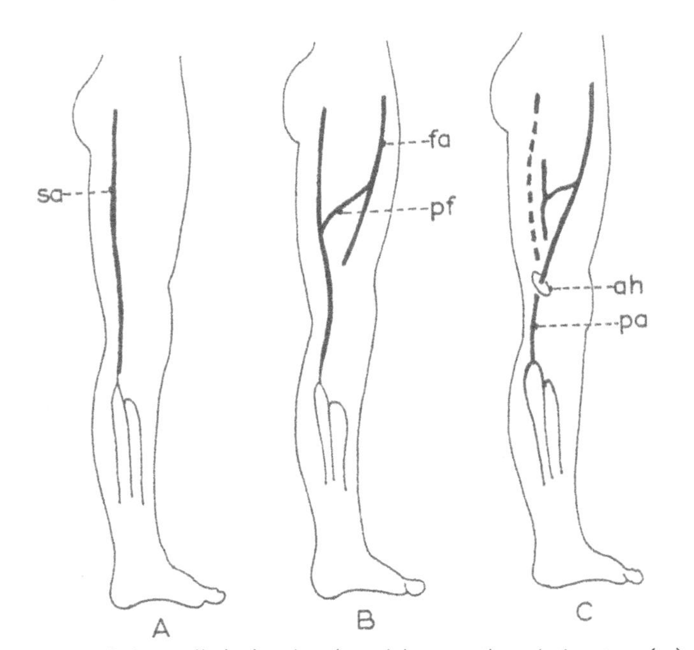

Figure 3.31: A: Lower limb showing the axial artery, the sciatic artery (sa). **B:** Formation of the femoral artery (fa) joining the sciatic artery via the profunda femoris artery (pf). **C:** Disappearance of the sciatic artery leaving the profunda femoris artery supplying the posterior part of the thigh.
ah = adductor hiatus; pa = popliteal artery.

Development of Veins

Many studies of the development of the intra-embryonic venous system have been made and their developmental complexity has resulted in a degree of confusion. In the adult, the systemic venous system is grossly asymmetrical but in the early embryo, the vascular network and its preferential channels are symmetrical on the two sides. Clearly, in the establishment of asymmetry, some channels will persist while others will disappear to achieve this. Thus, it is here proposed to treat the development separately on the two sides and in the most elementary way possible.

The veins of the developing embryo may be classified as follows:

- **Extra-embryonic**: Vitelline and umbilical veins;
- **Intra-embryonic**: Cardinal, subcardinal, supracardinal systems.

Since the heart is the central site for the venous return, the vascular networks may be divided into precardiac and postcardiac groups. The precardiac group is related to the head, neck, upper limbs and thorax and is drained by the **anterior cardinal veins.** The postcardiac network is related to the liver, mesonephroi, metanephroi and lower limbs and is drained by the **posterior cardinal**, **subcardinal** and **supracardinal** systems. The vitelline and umbilical veins are dealt with later.

The development of the trunk veins may be simplified by viewing their development in stages. The accompanying diagrams (Fig. 3.32A-G) attempt to illustrate the stages:

Stage 1: The blood in the cranial part of the embryo (cranial to the heart) is drained by the anterior cardinal veins (right and left); the blood in the caudal part of the embryo (caudal to the heart) is drained by the posterior cardinal veins (Fig. 3.32A).

Stage 2: The anterior and posterior cardinal veins join to form the common cardinal veins (ducts of Cuvier) which enter the sinus venosus (Fig. 3.32B).

Stage 3: Now that there are two parallel venous channels along the length of the embryo, they are joined by 'cross-anastomotic' channels. The cranial channel will become the **left brachiocephalic vein**; the middle channels will form the **azygos channels** and the caudal channel will give rise to the **left common iliac vein** (Fig. 3.32C).

Stage 4: The formation of cross-channels is followed by the formation of **supracardinal veins** which drain the dorsal body wall of the embryo and which join the azygos channels to the common cardinal veins. The venous system of the embryonic trunk is still symmetrical (Fig. 3.32D).

Stage 5: With the development of the suprarenal glands (5 weeks pf), gonads (6 weeks pf), mesonephroi (6 weeks pf) and metanephroi (7 weeks pf), the **subcardinal plexus** of veins appears, to provide a venous drainage for these organs. This plexus forms an extensive network between these organs as well as with the gradually disappearing posterior cardinal veins (Fig. 3.32E).

Stage 6: Although the diagram (Fig. 3.32F) indicates the disappearance of specific parts of the embryonic venous system, this is not a sudden process. It takes place in a gradual way as various organs develop and degenerate (see mesonephros). The diagram indicates those parts of the venous system which disappear or are modified.

Stage 7: The disappearance of specific parts of the venous system leaves the 'average'adult pattern of the venous drainage (Fig. 3.32G).

It should be emphasised that there are numerous variations in the pattern of venous drainage in all parts of the body and this is not surprising in the light of the 'comings and goings' of numerous intermediate channels. One interesting anomaly, which is not infrequently found, is the presence of a **renal venous collar** around the abdominal aorta. This anomaly would be of interest to the vascular surgeon dealing with an aortic aneurysm.

Vitelline and Umbilical Veins

The vitelline veins develop in the mesoderm covering the umbilical vesicle and enter the embryo by passing along the vitello-intestinal duct to reach the sinus venosus of the embryonic heart (Fig. 3.33). The two umbilical veins emerge from the placenta and enter the embryo via the connecting stalk (later, part of the umbilical cord) to enter the sinus venosus (Fig. 3.33). Both these sets of veins pass through the **septum transversum** to reach the sinus. Since the liver is developing in the lower part of the septum transversum, branches from the vitelline and umbilical veins form the blood sinuses of the liver.

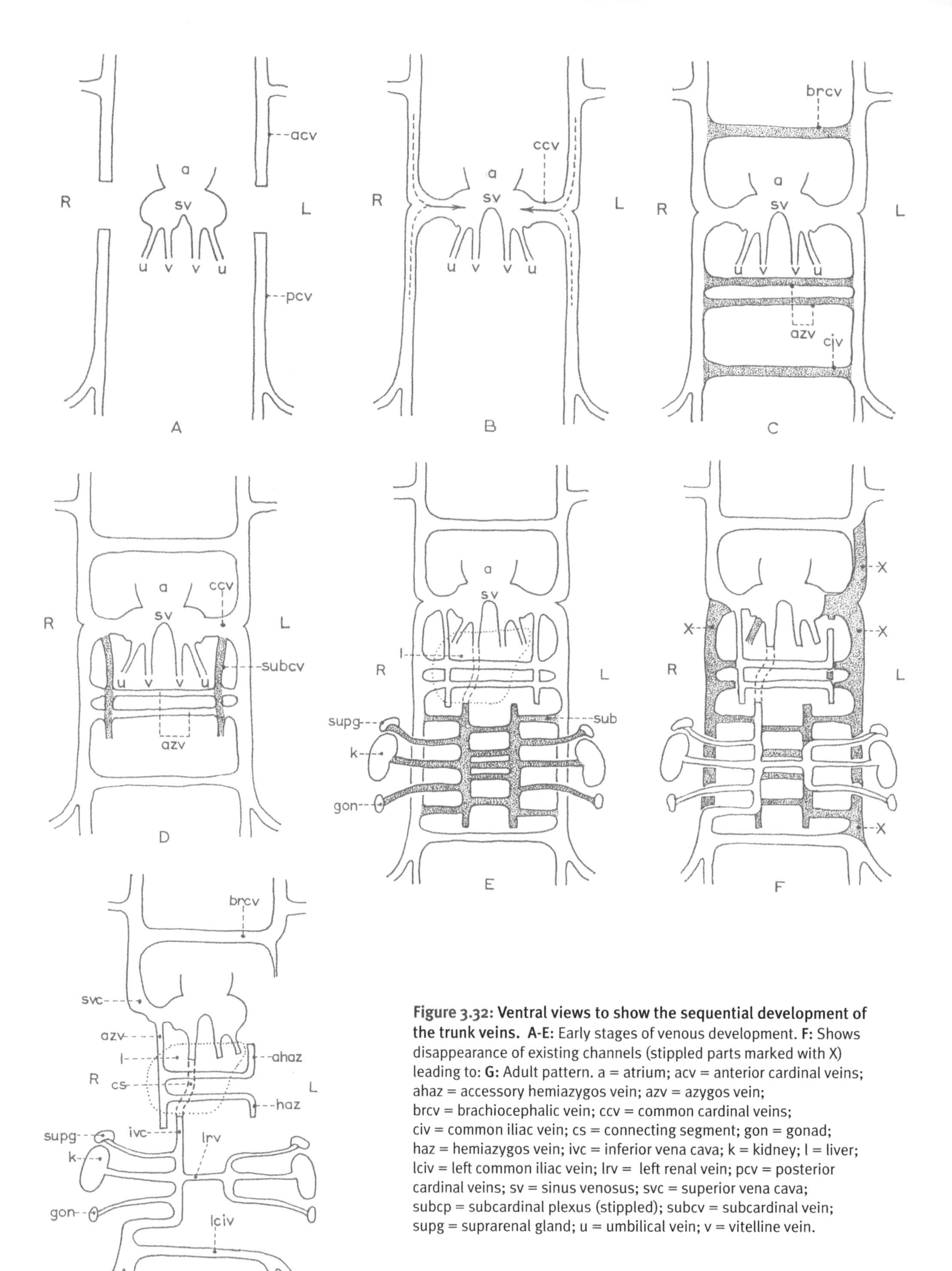

Figure 3.32: Ventral views to show the sequential development of the trunk veins. A-E: Early stages of venous development. **F:** Shows disappearance of existing channels (stippled parts marked with X) leading to: **G:** Adult pattern. a = atrium; acv = anterior cardinal veins; ahaz = accessory hemiazygos vein; azv = azygos vein; brcv = brachiocephalic vein; ccv = common cardinal veins; civ = common iliac vein; cs = connecting segment; gon = gonad; haz = hemiazygos vein; ivc = inferior vena cava; k = kidney; l = liver; lciv = left common iliac vein; lrv = left renal vein; pcv = posterior cardinal veins; sv = sinus venosus; svc = superior vena cava; subcp = subcardinal plexus (stippled); subcv = subcardinal vein; supg = suprarenal gland; u = umbilical vein; v = vitelline vein.

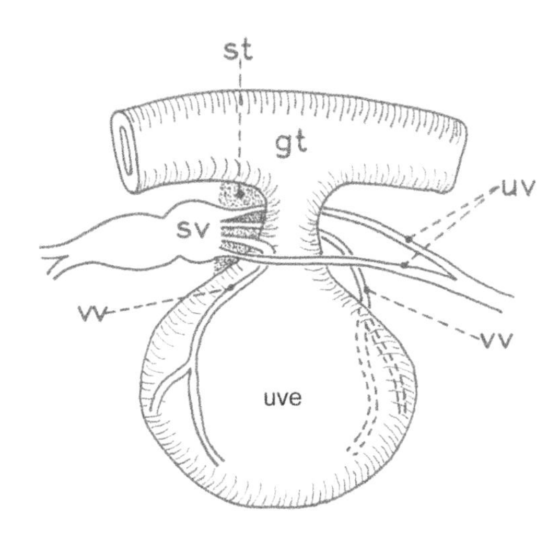

Figure 3.33: Lateral view of the vitelline veins (vv) and the umbilical veins (uv) passing through the septum transversum (st) to reach the sinus venosus (sv). gt = gut tube; uve = umbilical vesicle.

Formation of the Portal Vein

In the adult, the liver has a two-fold vascular supply, the hepatic artery and the portal vein. The portal vein drains the entire gastro-intestinal tract, and its purpose is to convey absorbed nutrients to the liver for metabolism. In the embryo, the vitelline veins branch in the septum transversum forming sinuses between the cords of liver cells (Fig. 3.34). The hepatic artery, which also supplies the developing liver, will give off branches which also open into the liver sinuses. At about this time, the vitelline veins form anastomotic 'cross-bars' around the gut tube, which lies between the two vitelline veins (Fig. 3.34C). When viewed from the ventral aspect (Fig. 3.35B,C), the stippled components of the vitelline veins and the cross-bars are the components which persist to form the portal vein, while the non-stippled parts will disappear. The portion of the vitelline vein on the right will persist, carrying venous blood into the liver (Fig. 3.35D).

Note too, that the right umbilical vein and the left vitelline vein have also disappeared.

In the embryo, when viewed from the ventral aspect, the venous channels loop around the gut tube (Fig. 3.35C), whereas in the fetus, the gut tube loops around the venous channel which follows a more direct route. This positional alteration is due to the rotation of the duodenal loop (Fig. 3.35D).

With the development of the liver, a massive capillary network forms in the septum transversum and this presents a considerable resistance to the

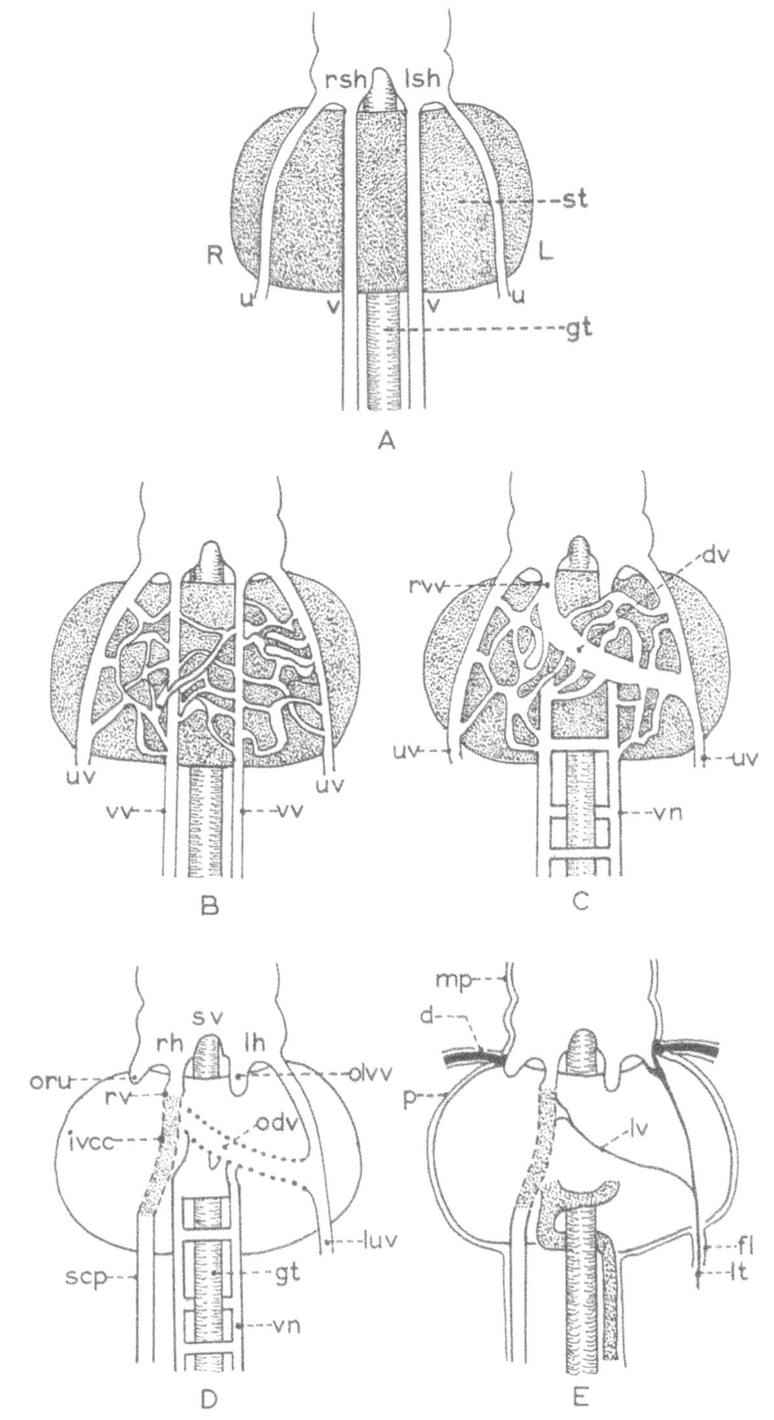

Figure 3.34: Ventral views to show the development of the ductus venosus, inferior vena cava and ligaments of the liver. d = diaphragm; dv = ductus venosus; fl = falciform ligament; gt = gut tube; ivcc = inferior vena caval connection; lh & lsh = left horn of sinus venosus; lt = ligamentum teres; luv = distal part of left umbilical vein; lv = ligamentum venosum; mp = mediastinal pleura; odv = obliterated ductus venosus, the original position of which is shown in interrupted lines; olvv = obliterated left vitelline vein; oru = obliterated right umbilical vein; p = peritoneum; rh & rsh = right horn of sinus venosus; rvv = proximal stump of right vitelline vein; scp = subcardinal plexus; st = septum transversum; sv = sinus venosus; u & uv = umbilical vein/s; v & vv = vitelline vein/s; vn = vitelline venous network.

flow of blood (cf. lungs). To overcome this, a shunt (short-circuit) develops between the left umbilical vein (now the only flow from the placenta) and the right vitelline vein (Fig. 3.34C), which will ultimately become the cranial part of the inferior vena cava. This important shunt is created by the opening up of a large channel (the **ductus venosus**), which diverts most of the oxygenated blood returning via the left umbilical vein directly to the proximal stump of the right vitelline vein. In this way a large proportion of the inflow to the heart is diverted away from the 'vascular resistance' (sinusoids) of the liver.

Note. At birth, with the severance of the umbilical cord, the pressure and blood flow in the left umbilical vein is drastically reduced (Fig. 3.34D,E). This results in collapse of the ductus venosus which undergoes fibrotic change to become the **ligamentum venosum** (Fig. 3.34E).

At about this time, the subcardinal plexus has formed in the lower abdomen of the embryo to drain all the abdominal organs except the gut tube. A connection develops between the upper end of the subcardinal plexus and the free (proximal) end of the right vitelline vein. This connection completes the formation of the major abdominal venous channel, the **inferior vena cava** (Fig. 3.34D). Thus, the upper part of the inferior vena cava develops from three separate vascular elements:

- the terminal part of the right vitelline vein,
- the right cranial part of the subcardinal plexus, and
- a connecting part developed between the cranial part of the right vitelline vein (septal or hepatic part) and the subcardinal plexus (Fig. 3.34D).

The student might find the study of Figure 3.34 somewhat complicated, as it is difficult to indicate depth in a two-dimensional diagram. The study of an actual liver together with Figure 3.36, reveals that the inferior vena cava has intra- and extra-hepatic parts which are on the posterior surface of the liver, while the portal vein enters the liver through the porta hepatis which is much anterior to the position of the vena cava.

A further point of interest becomes evident when the covering layer of the coelomic membrane is considered (Fig. 3.34E). On obliteration of the left umbilical vein (after birth), it becomes a fibrous cord, the **ligamentum teres**, but the extra-hepatic part is still covered with peritoneal (coelomic) membrane. This is called the **falciform ligament** (Fig. 3.34E).

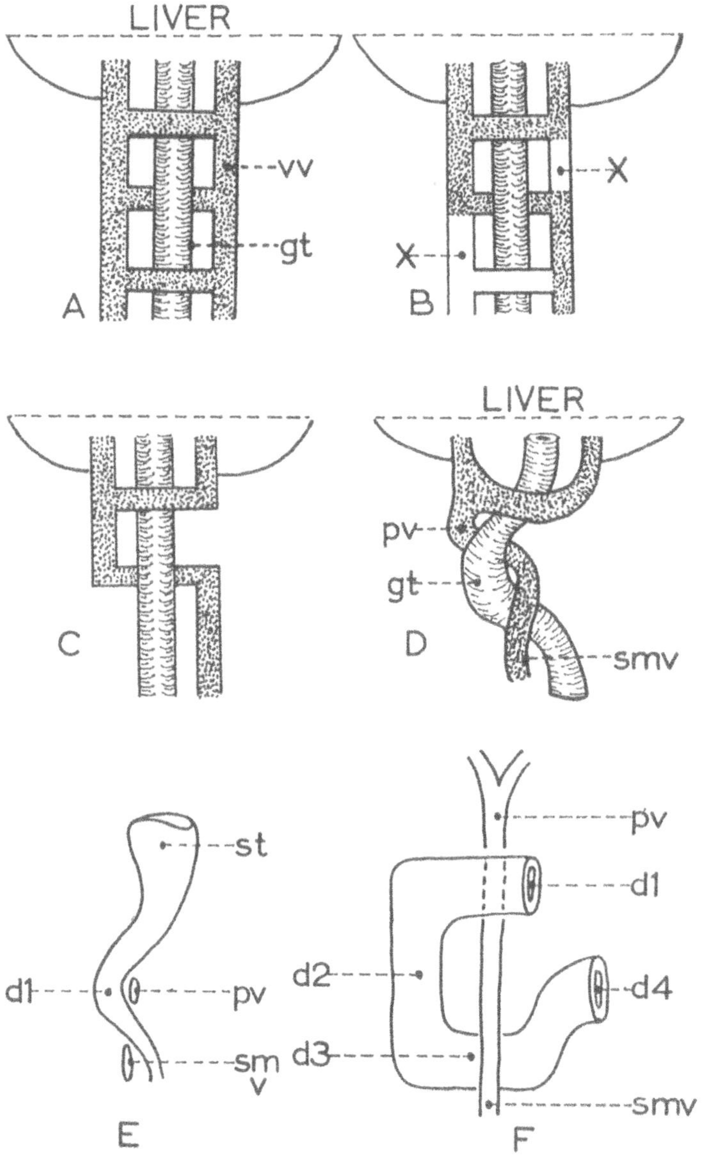

Figure 3.35: A-D: Ventral view of the formation of the portal vein.
Note. The regions marked X disappear leaving the pattern in C.
C: Positional changes result in D. **D and E:** Lateral views. **F:** Adult situation.
gt = gut tube; pv = portal vein; smv = superior mesenteric vein;
st = stomach; vv = vitelline veins; d1-d4 = parts of the duodenum.

Development of the Lymphatic System

Much as is found in the venous system, the early lymphatic system is symmetrical on the two sides of the embryo, and later by selective disappearance of some channels, the system becomes asymmetrical. Although there is some dispute about the

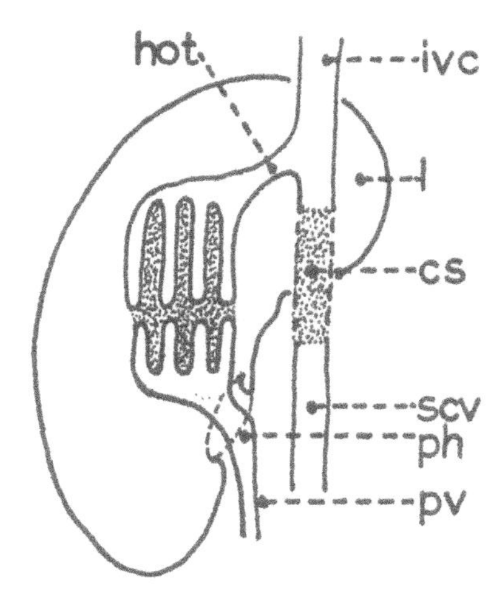

Figure 3.36: Lateral view of the relative positions of the portal vein and the inferior vena cava.
cs = connecting segment; hot = hepatic outflow tract; l = liver; ph = porta hepatis; pv = portal vein; scv = subcardinal vein.

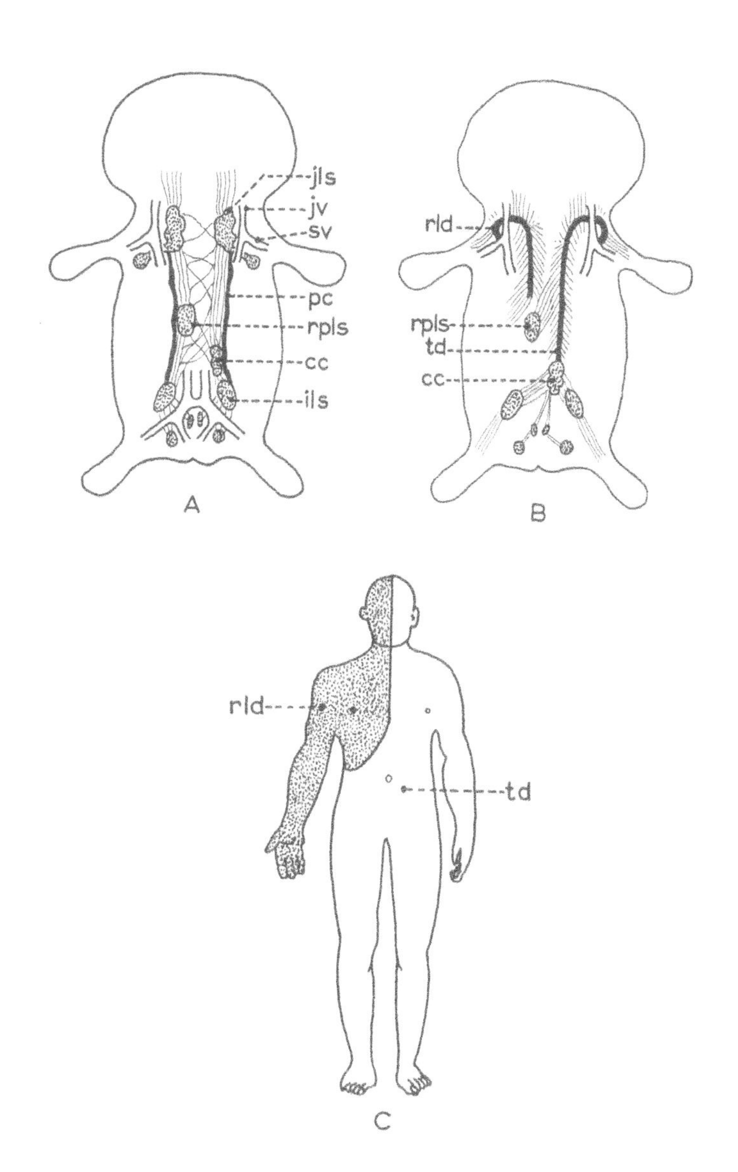

Figure 3.37: Development of the lymphatic system.
A and **B**: ventral views of the embryo. **C**: Anterior view of the adult.
cc = cisterna chyli; ils = iliac lymph sacs; jls = jugular lymph sacs; jv = jugular vein; pc = preferential channels; rld = right lymphatic duct; rpls = retroperitoneal lymph sac; sv = subclavier vein; td = thoracic duct.

origin of the lymphatic system, the generally accepted view is that it arises independently of both the arterial and venous systems.

In the early stages of development, fluid (lymph) collects within spaces between mesenchymal cells. These spaces become lined with endothelial cells to form lymphatic capillaries, resulting in a network lying more or less in the longitudinal centre of the embryo. This network has a main channel running along either side (**Fig. 3.37**). Within this network six lymph sacs arise:

- paired right and left **jugular lymph sacs** in the cervical region, related to the internal and external jugular veins (**Fig. 3.37**);
- paired **posterior lymph sacs**, related to the internal and external iliac veins;
- two abdominal lymph sacs, the **retrocoelomic sac** and the **cisterna chyli** which are related to the renal veins (**Fig. 3.37**).

Later, a preferential channel, the **thoracic duct**, is formed. It is a continuation of the cisterna chyli and lies at first in the midline of the embryo. The duct soon veers to the left, to ascend through the thoracic cavity and ends in the neck by opening into the junction between the left internal jugular and left subclavian veins.

On the right side, the **right lymphatic duct** is formed but it loses contact with the thoracic duct and exists as an independent channel. Its termination is identical to the thoracic duct, but on the right side (**Fig. 3.37**). This arrangement establishes the asymmetry of the system and the lymph drainage regions in the adult are indicated in Figure 3.37C.

All the lymph channels are replete with valves to ensure a unidirectional flow of their contents and there are valves at the terminations of the ducts to prevent the 'back-flow' of blood if the venous pressure should exceed that of the lymph pressure. **Lymph nodes** result from the invasion of the sacs with mesenchymal cells which form fibrous capsules and internal trabeculae. The interstices between the trabeculae become filled with haemocytoblasts which ultimately give rise to lymphocytes. These nodes act as 'filters' for any foreign material such as bacteria or cancer cells in the lymph.

The positions of the lymph sacs determine to a large extent the positions of the major lymph node groups of the adult body, namely, in the root of the neck, the axillae, the upper abdomen and mesentery of the bowel, the pelvis and the inguinal region.

The Conducting System of the Heart

At an early stage in the development of the heart, when the two cardiac tubes are undergoing fusion, the surrounding myoblasts exhibit a form of arhythmic (fibrillar) movement. As development proceeds, the movement becomes more regular and pulsatile so that the contained blood tends to move from the input (caudal) end to the output (cranial) end of the cardiac tube.

With the increasing complexity of the developing heart, this action would require a co-ordinating mechanism. The earliest evidence of this is a thickening on the dorsal wall of the atrio-ventricular junction. This is the **atrio-ventricular node** (Fig. 3.38A). The cells of the node, lying in the subendocardial plane of the heart extend along the dorsal wall of the ventricle as the **atrioventricular bundle** (of His), until they reach the cranial surface of the muscular interventricular septum (Fig. 3.38B,C).

At the cranial end of the septum, the bundle splits into **right** and **left bundle branches**, one supplying the right ventricle and the other the left ventricle. Much later, another node appears as a small knot of tissue situated at the junction of the superior vena cava and the atrial wall. This is the **sinu-atrial node**. With further development it comes to lie at the cranial part of the **sulcus terminalis**.

It has been suggested that there are specific interatrial pathways for the passage of impulses between the atria and the nodes. However, there seems to be no sound proof that these are actually present. Thus, it seems that impulses reach the atrio-ventricular node by passing along the ordinary atrial myocytes.

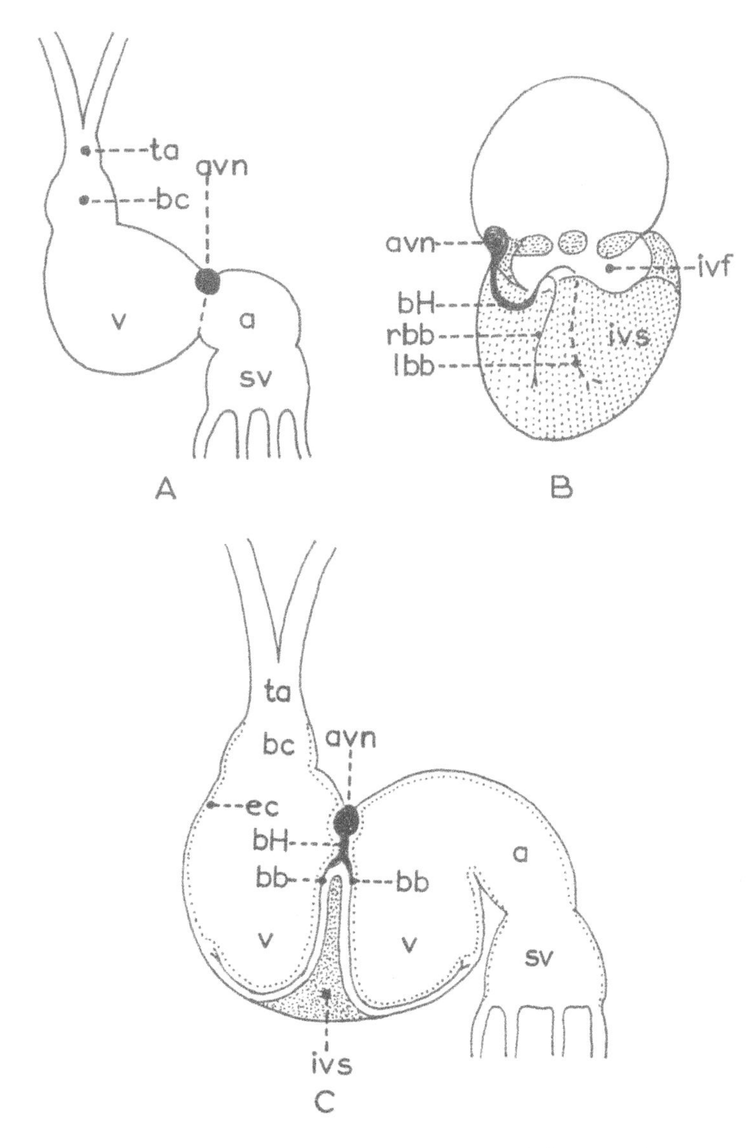

Figure 3.38: Diagrammatic representations (A and C) and sagittal section (B) to illustrate the development of the conducting system of the heart. a = atrium; avn = atrioventricular node; bb = bundle branches; bc = bulbus cordis; bH = bundle of His; ec = endocardium; ivf = interventricular foramen; ivs = interventricular septum; lbb = left bundle branch; rbb = right bundle branch; sv = sinus venosus; v = ventricle.

Anatomical and Physiological Changes to the Circulation at Birth

Humans are fundamentally terrestrial mammals and, therefore, live for most of their lives in a gaseous environment. However, nature has decreed that early human development shall take place in an aquatic environment, within the uterus of the mother.

The fetus is, therefore entirely dependent upon the placenta for its nourishment. Because of this, a number of anatomical adaptations are required to meet physiological necessities. At birth, these adaptations require re-adaptation or reversal to adjust to entry into a gaseous environment.

The embryonic (and fetal) adaptations are necessary for an aquatic environment where the placenta is responsible for oxygenation and the lungs are not. This situation is due to the presence

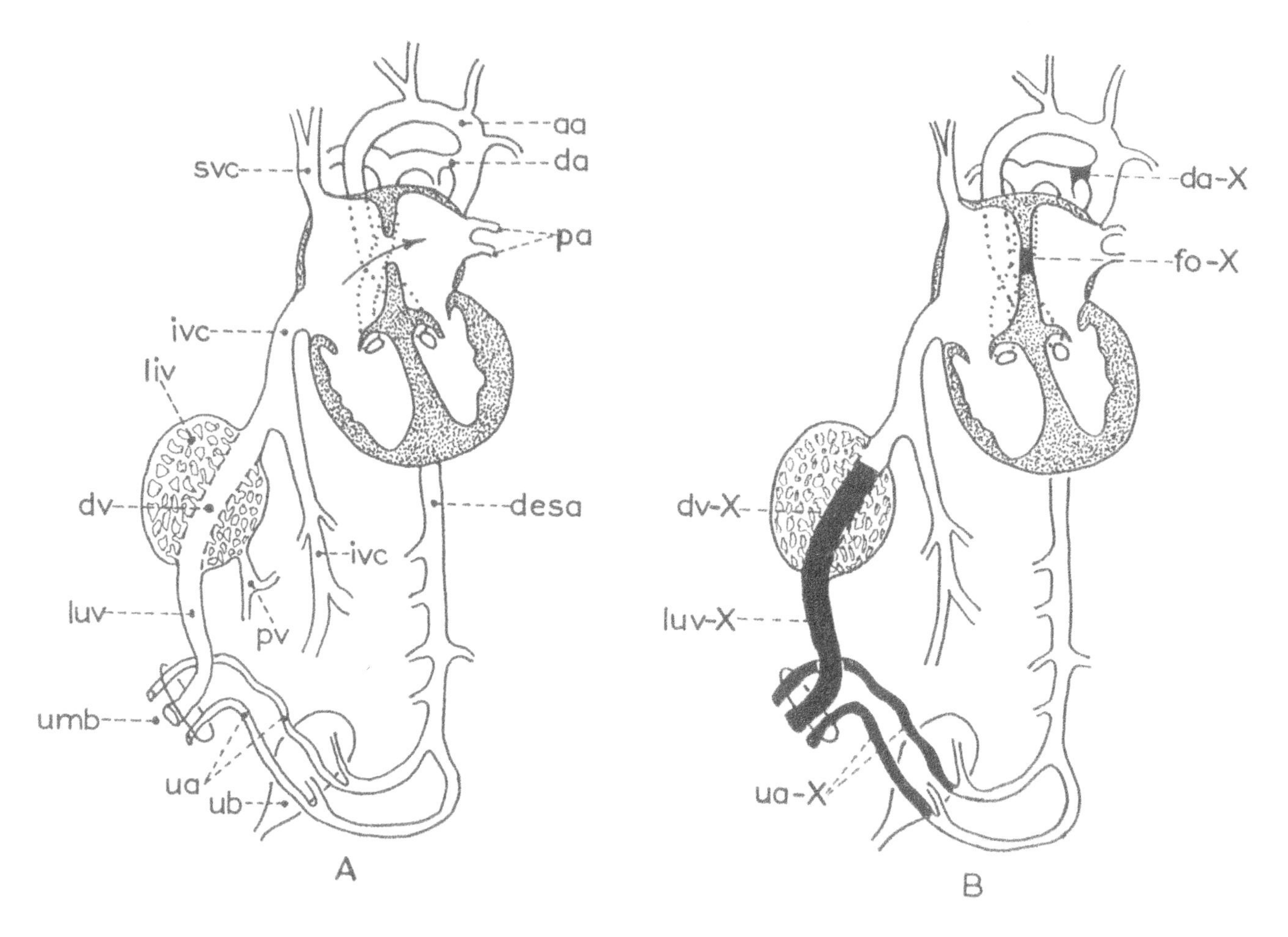

Figure 3.39: Circulation of blood in a fetus (A) and the changes which occur at birth (B). aa = aortic arch; arrow = foramen ovale; da = ductus arteriosus; desa = descending aorta; dv = ductus venosus; ivc = inferior vena cava; liv = liver; luv = left umbilical vein; pa = pulmonary veins; pv = portal vein; svc = superior vena cava; ua = umbilical arteries; ub = urinary bladder; umb = umbilicus;. **B:** Blackened vessels indicate closure of the shunts when the umbilical cord is ligated and the lungs become functional. Closure points are: da-X = ductus arteriosus; fo-X = fossa ovalis; dv-X = ductus venosus; luv-X = left umbilical vein; ua-X = umbilical arteries.

of a peripheral resistance in the lungs. A similar situation exists in the liver, but is concerned with blood flow rather than with oxygenation (**Fig. 3.39A**). The adaptations are:

(a) The presence of a **foramen ovale** between the right and the left atria so that oxygenated blood may reach the aortic arch without passing through the lungs which are inactive.
(b) The presence of a **ductus arteriosus** between the left pulmonary artery and the aortic arch so that any de-oxygenated blood entering the right ventricle may reach the aortic arch without passing through the lungs.
(c) The presence of a **ductus venosus** between the left umbilical vein and the inferior vena cava so that oxygenated blood from the placenta reaches the inferior vena cava without having to traverse the liver sinusoids.

Figure 3.39A represents the flow of blood through the fetus. The left ventricle pumps blood into the ascending aorta, aortic arch and descending aorta. The terminals of these vessels which carry de-oxygenated blood, enter the placenta as the umbilical arteries. The placenta is the maternal 'lung' within which the blood becomes oxygenated. This oxygenated blood returns to the fetus via the left umbilical vein, which enters the liver. Most of the oxygenated blood passes through the large calibre ductus venosus which acts as a shunt. A small quantity of the oxygenated blood enters the substance of the liver to provide for the liver's survival. After passing through the liver, this small quantity is de-oxygenated and is added to that passing through the ductus venosus to reach the inferior vena cava. Most of the highly oxygenated blood from the inferior vena cava will pass through the foramen ovale to the left atrium. From here the blood will enter the left ventricle and be pumped through the aortic circuit to eventually reach the placenta via the umbilical arteries.

Some of the blood from the inferior vena cava will mix with de-oxygenated blood returning via the superior vena cava from the head and neck. This blood will enter the right ventricle and leave via the pulmonary trunk. The greater part of this blood will enter the descending aorta via the ductus arteriosus to be distributed in the aortic circuit. A small amount of the pulmonary blood will enter the lungs to provide for their survival and this blood will be returned to the left atrium via the pulmonary veins.

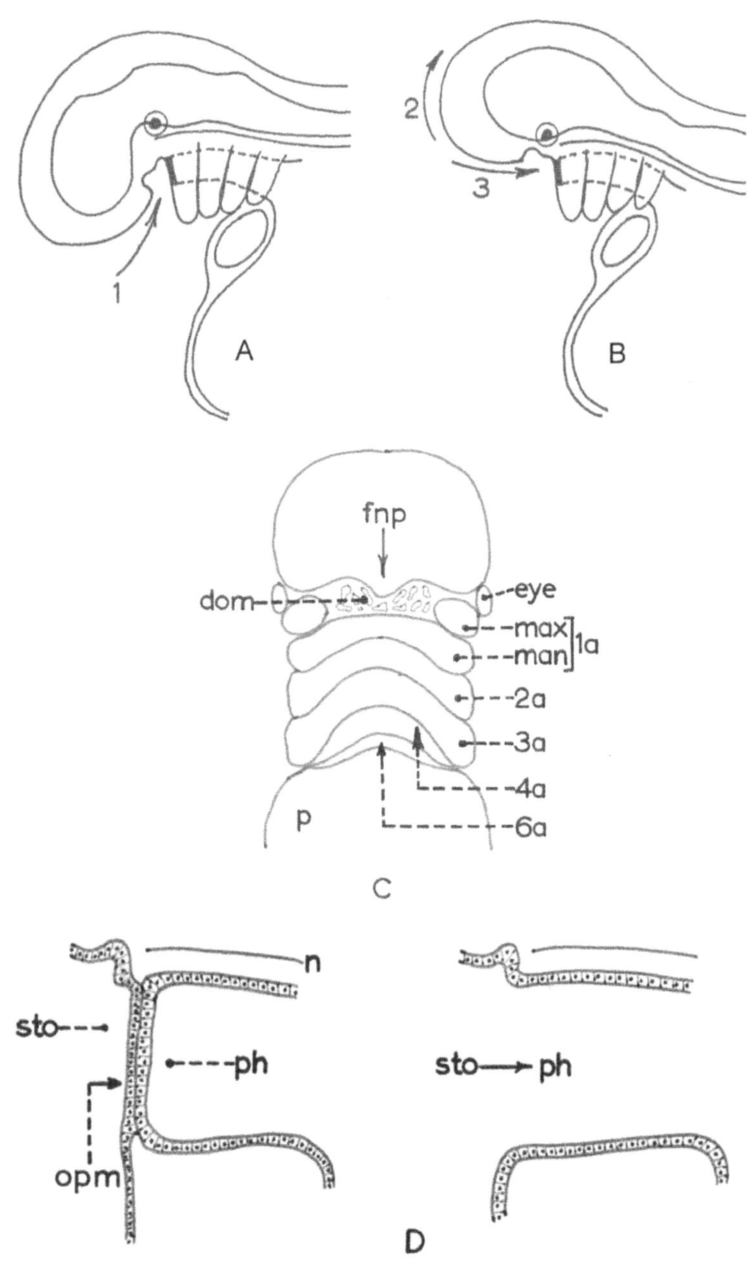

Figure 3.40: Development of the oropharyngeal membrane.
A and **B**: Lateral sections of embryo. **C**: Ventral view showing dissolution of oropharygeal membrane. **D**: Detailed view in lateral section of the breakdown of the oropharyngeal membrane. 1a – 6a = pharyngeal arches; 1 = stomodeum; 2 = extension of embryonic 'neck'; 3 = exposure of orophryngeal membrane; dom = dissolving oropharyngeal membrane; max = maxillary prominence; man = mandibular prominence; n = notochord; opm = oropharyngeal membrane; p = pericardium; ph = pharynx; sto = stomodeum.

In the adult, the right heart and its outflow tract are part of the pulmonary circuit, whereas the left heart and outflow tract are connected to the systemic circuit. Since in the fetus the lungs are uninflated, the increased capillary resistance is overcome by allowing the blood in the right atrium to leak away to the left atrium via an open foramen ovale. That blood which collects in the pulmonary artery from the right ventricle, leaks away through the ductus arteriosus to the aortic arch. These anatomical channels (foramen ovale and ductus arteriosus) may be regarded as physiological 'shunts'.

A similar system exists in the liver and the increased resistance to hepatic blood flow is overcome by the creation of a large calibre shunt between the umbilical vein (left) and the inferior vena cava. Lack of a shunt in this situation would result in the blood from the placenta passing through the liver sinusoids and its progress being retarded in reaching the heart.

When the child is born, it is imperative that it should breathe to fill the lungs with air, as the placenta will have ceased to supply oxygen to the newborn infant (**Fig. 3.39B**). When the umbilical cord is severed, the child must survive independently. The fetal adaptations are now no longer necessary and after a short period of time they begin to undergo reversal. With the expansion of the lungs, the capillary bed widens and the pulmonary vascular resistance diminishes. This allows the pulmonary blood flow to increase so that more blood reaches the left atrium with a consequent increase in pressure in that chamber. This, in turn, results in the septum primum being forced against the septum secundum to obliterate the foramen ovale. Due to their apposition, the two flaps become adherent and finally fuse, closing the foramen permanently.

With the reduction in pulmonary vascular resistance, the pulmonary arteries dilate so that blood from the pulmonary outflow tract enters the lungs more easily, and the pressure in the ductus arteriosus falls so that it undergoes constriction by action of its muscular wall. Gradually, the ductus closes completely, being converted to a knot of fibrous tissue, the **ligamentum arteriosum.**

Division of the umbilical cord stops the flow of blood in the two umbilical arteries and in the umbilical vein. The stasis (stillness) of the blood in these vessels results in it undergoing clotting and the clot propagates 'backwards' towards the infant until the first branch or tributary of the vessel is reached, where the blood is still flowing.

Note static blood anywhere in the body encourages clotting (coagulation); moving blood discourages clotting.

In the case of the arteries, the propagation stops at the superior vesical arteries (the last flowing branches before the blood clot). The clotted parts of the arteries will undergo fibrosis and become the two medial umbilical ligaments. In the case of the umbilical vein, the clot propagation will proceed (backwards) to where the vein enters the inferior vena cava. Within a short time, this clotted vessel too, is converted to a fibrous cord. The part within the liver becomes the **ligamentum venosum** and the part outside the liver is called the **ligamentum teres.** This latter ligament lies between the layers of the falciform ligament.

GASTRO-INTESTINAL TRACT AND RELATED STRUCTURES

Development of Structures of the Head and Neck

The intra-embryonic mesoderm is produced by the primitive streak and is restricted to the confines of the embryonic plate, at the edges of which it meets the extra-embryonic mesoderm **(Fig. 2.15 D,E,F** and **2.18)**. On either side, adjacent to the notochord, and around the cranial and caudal ends of the embryonic plate, this mesoderm forms a layer which extends along the trunk of the embryo. The plate differentiates into three longitudinal bands, the paraxial mesoderm, the intermediate mesoderm and the lateral plate mesoderm **(Fig. 2.15A,B,C)**. The intermediate mesoderm extends cranially only as far as the mid-cervical region, thereafter disappearing, so that the upper cervical and head regions contain only paraxial and lateral plate mesoderm **(Fig. 2.15C)**.

When the longitudinal 'bending' of the embryo takes place, the cranial end of the primitive umbilical vesicle is indented by the mesoderm surrounding the cardiogenic region and that of the septum transversum, resulting in the formation of a tubular cranial portion known as the foregut **(Fig. 2.16A,B,C)**. At this stage, the outer surface of the embryo is covered with surface ectoderm. As the cranial end of the embryo undergoes further acute flexion (the head-fold), a 'hollow' develops in the frontal region, next to the oropharyngeal membrane (formerly the prechordal plate). This is called the **stomodeum** (mouth region), which is also lined with surface ectoderm **(Fig. 2.16 A,B,C)**. The foregut, being a derivative of the primitive umbilical vesicle, is lined with endoderm. This leaves the stomodeum and foregut separated only by the oropharyngeal membrane **(Fig. 3.40A)**.

If we were able to extend the 'neck' of the embryo **(Fig. 3.40B)**, we should see the features shown in Figure 3.40C. During the 4th week pf, the oropharyngeal membrane disintegrates, allowing the stomodeum and the foregut to become continuous, not only spatially but also allowing continuity between ectoderm and endoderm **(Fig. 3.40D)**.

In the fetus (or adult), the position of the oropharyngeal membrane is thought to be at the anatomical oropharyngeal isthmus bounded by the lymphatic ring (of Waldeyer) and consisting of the nasopharyngeal tonsil (adenoid), the palatine (faucial) tonsils and the lingual tonsil (posterior one-third of the tongue) **(Fig. 3.41)**.

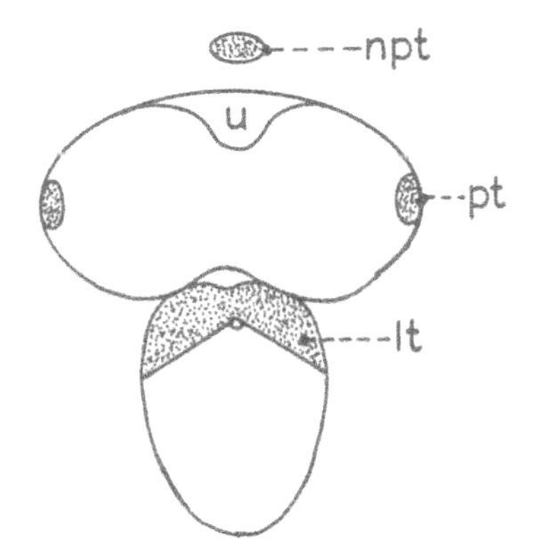

Figure 3.41: Transverse section to show 'presumed former' position of oropharyngeal membrane in the adult. lt = lingual tonsil; npt = nasopharyngeal tonsil; pt = palatine tonsil; u = uvula.

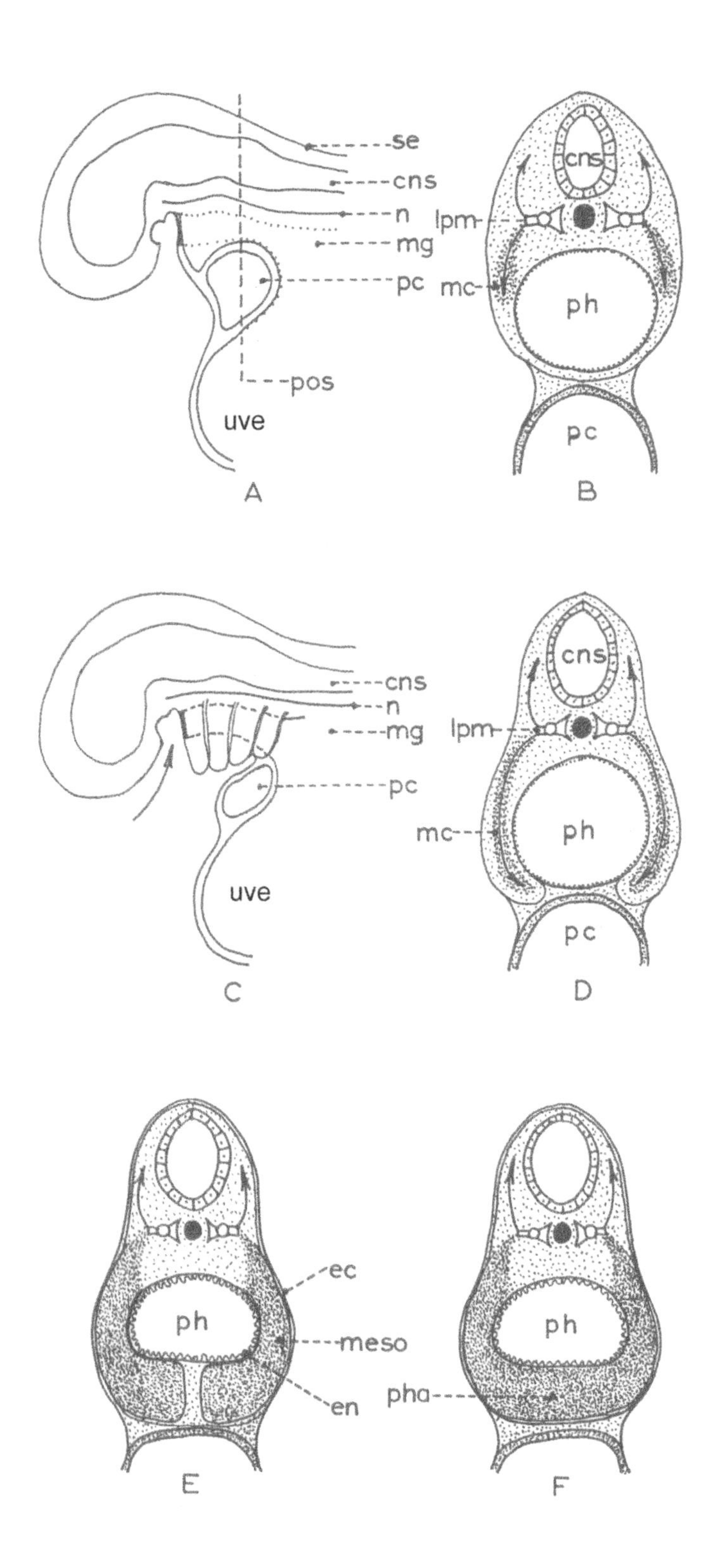

Figure 3.42: Mesodermal invasion of the pharynx. **A** and **C** (longitudinal sections) and **B, D, E,** and **F** (transverse sections) of pharyngeal region. **D, E** and **F** indicate the progressive migration of mesoderm (ectomesenchyme) around the pharynx. cns = central nervous system; ec = ectoderm; en = endoderm; lpm = lateral plate mesoderm; mc = mesodermal cells; meso = mesoderm; mg = midgut; n = notochord; pc = pericardial coelom; ph = pharynx; pha = pharyngeal arch; pos = position of sections; se = surface ectoderm; uve = umbilical vesicle.

At this time, the surface ectoderm of the neck and the endoderm of the foregut are very close together, and in the 'neck' region they become separated by the invasion, partly by splanchnic mesoderm and partly by neural crest mesenchyme. This invasion takes place by the proliferation of neural crest cells in the form of 'finger-like' processes passing between and separating ectoderm and endoderm (**Fig. 3.42A,B,C,D**). These mesodermal 'fingers' create 'ridges' on the outside and inside of the cranial part of the foregut (**Fig. 3.42C,D**) and which, in the fetus (or adult) becomes the anatomical pharynx. Because the finger-like processes are bilateral, they pass around the ventral part of the foregut and fuse in the ventral midline forming 'arches' (**Fig 3.42E,F**). In embryological terminology, these are known as the 'pharyngeal arches'.

There are said to be six such arches in the human embryo numbered I to VI from the cranial to the caudal end of the pharynx. However, there is serious doubt if arch V exists as a proper arch, and thus it is usually omitted, the numbering going from I to IV and then to VI. Numerous structures are developed in and around these arches, and these are dealt with later.

While this is happening, the nervous system has begun to develop in the central part of the ectoderm. The so-called neural ectoderm (neurectoderm) forms the neural plate. At the edges of the plate, two neural folds with a neural groove between them develop (**Fig. 3.43A**). At the summit or crest of each fold, a group of special cells differentiates (**Fig. 3.43B**). These are the **neural crest cells** and they are special for two reasons:

1. They are able to migrate throughout the embryo, and
2. they have the ability to transform into cells which form numerous special structures. In other words they are pluripotent.

These cells migrate out of the neural folds (**Fig. 3.43C**) and along special pathways in the embryonic body to form parts of organs in remote regions of the body. In the head and neck, they migrate into the ventral part of the head and into the pharyngeal arches and thus, take part in the formation of structures in these regions. The mesoderm of the head and neck may, therefore, be said to be of 'mixed' origin, partly from ordinary intra-embryonic mesodermal cells and partly from special neural crest mesenchymal cells. The latter mesenchyme is called **ectomesenchyme**.

Having considered the basic formation of the structures in the head and neck, we may now consider the more specific structures which are developed in that region. The first formation which we would encounter is the face.

Development of the Face

The development of the face takes place between the 4th to 8th week pf. It is only at about the 8th week that the face takes on a 'human' appearance.

By extending the 'neck' of the embryo and viewing the ventral region of the head, the structures surrounding the stomodeum are brought into view (Fig. 3.40C). These are:

- the **frontonasal process** (or **prominence**),
- the **maxillary processes** (or **prominences**), and
- the **mandibular processes** (or **prominences**).

The maxillary and mandibular processes will together constitute the 1st pharyngeal arch.

These processes are covered externally by ectoderm and are filled within by neural crest mesenchyme. The processes/prominences are bulges caused by the proliferation of the subectodermal ectomesenchyme and are demarcated externally by depressions or troughs.

In the depths of the stomodeum lies the **oropharyngeal membrane** where the ectoderm of the exterior abuts directly on the endoderm of the foregut (Figs. 3.40 and 3.44A). The oropharyngeal membrane will begin to perforate at about the 4th week pf, allowing continuity

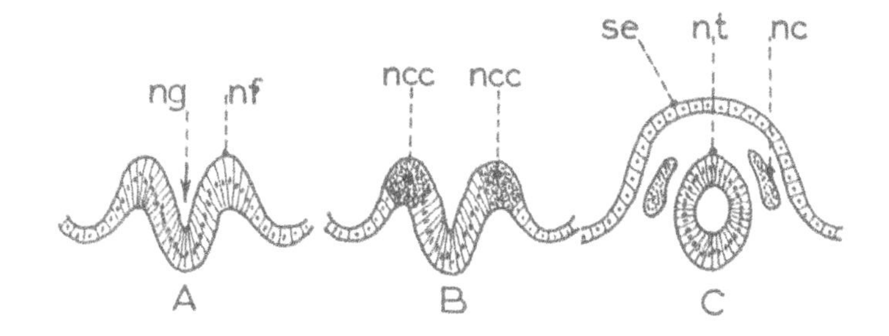

Figure 3.43: Transverse sections of the dorsal region of the embryo to show the development of the neural crest. nc = neural crest; ncc = neural crest cells; nf = neural fold; ng = neural groove; nt = neural tube; se = surface ectoderm.

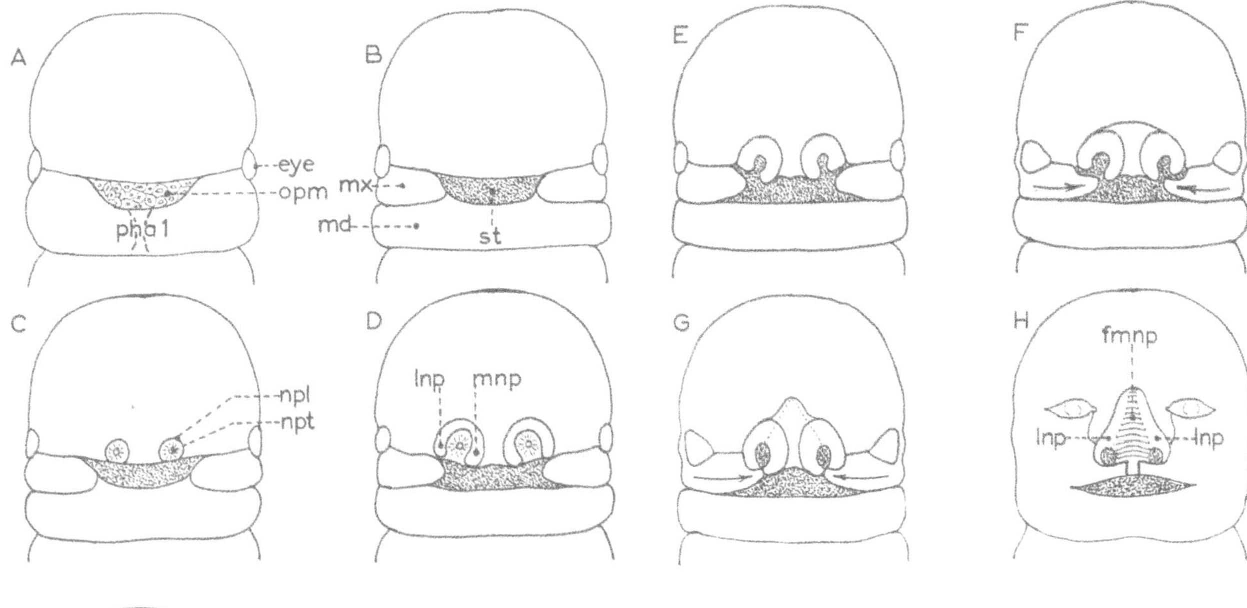

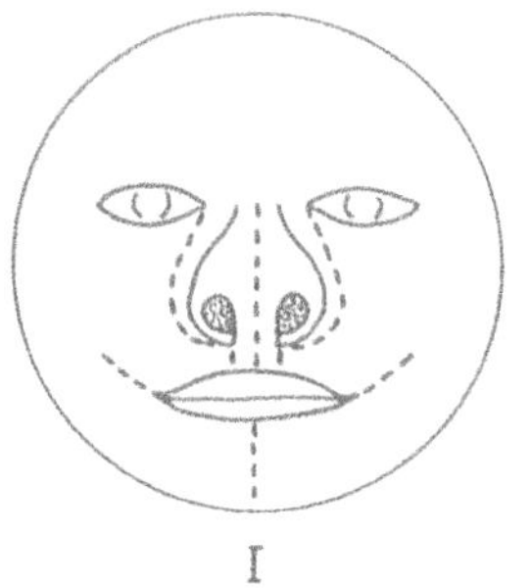

Figure 3.44: Ventral view of the developing face (A-H). I: Lines of closure. arrows = direction of growth of maxillary prominences; fmnp = fused medial nasal prominences (forming *median* nasal prominence); lnp = lateral nasal prominence; md = mandible; mnp = medial nasal prominence; mx = maxilla; npl = nasal placode; npt = nasal pit; opm = oropharyngeal membrane; pha 1 = pharyngeal arch 1; st = stomodeum.

between the stomodeal cavity (lined by ectoderm) and the pharyngeal region (lined by endoderm). Failure of perforation of the oropharyngeal membrane is rare (cf. cloacal membrane).

From about the 4th week, development proceeds as follows:

- The mandibular process acquires a ventral indentation producing the more cranially placed maxillary process (Fig. 3.44B).
- The mandibular processes merge in the midline to form the arch of the lower lip and jaw.
- On either side of the midline of the frontonasal prominence, a nasal placode appears (Fig. 3.44C); this is a thickening of the ectoderm induced by the underlying olfactory nerve.
- As the edges of the nasal (olfactory) placodes undergo thickening, due to proliferation of the ectomesenchyme underlying the ectoderm, the centres of the placodes deepen forming the nasal (olfactory) pits (Fig.3.46A), which will open postero-inferiorly into the stomodeum but are presently limited by the oronasal membranes (Fig. 3.46B,C,D). These membranes will degenerate by the 7th week pf.
- The thickened edges of the nasal pits form the medial and lateral nasal processes (or prominences) (Fig. 3.44D). Following the formation of the lateral and medial nasal processes, the frontonasal process is referred to as the frontal process.
- The medial nasal processes (which are 'crowded' towards the midline by the developing maxillary prominences) (Fig. 3.44E,F,G,H) merge to form the median nasal prominence which gives rise to the bridge of the nose, nasal septum, columella, philtrum, the median part of the upper lip and the frenulum (on the inside of the lip) (Fig. 3.44H).
- The maxillary prominences grow medially and make contact with the lateral nasal processes, pushing them and the medial nasal processes medially towards the midline. The upper lip and jaw are formed from the fusion of each maxillary process laterally with the median nasal process (Fig. 3.44F,G,H).
- The upper and lower lips become separated from their respective jaws by the formation of a sulcus (on the inside of the mouth and external to the jaws), the **labiogingival sulcus**, except in the midline where a frenulum will persist to attach the respective lip to the underlying jaw. The lower jaw will, similarly, become separated from the developing tongue by the formation of a **labiolingual sulcus**.
- The lens placode of the developing eye is set in the lateral line of junction of the frontonasal and mandibular processes, but is placed much further dorsally in early development. With further development, the eye moves ventrally and medially to come to lie opposite the bridge of the nose so that a 'trough' occurs between each lateral nasal prominence and the maxillary prominence (Fig. 3.44H); this is the **nasolacrimal groove** which ultimately closes, forming the **nasolacrimal duct**.
- The maxillary prominences merge with the mandibular prominences laterally to adjust the width of the opening into the stomodeum (Fig. 3.44G,H,I) i.e. restricting the opening of the mouth.

The lines of mergence/fusion/closure are indicated in Figure 3.44I.

Facial processes close by two mechanisms:

(a) Laterally and also in the mandibular midline, the 'fusion or merging' of processes comes about by the elevation of the ectoderm of the troughs due to proliferation of the underlying ectomesenchyme, i.e. the trough thus becomes level with the surroundings (Fig. 3.45A,B,C).

(b) Later in development, the maxillary processes are separated from the median nasal process by actual spaces, so that closure must take place by breakdown of epithelium and subsequent proliferation and confluence of ectomesenchyme (Fig. 3.45D,E,F).

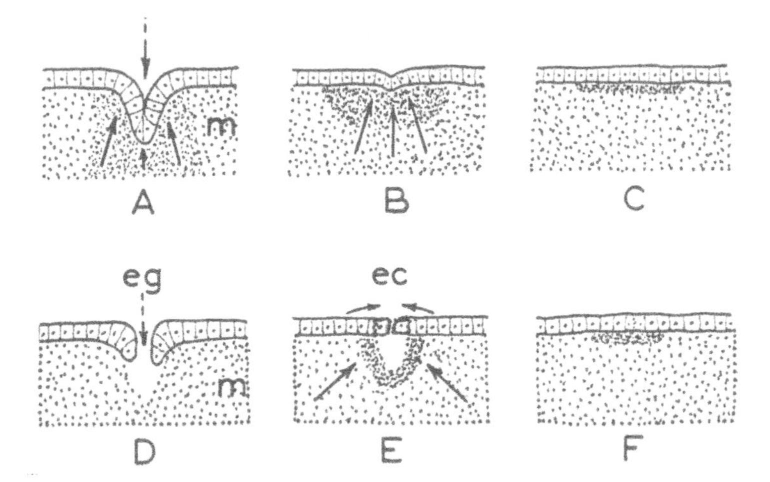

Figure 3.45: Diagrammatic representation to explain the difference between 'merging' and 'fusing'. A, B and C = merging. D, E and F = fusing. ec = epithelial closure; eg = epithelial gap; large arrows in E indicate trough.

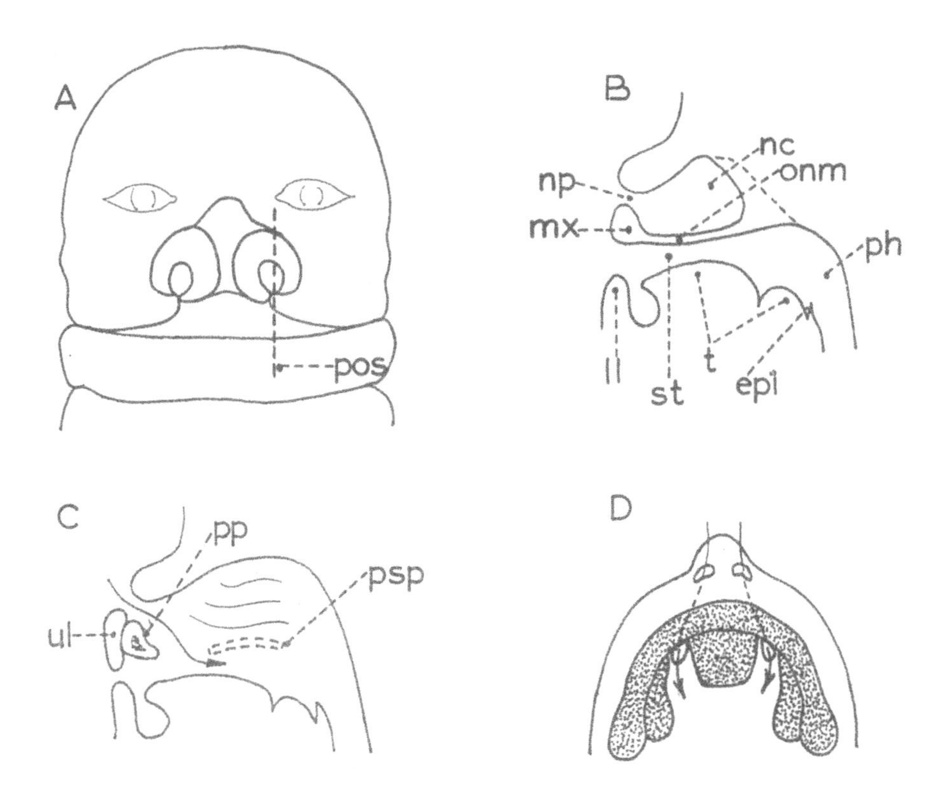

Figure 3.46: Development of nose and palate. A: Ventral view. **B** and **C:** Sagittal sections. **D:** Viewed inferiorly. arrows = channel leading from nose to mouth; epi = epiglottis; ll = lower lip; mx = maxilla; nc = nasal cavity; np= nasal pit; onm = oronasal membrane; ph = pharynx; pp = primary palate; pos = plane of sections B and C; psp = position of palate; st = stomodeum; t = tongue; ul = upper lip.

The structures of the face are thus derived as follows:

- The frontal prominence will give rise to the forehead and the root of the nose.
- The lateral nasal prominences give rise to the lateral walls and the alae of the nose. Where the lateral nasal prominences merge with the maxillary prominences laterally, they will contribute to the formation of the cheek. The merging of the maxillary and mandibular processes laterally will also contribute to the formation of the cheeks (Fig. 3.44I).
- The mandible forms in the merged mandibular processes. A cartilaginous bar, purely of neural crest origin, the **first pharyngeal arch cartilage** (**Meckel's cartilage**), develops in each mandibular prominence. The cartilage will degenerate but will leave some derivatives. Its dorsal tip will persist to give rise to the **malleus** and **incus** of the middle ear, and its ventral tip will give rise to a **mental ossicle** which will later become incorporated into the developing mandible. The perichondrium of Meckel's cartilage will form the **sphenomandibular ligament** and the **anterior ligament of the malleus**. The mandible is not derived from Meckel's cartilage, but will arise initially as two separate bones, on the lateral surface of each degenerating Meckel's cartilage. The site of the primary ossification center for each half of the mandible is at the level of bifurcation of the inferior alveolar nerve into its mental and incisive branches. The body and part of the ramus of the mandible will develop by **intramembranous ossification**, beginning at the 6th week pf. Later in development, the mental ossicles, which by now have undergone endochondral ossification, will be incorporated into the ventral ends of the mandibular body. Fusion of the two halves of the mandible occurs by the end of the first year of life. The condylar and coronoid processes will similarly develop from endochondral ossification between the 10th and 14th weeks pf. Alveolar bone will only form in response to the development of the teeth.

- A series of bulges occurs around the 1st pharyngeal groove from the 1st and 2nd pharyngeal arches. These bulges will ultimately merge to form the **pinna** of the ear. The pinna will move to a more lateral and cranial position due to the development of the mandible.

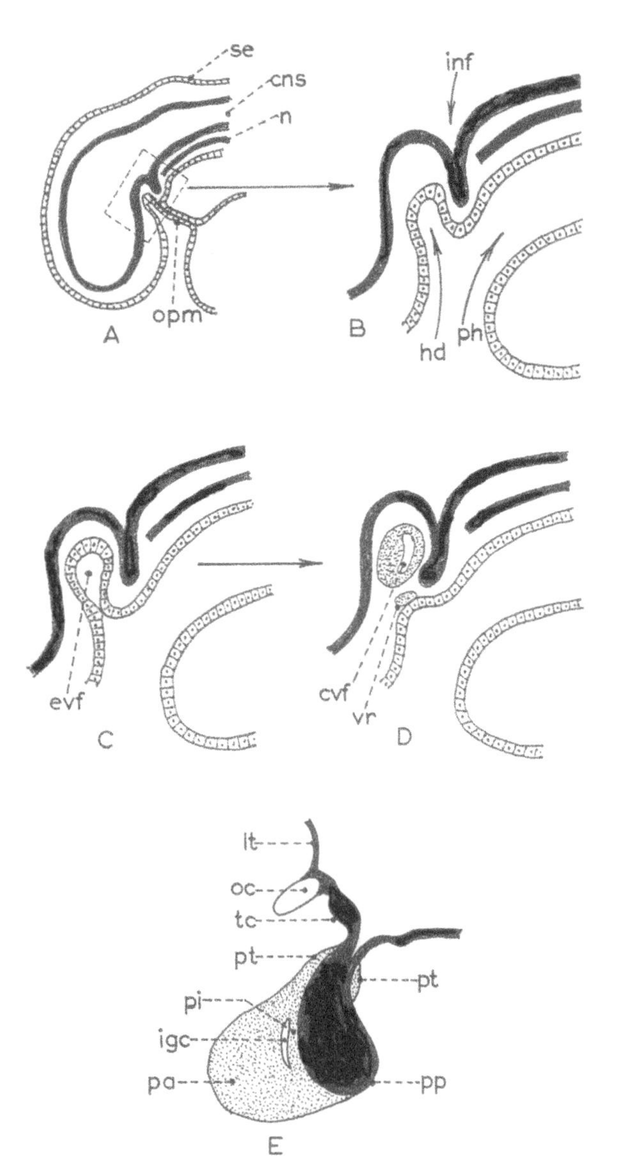

Figure 3.47: A-D: Longitudinal sections of the development of the hypophysis cerebri. E: Adult. cns = central nervous sytem; cvf = complete vesicle formation; evf = early vesicle formation; hd = hypophysial diverticulum; inf = infundibulum; igc = interglandular cleft; lt = lamina terminalis; n = notochord; oc = optic chiasma; opm = oropharyngeal membrane; pa = pars anterior; ph = pharynx; pi = pars intermedia; pp = pars posterior; pt = pars tuberalis; se = surface ectoderm; tc = tuber cinerium; vr = remnant of hypophysial diverticulum.

Congenital anomalies commonly occur at the lines of closure and may be listed as follows:

- Failure of merging of the mandibular processes results in cleft chin (**gnathoschisis**);
- **micrognathia** (underdevelopment) and **agnathia** (total lack of development) of the lower jaw. **Incorrect ear position** may result from a lack of development or maldevelopment of the lower jaw.
- Failure of the maxillary prominence (or prominences) to fuse with the median nasal prominence may result in **unilateral** or **bilateral cleft-lip** (more common in males); the left unilateral cleft is the most common.
- Failure of formation of the medial nasal prominences may result in a central cleft of the upper lip called a **median cleft of the lip** (rare), which may extend into the nose causing a cleft nose;
- Failure of the maxillary prominence/s to fuse with the lateral nasal prominence/s results in the formation of an **unilateral** or **bilateral oblique facial cleft**;
- Uneven or unequal fusion between maxillary and mandibular prominences may result in a **macrostomia** (large oral opening) or **microstomia** (small oral opening); rarely, if the maxillary and mandibular processes fuse completely a closed mouth, **astomia** results.
- **Cyclopia** in which the two eyes are fused (to a variable degree) in the centre of the face. The nose often occurs as a proboscis above the eye. The underlying cause of cyclopia is said to be the condition of holoprosencephaly, which is a failure of the forebrain to separate into two (to variable degrees).
- **Median cleft of the face** which is due to non-consolidation of the median parts of the face, commonly in the forehead region, and resulting from insufficiencies of neural crest cells in this region.
- **Hypertelorism** is a condition when the eyes are widely separated.

Many of these anomalies are thought to be associated with peculiar migration or proliferation of neural crest cells.

Derivatives of the Stomodeum

A number of important structures develop from the ectoderm of the stomodeal cavity. These are the **adenohypophysial** component of the

hypophysis cerebri, the **enamel** of teeth and the epithelium of the **parotid** salivary gland. The development of teeth and the salivary glands are dealt with elsewhere.

Development of the Hypophysis Cerebri (Pituitary Gland)

Shortly before the oropharyngeal membrane distintegrates, a small evagination develops in the ectodermal roof of the stomodeum (Fig. 3.47A,B,C,D). This deepens to form a pouch called the **hypophysial diverticulum** (of Rathke) which ultimately closes off to form a vesicle. The cells of this vesicle will differentiate into the various cells of the **adenohypophysis**. This vesicle will soon lose its attachment to the stomodeum and the hypophysial cavity will become partially or fully obliterated. The cells of the adenohypophysis are supported by an ingrowth of vascular mesenchyme from the region anterior to the termination of the notochord, and the mesenchyme surrounding the hypophysial cartilages, which ultimately become the body of the sphenoid bone. The adenohypophysis will ultimately gives rise to the **pars distalis, pars intermedia and pars tuberalis** of the pituitary gland.

The **neurohypophysis** is formed by a downgrowth of the diencephalon of the brain, called the **infundibulum** (Fig. 3.47A,B,C,D), and lies immediately behind the hypophysial diverticulum. The neurohypophysis is composed of a **pars nervosa** and an **infundibular stalk**. The neurohypophysis and adenohypophysis will together form the **hypophysis cerebri** or pituitary gland (Fig.3.47E). Ultimately, the hypophysis is surrounded by the bone of the body of the sphenoid and lies in the hypophysial fossa of the bone. It becomes covered by a membrane of dura mater, the **diaphragma sellae** in which there is an aperture through which the infundibulum (pituitary stalk) of the hypophysis cerebri passes.

In man, a remnant of adenohypophysial tissue is nearly always present under the mucous membrane of the upper nasopharynx and is probably a derivative of the hypophysial diverticulum. This is referred to as the '**pharyngeal' hypophysis**. It is capable of responding to blood-borne hormones and may be a source of pituitary hormones, especially in postmenopausal women.

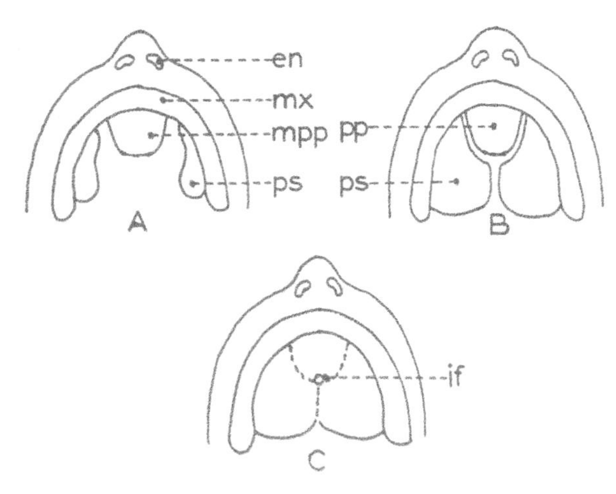

Figure 3.48: Development of the palate viewed from an inferior aspect. en = external nares; if = incisive foramen; mx = maxilla; mpp = median palatal process; pp = primary palate (median palatal process); ps = palatal shelf (lateral palatal processes).

DEVELOPMENT OF NOSE AND PALATE

In the adult, the nasal cavities are of considerable size and open to the exterior through the **external nares** (nostrils) and to the interior through the **posterior nares** (choanae). They are separated from one another by the nasal septum and from the mouth by the hard and soft palate. In the embryo, the development of the nasal cavity begins with the deepening of the nasal pits which burrow superiorly and posteriorly, ventral to the developing brain (Fig. 3.46A,B). At first, the primitive single nasal cavity is separated from the stomodeum by the oronasal membranes (Fig. 3.46B) but these break down in the 7th week pf to establish continuity between the primitive nasal and oral cavities which constitute the stomodeal cavity at this stage (Fig. 3.46C,D).

At the same time the maxillary prominences are approaching the midline of the face to merge with the median nasal process to complete the formation of the upper lip and jaw. The ectomesenchymal (neural crest) contents of the median nasal process form a triangular mass immediately posterior to the upper lip (Fig. 3.48A). This is called the **median palatal process**. At about the 5th week pf, the anterior part of the tongue is developing in the posterior part of the stomodeal cavity, and is pouting into the enlarging nasal cavity (Fig. 3.49A). Each maxillary prominence will form projections or 'shelves' called **lateral palatal** processes (Fig. 3.48A,B), which will ultimately contribute to the palate but at the present time, the projections are trapped in a vertical position, on either side of the protruding tongue (**Fig. 3.49A**). When the stomodeal cavity has enlarged sufficiently to allow the tongue to recede from the nasal cavity, these lateral (**vertical**) palatal processes become elevated to form the **horizontal** palatal shelves (**Fig. 3.49B**). The lateral palatal processes (now horizontal) join the median palatal process and then fuse with one another, from anterior to posterior, to form the **definitive palate** (**Fig. 3.48B,C**). The fusion of the three shelves occurs along a Y-shaped line, which has a deficiency at the junction of the median palatal process and the lateral palatal processes, called the **incisive foramen**, which transmits nerves and blood vessels (**Fig. 3.48C**). The fusion stops short of the posterior pharyngeal wall, thus leaving an opening or communication between nose and pharynx (**Fig. 3.46C**).

Ossification of the palate proceeds from primary and secondary ossification centres in the median palatal process appearing at approximately the 6th week pf, which gives rise to the '**premaxilla**', and from one primary ossification centre in each of the lateral palatal processes at the 6th week pf, which give rise to **the horizontal parts of the maxillary bones.** An additional primary ossification centre appears at the 8th week pf in the posterior region of each lateral palatal process and will give rise to the **horizontal process of the palatine bones.** The 'premaxilla' is said to be overgrown superficially by the maxillary bone, therefore no evidence of a premaxillary bone is evident after the 3rd month pf. Ossification does not occur in the posterior extremity of the palatal processes, thus resulting in the **soft palate** and **uvula.**

A number of theories exist about what causes the elevation of the lateral palatal processes from a vertical to a horizontal position. In addition it is known that the epithelium covering these processes develops a 'sticky' glycocalyx following elevation of the shelves. This would allow the shelves to remain in contact with each other until fusion occurs. 'Fusion' is brought about by *apoptosis* (programmed cell death) of the median edge epithelium and merging of the ecto-mesenchyme between the processes. The fusion of the ectomesenchyme results in a midline raphe in the mucous membrane of the palate.

During the development of the palate, the nasal septum will begin to grow inferiorly from a superior position and join with the fused palatal processes. The **ethmoid bone** (perpendicular plate) will ossify in the upper part of the septum and the **vomer bone** will ossify from two ossification centres in the mesenchyme in the floor of the septum. In addition a median cartilage will form in the septum. The fusion of the nasal septum with the palate (**Fig. 3.49C**) will complete the division of the nasal cavity into right and left cavities and the fusion of the palate will separate the nasal cavity from the oral cavity.

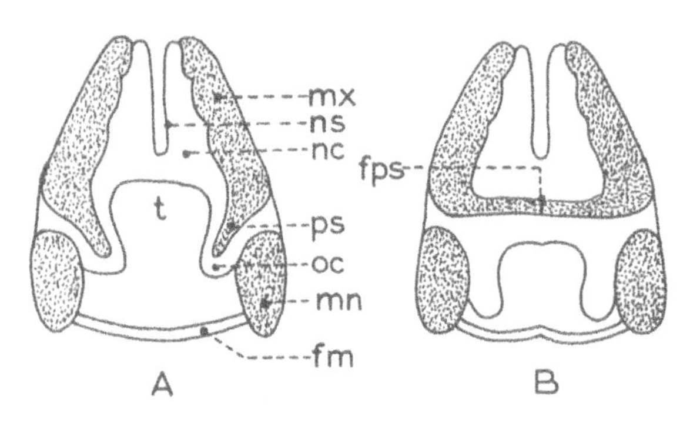

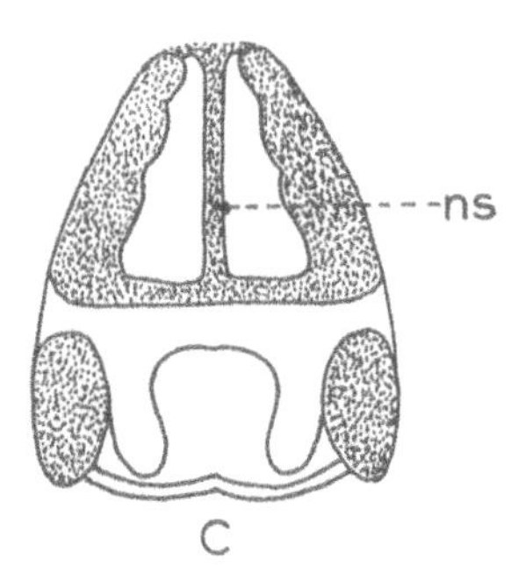

Figure 3.49: Development of the palate and the nasal septum in coronal section. fm = floor of mouth; fps = fused palatal shelves; mn = mandible; mx = maxilla; nc = nasal cavity; ns = nasal septum; oc = oral cavity; ps = palatal shelf (in vertical position); t = tongue.

COMMON ABNORMALITIES

- **Clefts of the palate** (uranoschisis) may occur along the normal lines of fusion.
- A **median** cleft will occur if the two lateral palatal processes fail to fuse. This cleft may extend anteriorly on one or both sides and result in a **unilateral** or **bilateral** cleft of the palate. These clefts may also be associated with clefts of the upper lip.
- **Bifid uvula** results from non-fusion of the most posterior parts of the lateral palatal processes (soft palate).

Numerous environmental agents are known to cause clefting of the palate. These include rubella, excess vitamin A; drugs such as amphetamines and barbiturates, epanutin, aminopterin, the benzodiazopines, thalidomide and others; and common habit forming drugs such as alcohol and smoking.

DEVELOPMENT OF THE TEETH

The development of the teeth is a complex matter and to understand it fully, the student should recall the following general facts:

(a) There are normally 20 teeth in the deciduous dentition and 32 teeth in the permanent dentition. In the infant and child, the first or **deciduous** dentition (primary or 'milk teeth') consists of two incisors, one canine and two molars in each of the four dental quadrants of the jaws (**Fig. 3.50A**). In childhood, the deciduous dentition is completely lost between the ages of 12-14 years and is replaced by the **permanent** dentition. The permanent dentition consists of two incisors, one canine, two premolars and three molars in each quadrant of the jaws (**Fig.3.50B**).

(b) Each tooth has a **crown**, a **neck** and one or more **roots** (**Fig. 3.51A**). The general structure of each tooth is of a laminar nature - **enamel, dentine, cementum** and **pulp** (**Fig. 3.51B**). The enamel is exceptionally hard, the dentine and cementum are also hard, while the pulp cavity is filled with soft vascular connective tissue containing nerve fibres derived from the trigeminal nerve. This nerve is the embryonic nerve of the 1st pharyngeal arch from which the dental arches are developed. Note that there is a degree of overlap in the laminar structure with the enamel being the main part which protrudes into the oral cavity.

(c) Each tooth is anchored at its root (or roots) to the surrounding **alveolar bone** of the maxilla or mandible by the **periodontal ligament** (**Fig. 3.51B**). The periodontal ligament extends from the **cementum** of the root to the alveolar bone.

(d) The pharyngeal arches are heavily infiltrated with neural crest cells and in the case of the 1st arch (maxilla and mandible), these cells have a profound influence on the development of the **dental lamina** and other dental structures.

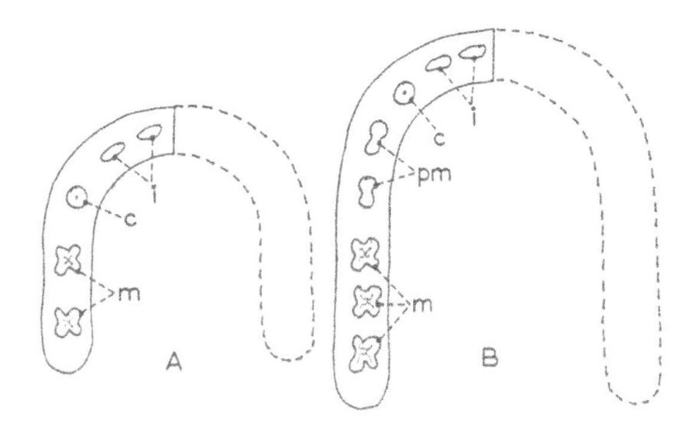

Figure 3.50: The arch of the jaw in a child (A) and an adult (B). c = canine; i = incisor; m = molar; pm = premolars.

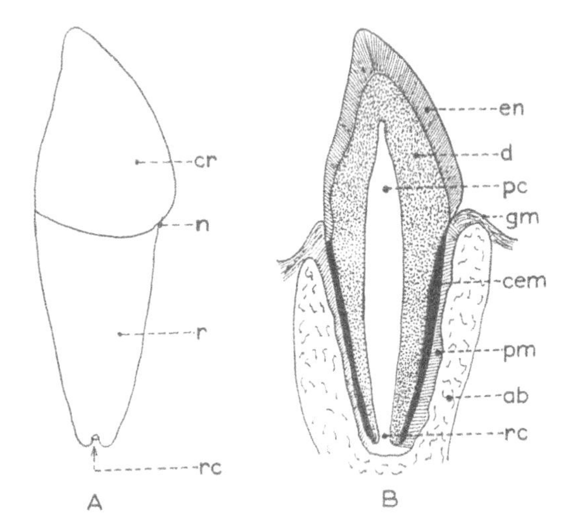

Figure 3.51: General structure of an adult tooth. ab = alveolar bone; cem = cementum; cr = crown; d = dentine; en = enamel; gm = gingival margin; n = neck; pc = pulp canal; pm = periodontal membrane (ligament); r = root; rc = root canal.

At a very early stage in the development of the face, it is possible to identify, in a section through the developing upper or lower jaw, a thickening of the stomodeal (oral) ectoderm, which is invaginating into the underlying ectomesenchyme. This thickening is the **primary epithelial band** (**Fig. 3.52A**). The primary epithelial band follows the curvature of the maxillary and mandibular arches (**Fig. 3.52B**). The primary epithelial band very soon gives rise to two laminae, the **vestibular** and **dental** laminae (**Fig. 3.53A,B**). The vestibular lamina develops on the labial surface of the dental lamina. The vestibular lamina will deepen and as a result of apoptosis of its cells, will give rise to the labio-gingival sulcus (**vestibular sulcus**) (**Fig. 3.53C**). The latter sulcus ultimately separates the cheek from the outer gingival surface, thus forming the **vestibule** of the mouth.

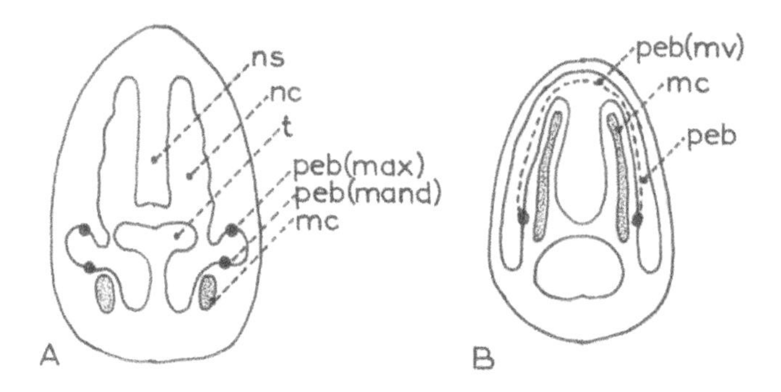

Figure 3.52: A: Coronal section of the upper and the lower jaws depicting the position of the primary epithelial band (peb). **B:** Horizontal section of the lower jaw indicating the direction of development of the primary epithelial band (pebmv). mc = 1st pharyngeal arch cartilage; nc = nasal cavity; ns = nasal septum; peb(mand) = primary epithelial band of mandibular prominence; peb(max) = primary epithelial band of maxillary prominence; t = tongue.

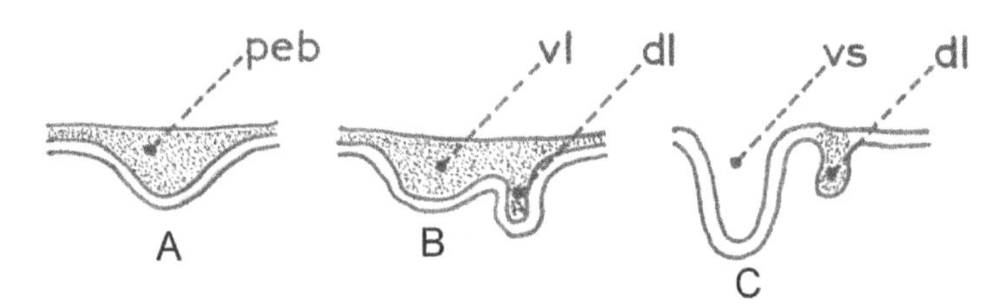

Figure 3.53: Transverse section of the early development of the stomodeal epithelium to show the development of the vestibular and dental laminae. A: Initially the primary epithelial band (peb) occurs. **B:** The primary epithelial band then forms two components, the vestibular lamina (vl) and the dental lamina (dl). **C:** The vestibular lamina undergoes apoptosis to form the vestibular sulcus (vs) and the dental lamina begins to form a tooth bud (tb) at specific points.

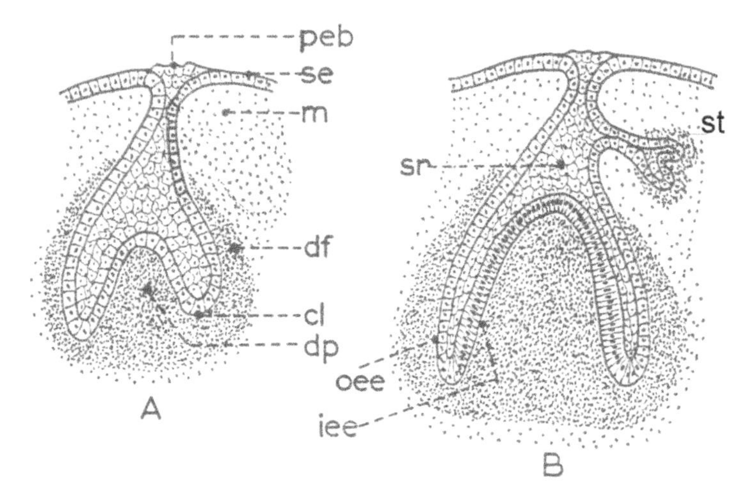

Figure 3.54: The cap and bell stage of tooth development. cl = cervical loop; df = dental follicle; dp = dental papilla; iee = inner enamel epithelium; m = mesoderm; peb = primary epithelial band; oee = outer enamel epithelium; se = surface ectoderm; sr = stellate reticulum; st = successional tooth.

Previously, the ectomesenchyme had condensed at five sites in each quadrant of the jaws subjacent to the primary epithelial band. This condensation of ectomesenchyme induces the formation of the **dental lamina** from the primary epithelial band (Fig 3.53B,C). The teeth are formed by the invagination of the dental lamina into the sites of condensation of ectomesenchyme of the arches (Fig. 3.53D,E). From about the 4th to the 5th week pf, the teeth develop by a series of fairly well-defined stages, namely the bud stage, the cap stage and the bell stage:

(a) **The Bud Stage**: The cells of the dental lamina invade the maxillary and mandibular arches, forming rod-like structures in the position of the future teeth. Each rod soon becomes bulbous at its deep invading end, forming the **tooth bud** (Fig. 3.53D). The bud is filled with oral ectodermal cells from the dental lamina and is separated from the underlying condensation of ectomesenchyme by its basement membrane.

(b) **The Cap Stage**: Due to increased mitosis of the cells of the lateral surfaces of the bud, the inner surface of the bud appears to deepen and changes its shape from a bud to a cap. Development of the cap results in it having two layers of epithelium – an outer layer of cuboidal epithelium called the **outer enamel epithelium**, and an inner layer of columnar epithelium called the **inner enamel epithelium** (Figs. 3.53 and 3.54A). At the junction of the outer and inner layers of epithelium, a blunt angle is formed and this is known as the **cervical loop**. The remaining cellular elements of the stomodeal ectoderm lying between the outer and inner enamel epithelia become the cells of the **stellate reticulum** (Fig. 3.54A). The inner and outer enamel epithelia and the stellate reticulum are referred to as the **enamel organ**, as this structure will eventually be responsible for the production of enamel. Simultaneously, the ectomesenchyme deep to and around the bud undergoes further condensation. That part deep to the bud is called the **dental papilla** while the part around the bud is called the **dental follicle** (Figs. 3.53E and 3.54A). The dental papilla indents the deep surface of the bud so that the bud overlies the papilla in the form of a 'cap'.

During the cap stage, a condensation of cells appears between the inner enamel epithelium and the stellate reticulum. This is the **enamel knot** and it is known to be an important signalling centre in the development of the tooth. While a single primary enamel knot is found in teeth at the cap stage, this will undergo apoptosis. Secondary enamel knots will later appear at the tips of the future cusps in molars.

As development proceeds, the deep surface of the 'cap' undergoes further deepening so that the cervical loop elongates and its angle becomes more acute. The inner and outer layers of the enamel epithelia thus become more closely apposed in the region of the cervical loop (Fig. 3.54A), giving the appearance of a 'bell'.

(c) **The Bell Stage**: At this time, the ectomesenchyme of the dental papilla induces the inner enamel epithelium to reverse its polarity. The nuclei of the inner enamel cells move from the basal pole (basement membrane side) to the opposite pole of the cells (Fig. 3.55). The reason for this action is that the cells will then be able to secrete enamel from their deep surface. A cellular condensation occurs between the stellate reticulum and the inner enamel epithelium, and is called the **stratum intermedium** (Fig. 3.56). The full function of the stratum intermedium is not clearly known but it is thought to assist in the formation of enamel.

The repolarised cells of the inner enamel epithelium now induce the adjacent cells of the ectomesenchyme of the dental papilla to become aligned in a layer under the inner enamel epithelium (Fig. 3.56A). These transformed ectomesenchymal cells now begin to secrete **dentine** and are thus known as **odontoblasts** (Fig. 3.56B). The formation of dentine, in turn, induces the inner enamel epithelium to secrete enamel and these cells may now be called **ameloblasts** (Fig. 3.56B).

It is now clear that a series of **inductive epithelio-mesenchymal interactions** have taken place to initiate **histodifferentiation** (tissue formation) and **morphodifferentiation** (shape) of the crown of the tooth. As more and more enamel and dentine are formed, the thickening layers cause the ameloblasts and the odontoblasts to be pushed away from one another. The formation and mineralisation of enamel (amelogenesis) is a complex matter which results in an extremely hard crystalline material. In the formation of the dentine (dentinogenesis), the cellular processes of the odontoblasts remain trapped in the dentinal substance. The odontoblasts continue to lay down dentine around these processes, forming **dentinal tubules**. The mineralisation of the dentine contributes to its strength and hardness.

Dentine is laid down until the dental papilla is reduced to a narrow cavity, the **pulp cavity**, which contains connective tissue as well as blood vessels and nerves which supply the tooth.

Prior to the formation of enamel and dentine, the ameloblasts were nourished by a process of diffusion from blood vessels in the ectomesenchyme of the dental papilla. The continued deposition of dentine and enamel results in a thickened layer between the dental papilla and the

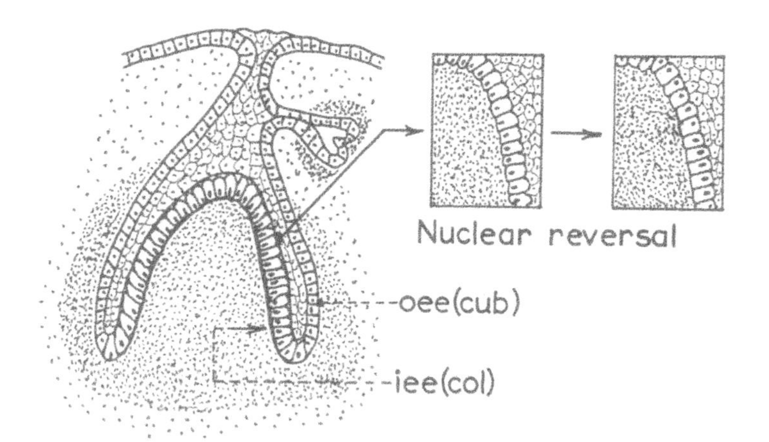

Figure 3.55: Nuclear reversal in the inner enamel epithelium during induction. iee = columnar inner enamel epithelium; oee = cuboidal outer enamel epithelium.

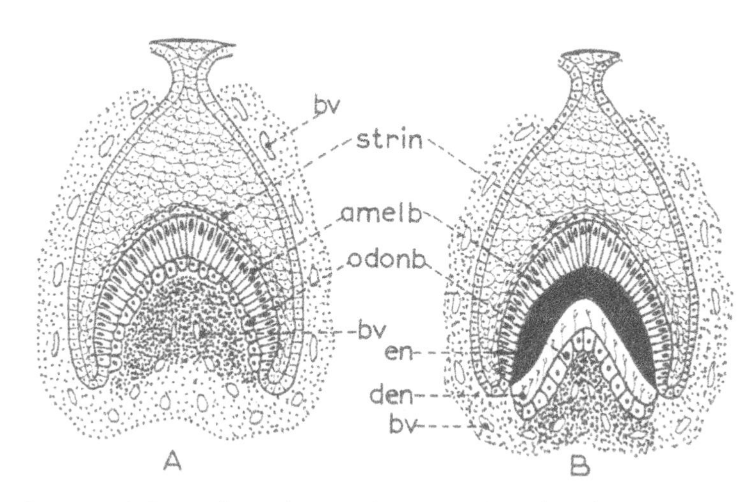

Figure 3.56: Crown formation. amelb = ameloblasts; bv = blood vessels; den = dentine; en = enamel; odonb = odontoblasts; strin = stratum intermedium.

ameloblasts, thus reducing the diffusion gradient. Since nourishment must continue for the ameloblasts to remain viable, continued nourishment is supplied by vessels in the dental follicle. This is achieved by the attenuation of the stellate reticulum so that the outer enamel epithelium becomes crenated (Fig. 3.57A,B). In this way blood vessels in the dental follicle are 'brought closer' to the ameloblasts, thus facilitating diffusion of nutrients. The collapse of the enamel organ results in the formation of the **reduced enamel epithelium**, which will fuse with the oral epithelium during eruption and will give rise to the **dentogingival junction** in the erupted state.

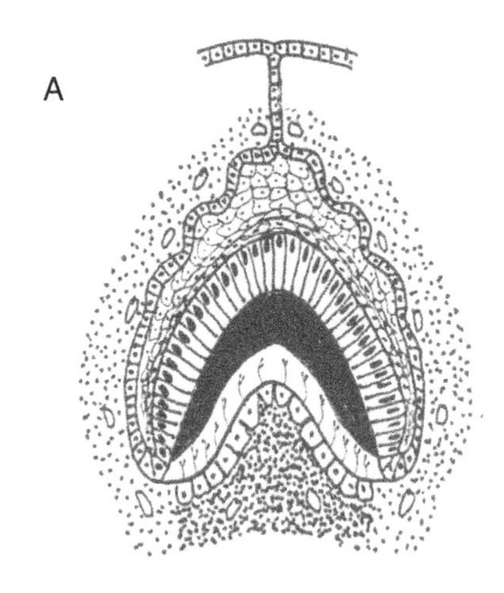

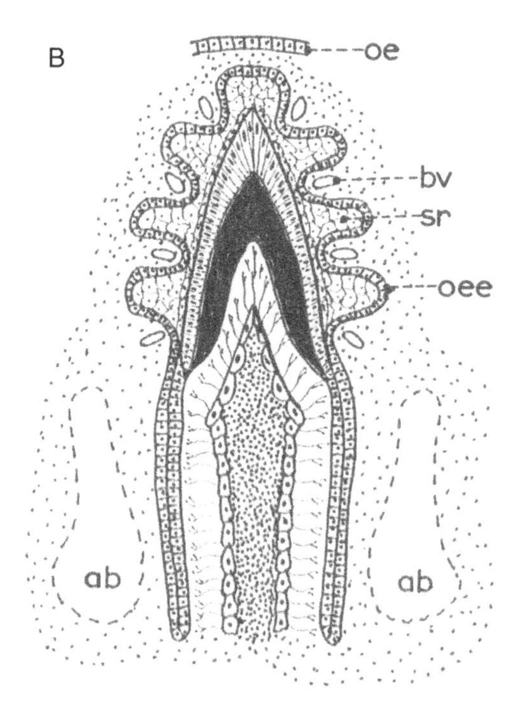

Figure 3.57: 'Collapse' of the outer enamel epithelium (oee). ab = alveolar bone; bv = blood vessel; oe = oral epithelium; sr = stellate reticulum.

During the bell stage, the invading dental lamina gives an 'offshoot' on its *lingual side,* which will form an adjacent tooth bud that will give rise to a permanent (successional) tooth (Fig. 3.54B). Note that the original buds form the deciduous (milk) teeth only. The attachment of the developing tooth to the surface oral epithelium is lost by the degeneration of the dental lamina during the bell stage (Fig. 3.58).

At the edges of the enamel organ, the inner and outer enamel epithelia of the cervical loop continue to elongate into the surrounding ectomesenchyme. The layers are in close apposition and form the **epithelial root sheath** (of Hertwig) (Fig. 3.59). No stratum intermedium is present between the layers of the inner and outer enamel epithelia of the root. The inner enamel epithelium of the root sheath now induces the formation of a layer of odontoblasts in the underlying ectomesenchyme of the dental papilla. These odontoblasts secrete dentine which will form the root of the tooth. When the dentine has formed, the root sheath begins to undergo patchy degeneration (Fig. 3.60A,B). This degeneration results in openings in the root sheath through which surrounding vessels, accompanied by ectomesenchymal cells from the dental follicle, will penetrate. The remaining cells of the root sheath form the **cell rests** (of Malassez), which are found in the periodontal ligament of the erupted tooth. The ectomesenchymal cells of the dental follicle come into contact with the dentine of the root and are induced to form **cementoblasts**, which secrete **cementum** (cementogenesis) on to the surface of the dentine of the root (Fig. 3.60B).

The number of roots of a tooth is determined by the ingrowth of a transverse diaphragm/s from the epithelial root sheath. This will also cause narrowing of the apical foramen of each root.

As the cervical loop grows to form the root sheath, ectomesenchyme of the dental follicle will give rise to the **periodontal ligament** (Fig. 3.60B) and to the **alveolar bone** of the mandible and maxilla. Alveolar bone deposition will only occur in response to the presence of tooth buds. The cell rests are located between the fibres of the periodontal ligament, which are anchored on one side to the surrounding alveolar bone and on the other side to the cementum. The fibres inserting into the bone and the cementum are called **Sharpey's fibres**.

Although there are a number of theories concerning the eruption of teeth, it is thought that the most likely is due to forces exerted by the periodontal ligament.

POSSIBLE ABNORMALITIES

The complexity of the development of the teeth may lead to a large number of possible anomalies of site, number, size and shape. Anomalies may occur in the enamel, dentine and in root formation.
Some common anomalies are:
- Anodontia
- Macrodontia
- Microdontia
- Amelogenesis imperfecta
- Dentinogenesis imperfecta

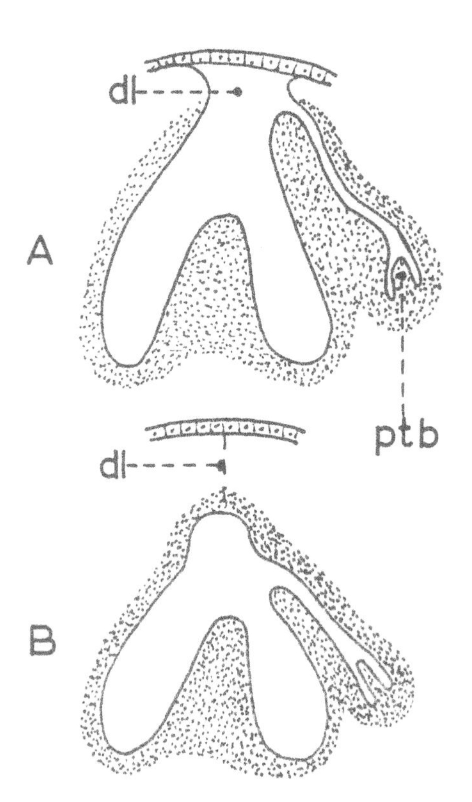

Figure 3.58: Degeneration of the dental lamina (dl). ptb = permanent tooth bud.

THE PHARYNX AND ITS DERIVATIVES

Development of the Pharyngeal Arches

The developing pharynx lies caudal to the oropharyngeal membrane. When the oropharyngeal membrane dissolves, the stomodeal cavity (future mouth) and pharynx, become continuous. Thus, the ectoderm of the mouth becomes continuous with the endoderm of the gastrointestinal tube **(Fig. 3.40D)**.

The external surface of the embryo is covered by ectoderm (which gives rise to skin) and is continous with the ectoderm of the oral cavity (lining of the first pharyngeal arch). The pharynx is lined with endoderm. Both the internal ectoderm and endoderm will give rise to a common mucous membrane. This is surrounded by mesoderm which has arisen from the lateral plate mesoderm **(Fig. 3.42A,B,C)**. This lateral plate mesoderm becomes heavily invaded by the migration of neural crest cells (ectomesenchyme) during the 4th week of development. As a result, the greater number of the mesenchymal cells in the pharyngeal region are actually derived from the neural crest. *Hox* genes are important in specifying the position of the neural crest cells within the pharyngeal arches. At selected places, the mesenchyme surrounding the pharynx proliferates and condenses into 'bars' or

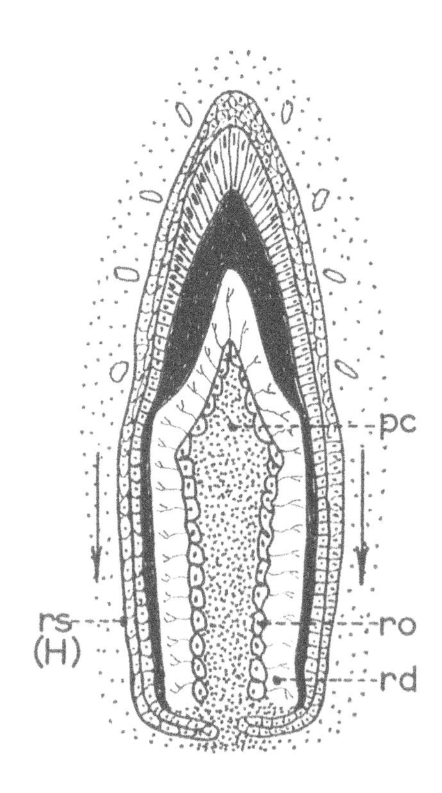

Figure 3.59: Root formation. arrows = penetration of cervical loop to form epithelial (Hertwig's) root sheath (rs[H]); pc = pulp cavity; rd = root dentine; ro = root odontoblasts.

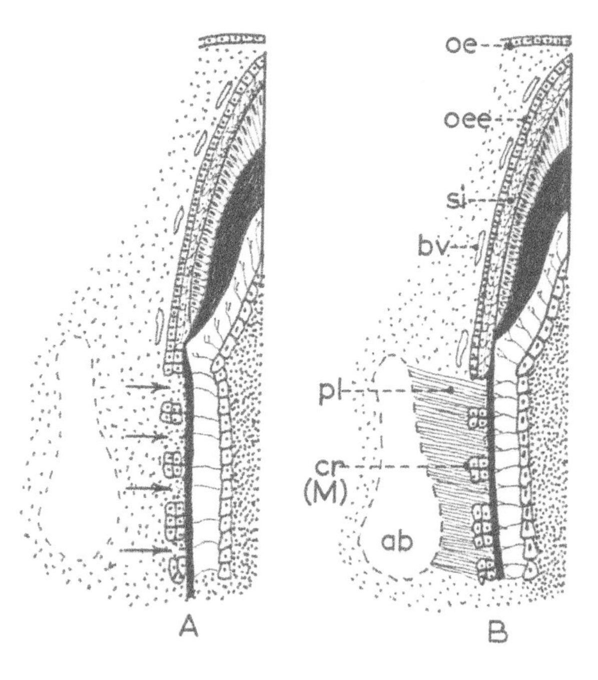

Figure 3.60: Formation of the root and cementum. Arrows indicate direction of penetration of vascular mesenchyme. ab = alveolar bone; bv = blood vessel; cr (M) = cell rests (of Malassez); oe = oral epithelium; oee = outer enamel epithelium; pl = periodontal ligament; si = stratum intermedium.

arches on either side, with 'depressions' or troughs between them **(Fig. 3.42D)**. These arches, known as **pharyngeal arches**, extend from a position lateral to the developing hindbrain, around the walls of the pharynx into the ventral region where they grow towards their opposite member in the midline **(Fig. 3.42)**. They give rise to six curved, cylindrical thickenings which form a collar around the pharynx. Externally, the development of these arches separates the stomodeum from the developing pericardial cavity. The arches are subject to the law of 'cranio-caudal development' in that those in the upper part of the pharynx are developmentally in advance of the more caudal arches. Thus, they are never all present at the same time in the form of arches. In this text, for the sake of simplicity and understanding, let us consider them as all being present at the same time.

The **1st (mandibular) arch** has an additional process called the **maxillary** process. The **2nd (hyoid) arch** later grows caudally and covers the succeeding arches. The mandibular processes of the 1st arch merge in the midline as do the hyoid arches. The ventral ends of the succeeding arches fail to reach the midline and are separated by a median swelling of the pharyngeal floor (the **hypopharyngeal eminence**). In the human embryo, the **5th arch** is only transiently present and leaves no derivatives.

The depressions between the arches, known as **pharyngeal grooves** or **clefts** (externally), are covered by ectoderm. Internally the depressions or **pouches** between the arches are lined with endoderm. Mesenchyme intervenes between the ectoderm (of the groove) and the endoderm (of the pouch) forming a 'closing' membrane **(Fig. 3.42D)**. It is noteworthy that the closing membranes in fish become perforated to form the gill slits.

Look at Figure 3.61A, which is a lateral view of the developing embryo and note the presence of the protruding brain with the pharyngeal arches below. Inferior to the third arch is the bulge of the pericardium containing the heart and below that, the developing umbilical cord. Now imagine that we make a section through the embryo as indicated in Figure 3.61A such that the arches are cut from side to side. If we remove the ventral part of the embryo and look at it from its posterior aspect, we see the 'back of the front of the pharynx' **(Fig. 3.61A,B,C and 3.62)**. On the cut sections we see ectoderm (skin) on the outside, endoderm (mucous membrane) on the inside and ectomesenchyme between these layers. In this ectomesenchyme, condensations of cartilage will occur and some of these will ultimately ossify. In addition to the formation of cartilage, the surrounding ectomesenchyme will differentiate into muscle and blood vessels and will also be invaded by cranial nerves.

In all vertebrates (animals which possess a vertebral column), the developing pharyngeal arches have the same **basic plan** **(Fig. 3.63)**. This plan takes the form of the following:

(a) Each arch is a mass of ectomesenchyme, covered externally by ectoderm and lined internally by endoderm (except in the case of the 1st arch, which is lined internally by stomodeal ectoderm).
(b) Ectomesenchyme condenses in each arch to form a cartilaginous bar (this is purely neural crest in origin) and is known as the **pharyngeal arch cartilage**.
(c) Some of the mesenchyme of each arch differentiates into **striated muscle** which is unlike most of the muscle of the gut (which is smooth muscle).
(d) A **pharyngeal arch artery** will develop caudal to the arch cartilage.
(e) A mixed, **motor** and **sensory nerve** grows into each arch from the hindbrain. In each

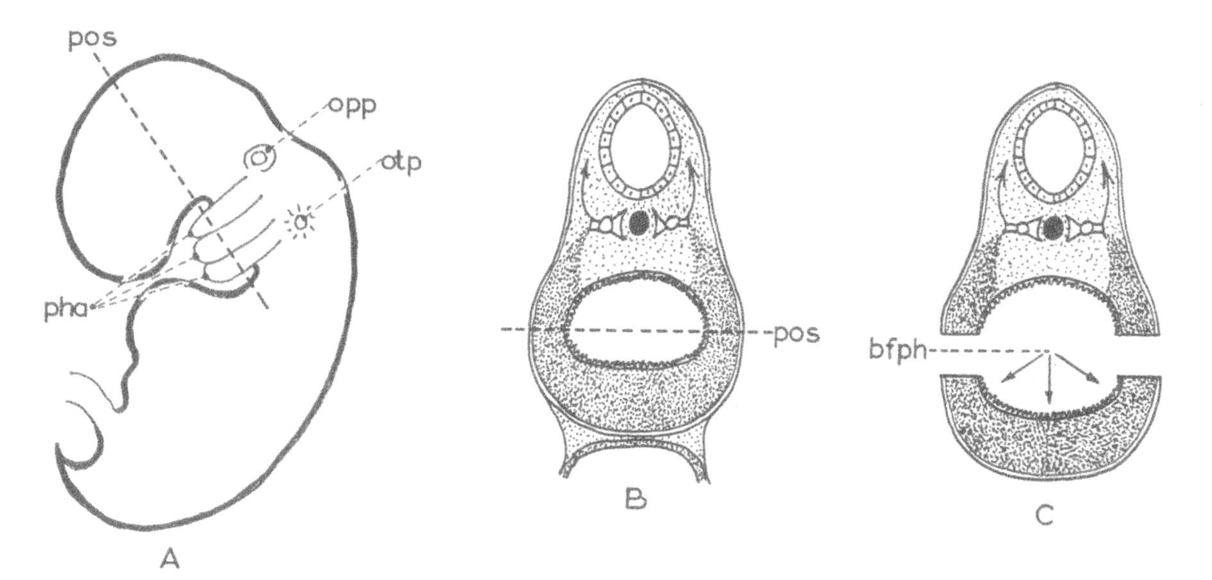

Figure 3.61: Development of pharyngeal arches. A illustrates the plane of section (pos) of **B** and **C**. The interrupted line in **B** shows a cut which would be taken to derive the view in **figures 3.62A** and **B**. bfph indicates a view of the 'back of the front of the pharynx' as depicted in figures **3.62A** and **B**. opp = optic placode; otp = otic placode; pha = pharyngeal arches.

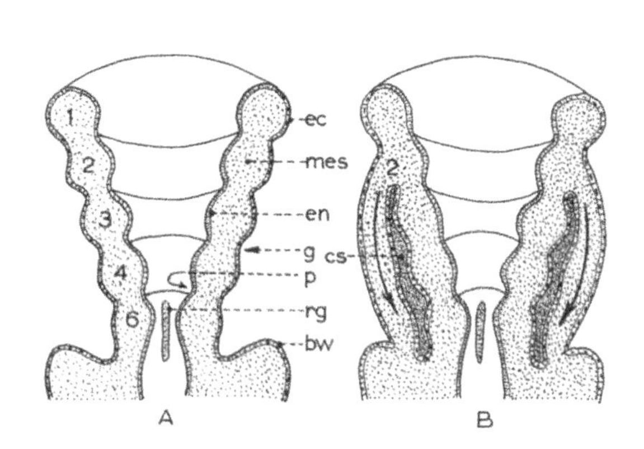

Figure 3.62: Section through pharyngeal arches, view indicated in figure 3.61C. 1, 2, 3, 4, 6 = numbered pharyngeal arches; arrow = overgrowth by 2nd pharyngeal arch; bw = body wall; cs = cervical sinus; ec = ectoderm; en = endoderm; g = groove; mes = mesoderm; p = pouch; rg = respiratory groove.

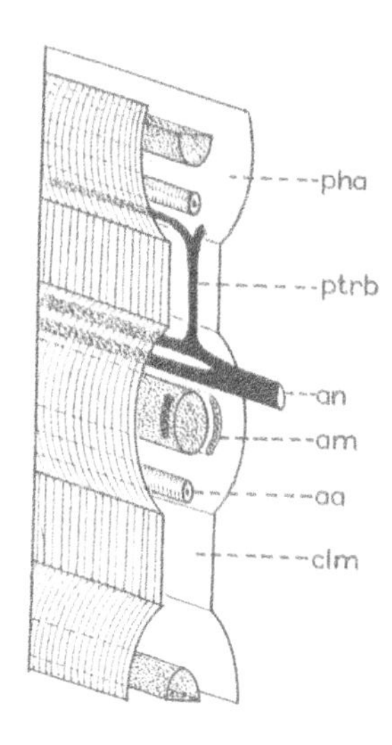

Figure 3.63: Basic vertebrate plan of pharyngeal arches. aa = arch artery; am = arch muscle; clm = closing membrane; pha = pharyngeal arch; ptrb = pretrematic branch of arch nerve (an).

arch, except the 6th, the nerve is cranial to the cartilage. In the 6th arch the nerve is caudal to the artery which means that it is also caudal to the cartilage. The motor component of the nerve innervates the striated muscle of that arch. The sensory component divides into **pre-** and **post-trematic** branches (trema– Greek = opening [or gill in fish]) separated by the pharyngeal groove and pouch. The post-trematic branch runs with the motor component of the nerve and supplies the epithelium on the cranial aspect of its own arch (Fig. 3.63). The pretrematic branch arches over the preceeding pharyngeal groove or pouch to supply the caudal epithelium of the preceeding arch. Thus each arch contains its own motor and post-trematic nerve plus the pretrematic branch of the nerve of the succeeding arch.

This basic plan is modified in man so that only two pretrematic nerves are known to persist: The facial nerve (nerve of arch 2) has a **chorda tympani branch** (pre-trematic branch) which

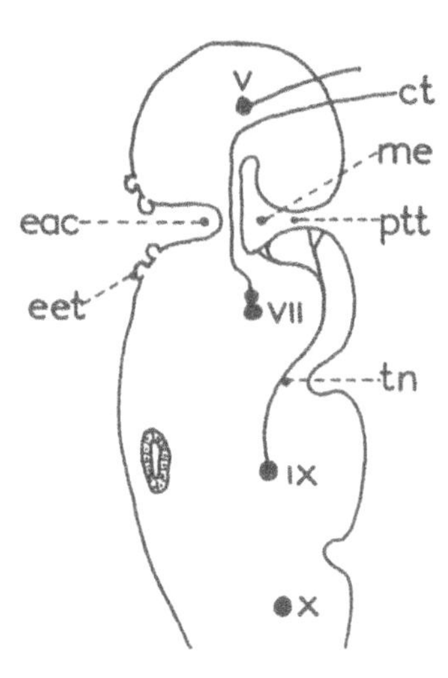

Fig. 3.64: Illustration to show how the pretrematic nerves reach their area of supply.
V = mandibular branch of trigeminal nerve; VII = facial nucleus; IX = glossopharyngeal nerve; X = vagus nerve; ct = chorda tympani; eac = external auditory canal; eet = external ear tubercles; me = middle ear; ptt = pharyngotympanic tube; tn = tympanic nerve.

supplies the anterior two-thirds of the tongue (arch 1) with taste sensation (**Fig. 3.64**).

The glossopharyngeal nerve (nerve of arch 3) has a **tympanic branch** (pre-trematic) which supplies the walls of the auditory tube and tympanic cavity (parts of pouches/arches 1 and 2).

These relationships are quite typical of pretrematic nerves as found in lower vertebrates.

While the nerve supply to the 4th and 6th arches is said to be the vagus nerve, it is in fact provided by the *cranial part of the accessory nerve which 'hitch-hikes' with the branches of the vagus.*

Each arch and its contents become modified during development to give rise to very specific structures in the adult. These are detailed in Table 3.1. Although a nerve is specified for each arch, in the case of the lower arches, these nerves may be acting as conduits for other nerves (e.g. the vagus will carry fibres from the cranial part of the accessory nerve to supply arches 4 and 6). In this way the striated laryngeal muscles are not supplied by the vagus nerve but by the cranial part of the accessory nerve.

The Pharyngeal Grooves

In the 5th week pf, four pharyngeal grooves are evident externally between the arches. Only the first groove will contribute to a definite structure during development. The first groove deepens to form the **external acoustic meatus.** A series of nodules form from the 1st and 2nd arches around the 1st groove. These nodules will fuse to become the external ear (pinna). The deepened first groove ends at the ear-drum (**tympanic membrane**) (**Fig. 3.64**), which, like the pinna is covered with stratified squamous epithelium.

The second arch expands laterally and caudally and grows down to join the tissue surrounding the upper thorax and thus overlies the 2nd, 3rd and 4th grooves. The cavity formed deep to it, is the **cervical sinus** (**Fig. 3.62B**). Thus, the 2nd, 3rd and 4th pharyngeal grooves make no normal contribution to the neck structures, as they are covered by the expanded 2nd arch. However, the expanded arch allows the neck surface to be smoothed out. The cavity created by the overlying 2nd arch is gradually obliterated and normally disappears. Occasionally, a pocket remains beneath the expanded second arch, and forms a '**pharyngeal cyst**' which is lined with stratified squamous epithelium (which is in fact skin). If this perforates to the exterior it will form a **pharyngeal fistula.**

The Pharyngeal Pouches

As already described, the **pharyngeal pouches** lie between the pharyngeal arches on the internal aspect of the developing pharynx and are, therefore, lined with endoderm. The pharynx is a dorso-ventrally compressed tube and the pouches arise as lateral extensions of the tube (**Figs. 3.62** and **3.66**) which grow towards each pharyngeal groove on the external surface of the pharyngeal arches. Thus a layer of endoderm, intervening ectomesenchyme and a layer of ectoderm separate each pouch from the respective groove (**Fig. 3.62A**).

Each pouch develops a **dorsal** and **ventral diverticulum** (**Fig 3.67** and **3.68**) except in the case of pouch 1 and pouch 2, which do not develop a ventral diverticulum as the tongue will be developing in this region. The development of the 1st pouch will be dealt with during the development of the ear.

The endoderm of the ventral diverticulum of pouch 2 proliferates to give rise to the **palatine tonsils** (**Figs. 3.67** and **3.68**). Each tonsil will retain continuity between its lumen and the cavity of the pharynx. Lymphocytes will differentiate in the tonsil at about 5 months pf.

Table 3.1: Derivatives of Pharyngeal Arches

Pharyngeal Arch	*Skeletal components (see Figures 3.65A,B)*	*Muscles*	*Pharyngeal Arch Artery*	*Nerve*
1st: **Mandibular** and **maxillary**: contributes to facial prominences and anterior 2/3rds tongue mucosa	**Maxillary component:** facial bones (maxilla, zygomatic, squamous part of temporal, palatine and vomer) **Mandibular component:** Meckel's cartilage, which gives rise to malleus and incus (dorsal end) and mental ossicle (ventral end) sphenomandibular ligament and anterior ligament of malleus (perichondrium of Meckel's cartilage), body of mandible, spine of sphenoid	Muscles of mastication (temporalis, masseter, medial and lateral pterygoids) anterior belly of digastric, tensor veli palatini, tensor tympani	Primitive maxillary artery	V – **Trigeminal** nerve
2nd: Hyoid	Stapes, styloid process, stylohyoid ligament, lesser horns and upper part of body of hyoid bone	Muscles of facial expression, posterior belly of digastric, stylohyoid, stapedius, auricular and epicranial muscles	Stapedial artery	VII – **Facial** nerve: the pathway of migration of the muscles of facial expression are mapped out by the branches of the facial nerve
3rd: contributes to the mucosa of the posterior 1/3 of the tongue	Greater horns and lower part of body of hyoid bone	Stylopharyngeus	Internal carotid artery	IX – **Glossopharyngeal** nerve
4th: contributes to the mucosa of the posterior 1/3 of the tongue	Upper part of thyroid cartilage	Levator veli palatini, uvular muscles, palatoglossus, cricothyroid, ? constrictors of the pharynx	Left: aortic arch Right: proximal part of right subclavian	X – The superior laryngeal branch, which carries fibres from the cranial part of the accessory nerve innervates this arch via the vagus.
5th	No derivatives	No derivatives	No derivatives	No derivatives
6th	Lower part of thyroid cartilage, arytenoids and corniculate cartilages, epiglottis	Intrinsic laryngeal muscles, ? constrictors of the pharynx	Pulmonary arteries, on the left – ductus arteriosus	X –The recurrent laryngeal branch which carries fibres from the cranial part of the accessory nerve innervates this arch via the vagus.

Each 3rd pouch develops a ventral and dorsal diverticulum. These will become solid due to the proliferation of the endodermal lining. The ventral diverticulum on each side will lose continuity with the pharynx and migrate inferiorly to join its counterpart from the opposite side to form the **thymus** gland. The gland will be carried inferiorly into the thorax to attain its fetal position (Figs. 3.67 and 3.68).

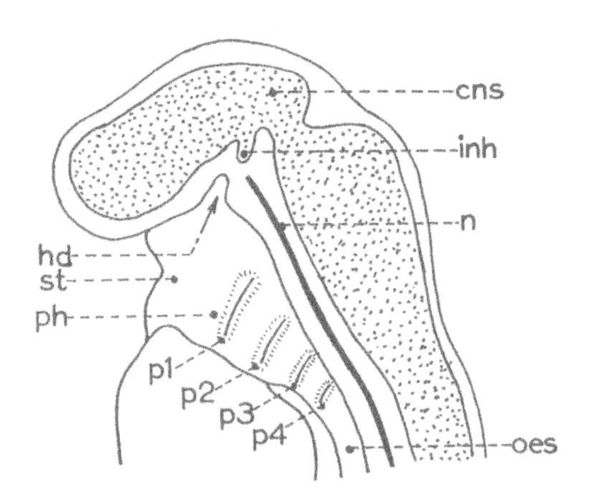

Figure 3.66: Median section through the embryo to show the pharyngeal pouches. cns = central nervous system; hd = hypophyseal diverticulum; inh = infundibulum of hypophysis; n = notochord; oes = oesophagus; p1, p2, p3, p4 = pouches; ph = pharynx; st = stomodeum.

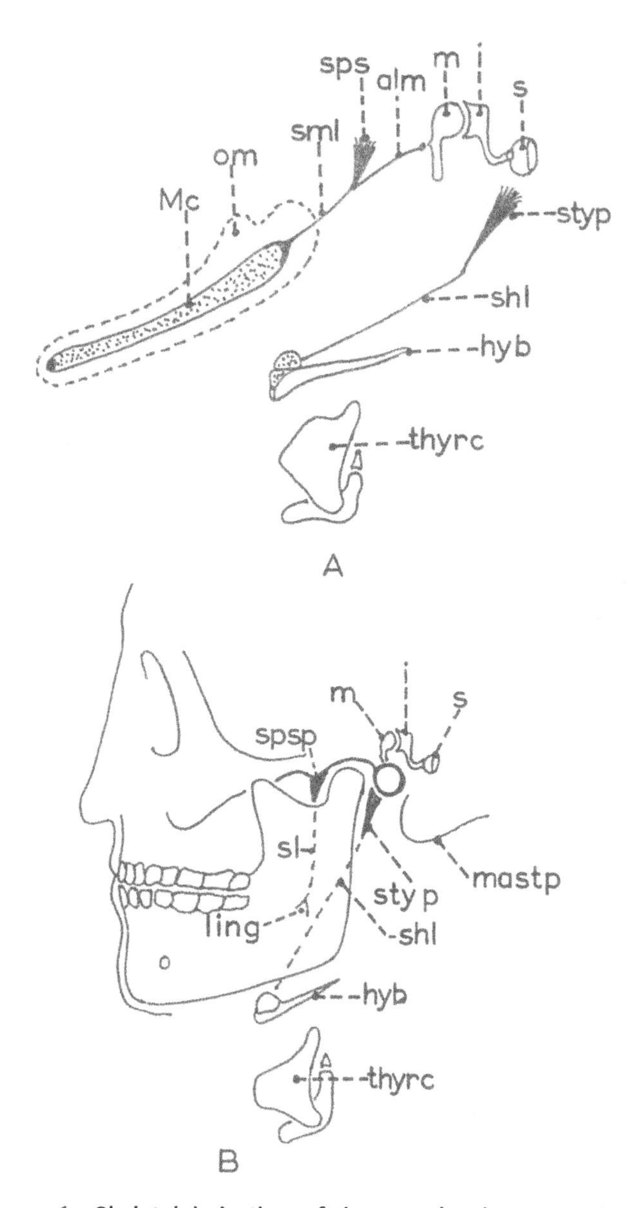

Figure 3.65: Skeletal derivatives of pharyngeal arches. A = embryo, **B** = adult. alm = anterior ligament of malleus; hyb = hyoid bone; i = incus; ling = lingula; m = malleus; mastp = mastoid process; Mc = Meckels'cartilage; om = ossification of mandible (interrupted lines); s = stapes; shl = stylohyoid ligament; sl and sml = sphenomandibular ligament; sps and spsp = spine of sphenoid bone; styp = styloid process; thyrc = thyroid cartilage.

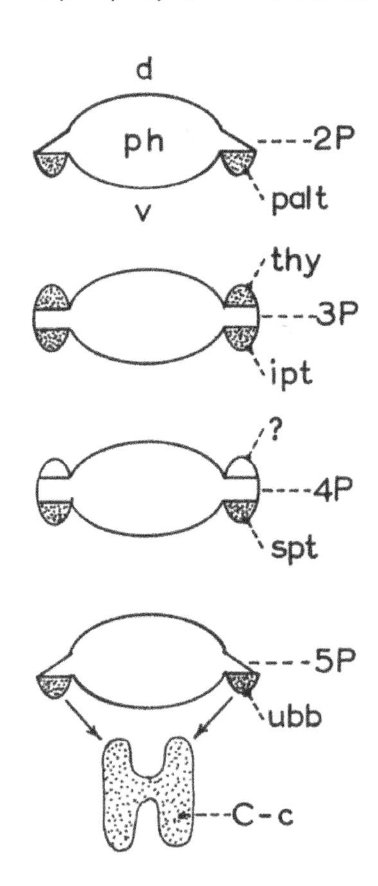

Figure 3.67: Transverse sections through the pharynx illustrating the development of the dorsal and ventral diverticula of pouches 2-5 (p2-p5). C-c = C cells of thyroid gland; d = dorsal; ipt = inferior parathyroid gland; palt = palatine tonsil; ph = pharynx; spt = superior parathyroid gland; thy = thymus gland; ubb = ultimobranchial body; v = ventrical..

The dorsal diverticulum of pouch 3 on each side will form an **inferior parathyroid gland** (parathyroid III), which loses connection to the pharynx and migrates inferiorly to attach to the inferior pole of the thyroid gland on each side (Figs. 3.67 and 3.68).

The 4th pouch on each side also develops a dorsal and ventral diverticulum. These too, become solid by proliferation, lose continuity with the pharynx and migrate inferiorly. It is not known what the ventral diverticulum on each side gives rise to in man, but it has been proposed that it may contribute to the thymus gland. The dorsal diverticulum of the 4th pouch gives rise to the **superior parathyroid gland** (parathyroid IV) on each side. These glands migrate inferiorly to attach to the superior pole of the thyroid gland on either side (Figs. 3.67 and 3.68).

The rudimentary 5th pouch is said to give rise to the **ultimopharyngeal body** in humans. This leaves no derivatives in the human but is believed to act as a resting place for neural crest cells that migrate to populate the thyroid gland at a later stage. These neural crest cells will give rise to the **C cells (parafollicular cells**) of the thyroid gland (Figs. 3.67 and 3.68).

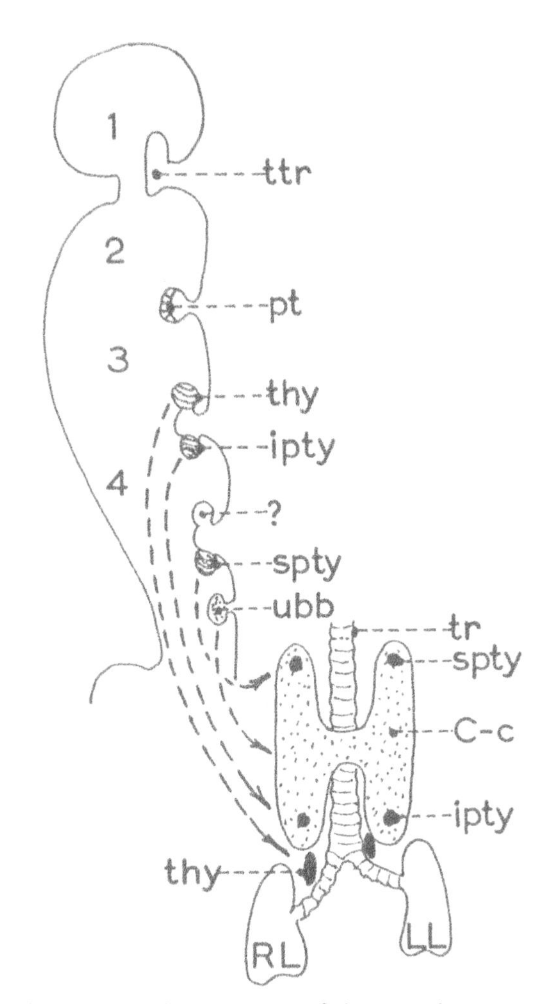

Figure 3.68: Migratory route of pharyngeal derivatives. C-c = C cells of thyroid; ipty = inferior parathyroid gland; pt = palatine tonsil; spty = superior parathyroid gland; thy = thymus gland; thyr = thyroid gland; tr = trachea; ttr = tubotympanic recess; ubb = ultimobranchial body; ? = undetermined.

POSSIBLE ABNORMALITIES

As the structures derived from some of the pouches 'migrate' to attain their final adult position, ectopic tissue may be found along the routes of migration or, as in the case of the parathyroid glands, they may migrate excessively to come to lie in the superior mediastinum.

The Development of the Tongue

The complicated innervation of the adult tongue can best be understood from its developmental history. The complex interlocking of a variety of derivatives to form the tongue gives rise to its intricate innervation.

The adult tongue is a membranous envelope filled with striated muscle, which has sensory (membranous) and motor (muscle) functions and can be divided anatomically into an anterior two-thirds and a posterior third. The demarcation between these two parts is a V-shaped line, the **sulcus terminalis**, the apex of which points posteriorly.

The tongue forms from a series of structures in the ventral floor of the pharynx (Fig. 3.69).

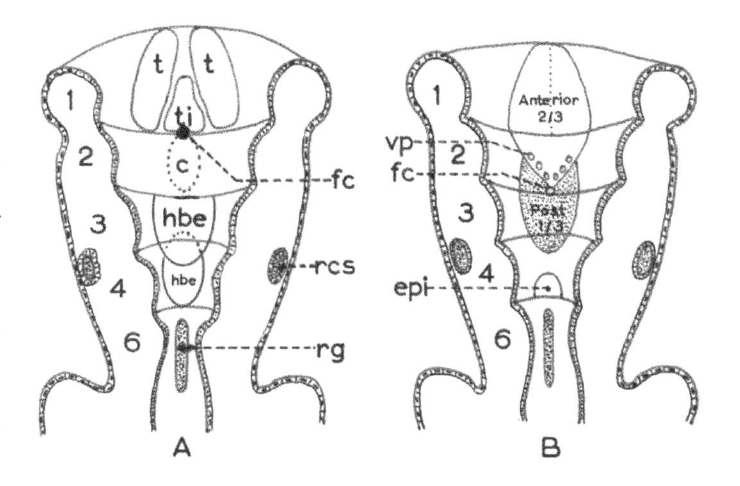

Figure 3.69: Formation of the tongue. 1, 2, 3, 4, 6 = pharyngeal arches; c = position of copula; epi = epiglottis; fc = foramen caecum; hbe = hypobranchial eminence; t = tongue; ti = tuberculum impar; rcs = remnant of cervical sinus; rg = respiratory groove; vp = circumvallate papilla.

In the 5th week pf three bulges appear in the middle of the ventral pharyngeal floor. These bulges are derived from the first pharyngeal arch and will be covered on their exterior by stomodeal ectoderm (as they lie anterior to the oropharyngeal membrane). They are filled with ectomesenchyme derived from the 1st pharyngeal arch. The three bulges are the **median tuberculum impar** and on either side of this, a **lateral lingual swelling**. The three bulges fuse to form the **anterior two-thirds** of the tongue, although the tuberculum impar does not grow as prolifically as the two lateral lingual swellings. The lateral lingual swellings will thus fuse anterior to the tuberculum impar. These bulges are derived from the 1st pharyngeal arch and will be supplied by a branch of the Vth cranial nerve (trigeminal) (Fig. 3.70). In addition, a (**pretrematic branch** – see Fig. 3.63, p.72) special sensory branch of the facial nerve, the **chorda tympani**, will also innervate this area (Fig. 3.70). The deepening of the 1st groove and the 1st pouch result in the formation of the **tympanic membrane**. This closing membrane is the only pathway through which the pretrematic branch of the VII cranial nerve can pass from the 2nd arch to supply the 1st arch. The chorda tympani supplies taste fibres to the anterior two-thirds of the tongue and in the adult must necessarily pass through the tympanic membrane (Fig. 3.70).

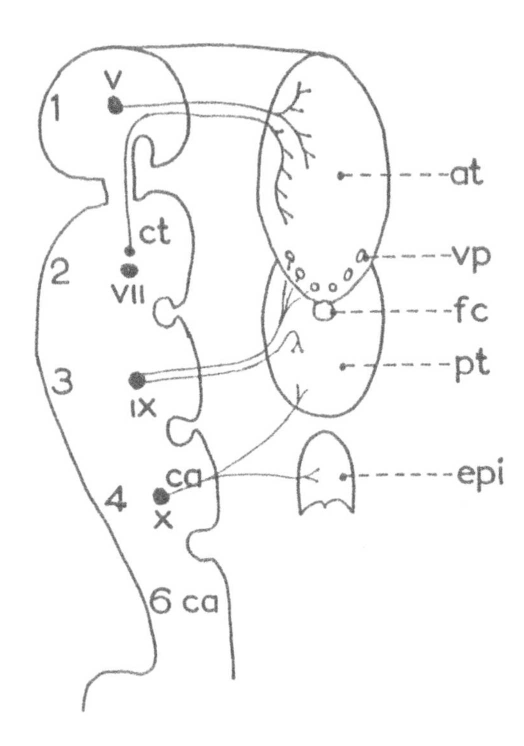

Figure 3.70: Sensory and autonomic nerve supply of the tongue and epiglottis.
1, 2, 3, 4, 6 = pharyngeal arches; V = trigeminal nerve; VII = facial nerve; IX = glossopharyngeal nerve; X = vagus nerve; at = anterior tongue; ca = cranial accessory via the vagus; ct = chorda tympani; epi = epiglottis; fc = foramen caecum; pt = posterior tongue; vp = circumvallate papilla.

Note there are three types of sensation:

- **general sensation** of pain, touch, temperature, vibration sense;
- **special sensation** of vision, audition, smell and taste;
- **propioceptive sensation** for the detection of position.

While the 2nd pharyngeal arch shows a swelling in its midline, known as the **copula** (meaning yoke, to 'span' or unite), this does not contribute to the adult tongue. The copula is submerged by proliferation of ectomesenchyme covered by endoderm, from the 3rd and 4th pharyngeal arches (Fig. 3.69). There is a greater contribution from the 3rd pharyngeal arch than the 4th arch to the posterior part of the tongue. The proliferation of the ectomesenchyme from the 3rd and 4th arches will interlock with the anterior two-thirds of the tongue. The contributions from the 3rd and 4th arches thus form the **posterior one-third** of the adult tongue. The innervation of the posterior one third of the tongue is thus derived from the branches of the cranial nerves to the 3rd and 4th arches, i.e. the IXth cranial nerve (glossopharyngeal) and Xth cranial nerve (superior laryngeal, which is actually the cranial part of the accessory nerve) respectively (Fig. 3.70).

The anatomical line along which the anterior two-thirds and posterior one-third of the tongue meet is known as the **sulcus terminalis**, in the centre of which is the **foramen caecum** which lies posterior to the tuberculum impar. The tissue surrounding the foramen will give rise to the thyroid diverticulum.

The **hypopharyngeal eminence** is said to arise from the fusion of the 3rd and 4th arches and its posterior part will give rise to the epiglottis.

The striated muscle of the tongue is said (from comparative vertebrate studies) to be derived from **occipital (postotic) myotomes** (muscle blocks). These migrate a consideable distance to reach the tongue and drag along with them the XIIth cranial nerve as their innervation (Fig. 3.70).

The tongue thus forms from an interlocking of stomodeal ectoderm, pharyngeal endoderm and ectomesenchyme, and occipital myotomes. Its complex innervation from diverse sources (cranial nerves V, VII, IX, X and XII) is thus explained. The sulcus terminalis forms a convenient anatomical division but does not act as a distinct boundary, as fibres from the glossopharyngeal nerve will cross it to supply the **circumvallate papillae**, which lie anterior to it.

Papillae and taste buds will arise on the dorsal surface of the tongue from the 7th week pf. The taste buds are said to be functional in the fetus and are used by the fetus to 'sense' the intra-amniotic environment.

POSSIBLE COMMON ABNORMALITIES

Aglossia:	lack of tongue
Macroglossia:	enlarged tongue
Microglossia:	tongue reduced in size
Ankyloglossia:	tongue tied down to ventral surface of oral cavity

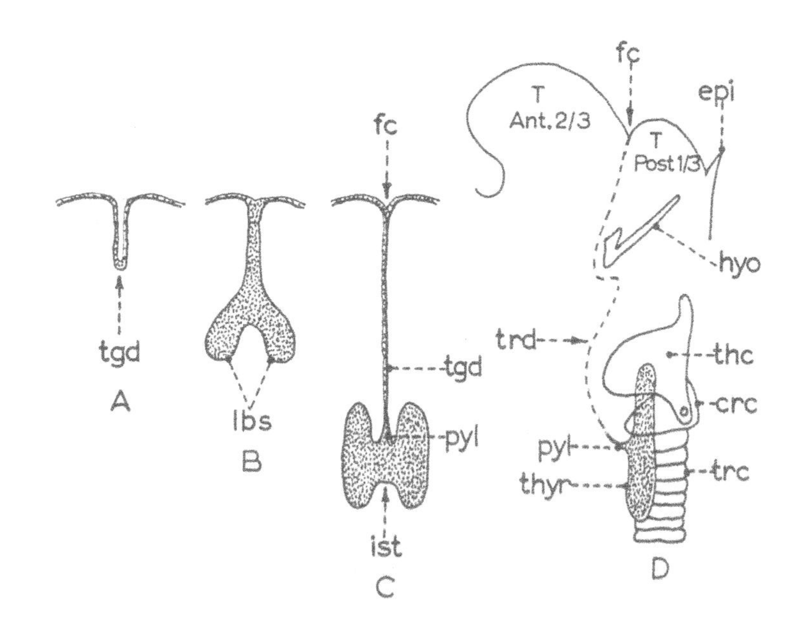

Figure 3.71: Development of the thyroid gland. A-C: ventral view; **D:** lateral view. crc = cricoid cartilage; epi = epiglottis; fc = foramen caecum; hyo = hyoid bone; ist = isthmus; lbs = lobes of thyroid; pyl = pyramidal lobe; thc = thyroid cartilage; tgd = thyroglossal duct; trc = tracheal cartilage; trd = track of thyroglossal duct; thyr = thyroid gland.

The Development of the Thyroid Gland

The thyroid gland develops in the midline of the ventral floor of the pharynx at the 4th week pf, caudal to the tuberculum impar as a ventral outgrowth from the foramen caecum (at the level between the 1st and 2nd pharyngeal pouches) **(Fig. 3.68 & Fig. 3.69)**. It is derived from endoderm. The evagination from the foramen gives rise to a tube, which elongates and extends caudally, ventral to the surface of the pharynx **(Fig. 3.71A,B)**. This tube is known as the **thyroglossal duct** as it will unite the developing thyroid gland to the developing tongue in the embryo. During its elongation the duct develops a bi-lobed tip which will give rise to the **thyroid gland (Fig. 3.71B,C)**. As it develops, it passes anterior to the cartilages of the 2nd and 3rd pharyngeal arches. The bifid tip forms the two lobes of the thyroid gland, which are joined by an isthmus. By the time the thyroid gland has reached its adult destination (level of the 2nd and 3rd tracheal cartilages) at about the 7th week pf, the thyroglossal duct will normally have degenerated **(Fig. 3.71D)**. The distal part of this duct may persist to give rise to the **pyramidal lobe** of the thyroid (this is sometimes accompanied by a muscle, the **levator thyroideae**, which may be a slip of muscle from the strap muscles of the neck). The foramen caecum persists as a small blind pit at the base of the tongue. If the thyroid gland does not 'descend' a **lingual thyroid** gland or **thyroid nodules**, will be present at the base of the tongue in the child. Aberrant thyroid tissue may arise from the ventral diverticulum of the 4th pharyngeal pouch.

By the 10th week pf, thyroid follicles, derived from endoderm, are observed in the thyroid gland and some time after, the gland will start to produce thyroglobulin. **Parafollicular cells (C cells** of the thyroid) are derived from neural crest cells, as is the connective tissue and capsule of the gland.

POSSIBLE COMMON ABNORMALITIES

- **Absence** of thyroid gland: leads to the clinical syndrome of cretinism (arrested physical and mental development).
- If the thyroglossal duct does not degenerate completely it may produce a **suprahyoid cyst** or an **infrahyoid cyst**. These arise as a result of secretion from the epithelium on the interior of the duct. These cysts are movable with the thyroid gland and are found in the midline of the neck.

Development of the Salivary Glands

In the adult all the salivary glands open into the oral cavity, for obvious functional reasons. It seems reasonable, therefore, to suppose that their origins and development would be related to the stomodeum. However, only the **parotid gland** (epithelium) is derived from stomodeal ectoderm, while the epithelium of the **submandibular** and **sublingual glands** is derived from the nearby pharyngeal endoderm. The area from which the glands arise is a junctional region between stomodeal ectoderm and pharyngeal endoderm. Extensive epithelio-mesenchymal interactions between the epithelium of each gland and the surrounding ectomesenchyme will result in a compound (branched) gland.

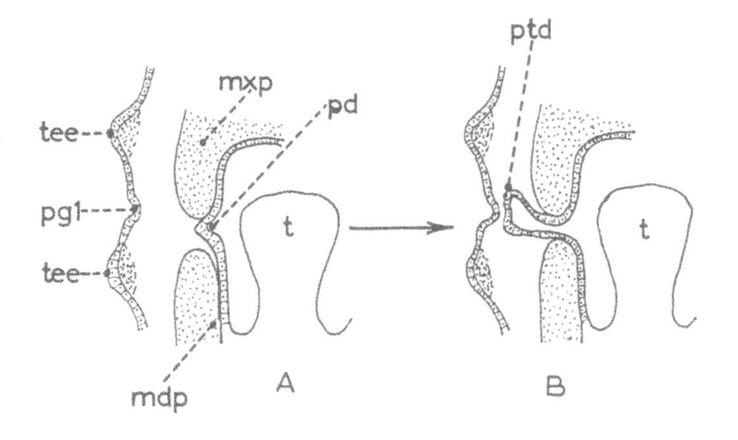

Figure 3.72: Development of the parotid gland seen in coronal section. mxp = maxillary prominence; mdp = mandibular prominence; pd = parotid dimple; pg1 = pharyngeal groove 1; ptd = parotid duct; t = tongue; tee = tubercle of external ear.

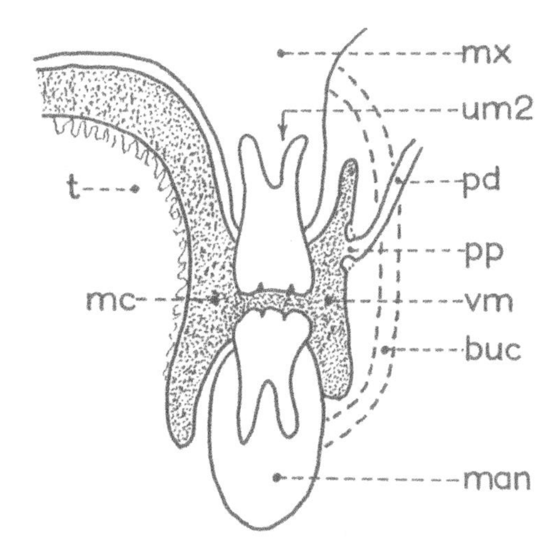

Figure 3.73: Coronal section of the adult oral cavity to show the opening of the parotid duct. buc = buccinator; mx = maxilla; pd = parotid duct; pp = parotid papilla; man = mandible; mc = mouth cavity; t = tongue; um2 = upper molar 2; vm = vestibule of mouth.

The **parotid gland** is the first to appear (±5 weeks pf) with the formation of a groove in the interior of the stomodeum between the maxillary and mandibular prominences. The groove deepens and closes over forming a tube (parotid duct) with an opening close to the angle of the stomodeum (**Fig. 3.72**). The deeper blind end of the tube burrows into the cheek to form the ductules and parenchyma of the gland. As the positions of the angles of the stomodeum are adjusted during the development of the face, the opening of the main duct recedes into the oral cavity to lie ultimately, opposite the upper second molar tooth (**Fig. 3.73**).

The **submandibular gland** is the next to appear (± 6 weeks pf) and is first visible as a groove emanating from the linguogingival region (**Fig. 3.74A**), which closes over to form the submandibular duct with an opening ultimately, under the tip of the tongue (**Fig. 3.74B**). The blind end of the duct burrows into the ectomesenchyme of the floor of the stomodeum, dividing progressively to form the ductules and substance of the gland.

The **sublingual gland** is the last of these glands to appear (± 8 weeks pf) and is the smallest of the three. It arises as a series of epithelial ingrowths in the floor of the stomodeum, close to, and to the lateral side of, the groove of the duct for the submandibular gland. By this time, this groove is closing over to form the submandibular duct. The epithelial ingrowths of the sublingual gland canalise and burrow into the ectomesenchyme of the floor of the stomodeum to form the ductules and the substance of the gland. The closeness of the epithelial ingrowths to the developing submandibular duct results in several of the sublingual ducts opening into the submandibular duct in the adult.

Three facts about the development of the salivary glands are noteworthy:

(a) They all have similar mechanisms of development related to branching morphogenesis.
(b) The presence of ectomesenchyme is essential for their induction and to give rise to the connective tissue of the glands.
(c) The parotid gland is of **ectodermal** origin; the submandibular and sublingual glands, from their origin behind the oropharyngeal

membrane are of **endodermal** origin. This is difficult to appreciate in the adult as massive positional changes have taken place in the development of the mouth and its contained structures.

Figure 3.75 is a composite diagram of the position of the three salivary glands in the adult in relationship to the mylohyoid muscle.

POSSIBLE ABNORMALITIES

A **ranula** results from the failure of one or more of the sublingual ductules to canalise. The ductule will thus form a cystic structure in the floor of the mouth. A ranula may 'plunge' into the neck by passing around the posterior free edge of the mylohyoid muscle.

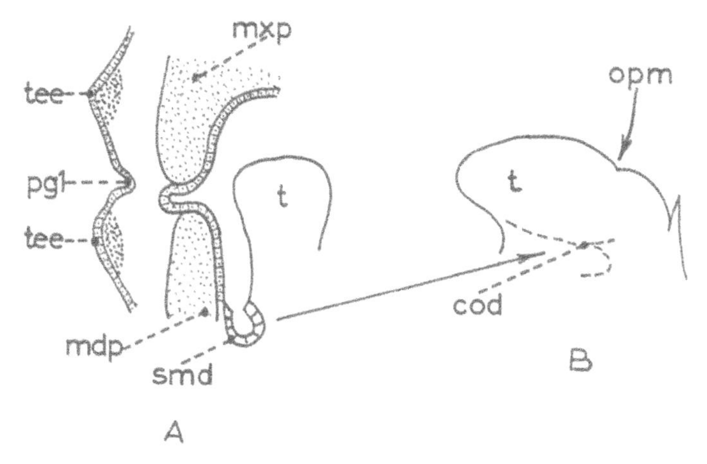

Figure 3.74: Development of the submandibular duct. A: embryo; **B:** adult. cod = combined opening of ducts; mxp = maxillary prominence; mdp = mandibular prominence; opm = position of oropharyngeal membrane; pg1 = pharyngeal groove 1; smd = submandibular duct; t = tongue; tee = tubercle of external ear.

The Development of the Gastro-intestinal Tract

The gastro-intestinal tract leads from the **oral cavity** to the **anus** and has a number of **associated glands**. The epithelium of the tube is mainly derived from endoderm with an ectodermal contribution at the stomodeal and proctodeal regions. Most of its connective tissue and muscle are derived from splanchnic mesoderm.

The development of the gastro-intestinal tract begins with the 'folding' of the embryo (Fig. 2.16A-C and 2.17B-F). When the embryo begins to convert from a flat structure to a tubular structure, the umbilical vesicle is drawn (pinched up) into the 'body' of the embryo. This occurs first in the head end of the embryo and then in the tail end, so that a tube (the gut tube), blind at either end but open widely ventrally in the middle, develops (Fig. 2.16C and 2.17A). Due to the folding, the opening between the tube and the umbilical vesicle gradually becomes narrowed to form the **vitello-intestinal duct**. The cranial part of the tube will form the **foregut**, the middle part which, at this stage, is still continuous with the umbilical vesicle forms the **midgut**, and the part which has folded into the caudal end of the embryo will become the **hindgut** (Fig 3.76).

The cranial termination of the tube is at the **oropharyngeal membrane** (the thinned prechordal plate). The part more rostral to the membrane is the **stomodeum** (part of which will form the mouth), which is lined with ectoderm (Fig. 3.40D). The oropharyngeal membrane, being

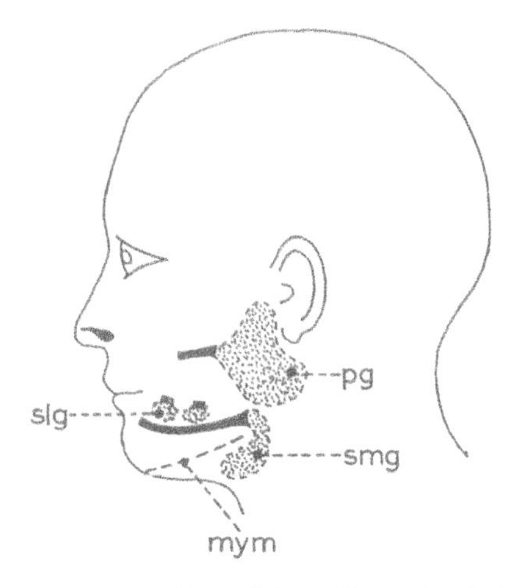

Figure 3.75: Position of the salivary glands in an adult to illustrate the relationship of the submandibular gland (smg) to the mylohyoid muscle (mym) (interrupted line). pg = parotid gland; slg = sublingual gland.

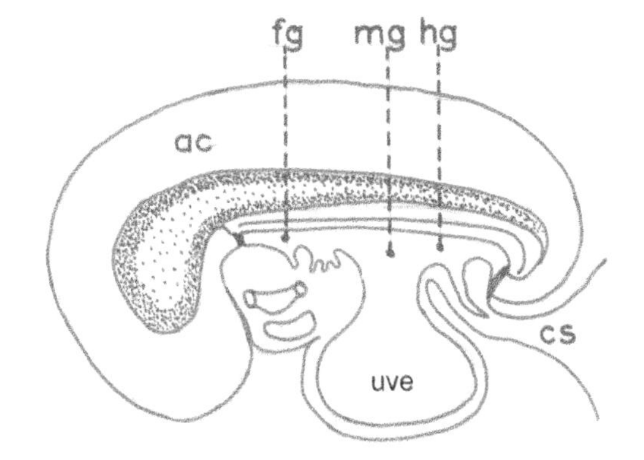

Figure 3.76: Divisions of the gut tube. ac = amniotic cavity; cs = connecting stalk; fg = foregut; hg = hindgut; mg = midgut; uve = umbilical vesicle.

devoid of mesoderm, is unstable, and breaks down at about the 4th week pf, resulting in continuity between the stomodeal cavity and the foregut tube. At the tail-end of the embryo, the hindgut is separated from the **proctodeum** by the **cloacal membrane** which, also being devoid of mesoderm, breaks down at about the 7th week pf.

Note the difference in time between the two disintegrations, this being indicative of the law of cranio-caudal development.

The vitello-intestinal duct will eventually close off from the midgut and the umbilical vesicle will degenerate. If the duct persists a **fistula** will open from the small intestine to the exterior at the level of the umbilicus.

> **COMMON ABNORMALITY**
>
> If only the proximal part of the vitello-intestinal duct persists, this will give rise to a pocket coming off the ileum which is known as the **diverticulum of the ileum** (Meckel's diverticulum).

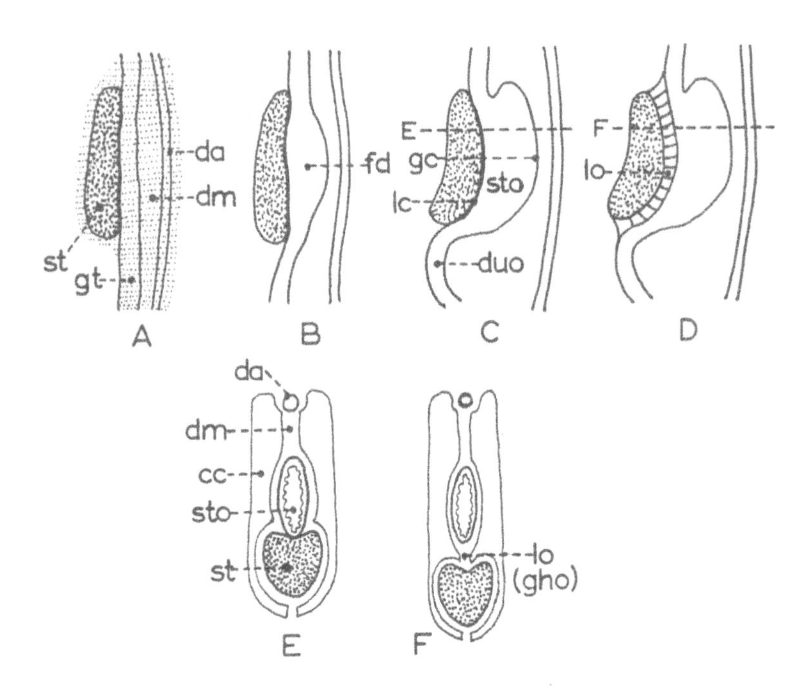

Figure 3.77: Lateral views (A-D) and transverse sections (E and F) of the developing stomach and the duodenum. Interrupted lines in figures C and D labelled E and F, indicate the plane of section of figures E and F respectively. cc = coelom; da = dorsal aorta; dm = dorsal mesentery; duo = duodenum; fd = fusiform dilatation; gc = greater curvature; gt = gut tube; lc = lesser curvature; lo = lesser omentum (gho = gastrohepatic omentum); st = septum transversum; sto = stomach.

The primitive gut tube may be divided into three parts which in the adult gives rise to specific regions of the gut: the foregut (from the tonsillar fossa to the middle of the second part of duodenum), the midgut (from the middle of the second part of the duodenum to the junction between the middle and lateral one-third of the transverse colon) and the hindgut (from the lateral third of the transverse colon to the ano-rectal junction). In the embryo, the junction between foregut and midgut is called the **anterior intestinal portal** and that between the midgut and hindgut, the **posterior intestinal portal**.

The **pharynx** is a dorso-ventrally compressed tube which will give rise to lateral pouches, the **pharyngeal pouches**. The development of these pouches is intimately related to the development of the pharyngeal arches. The **laryngotracheal diverticulum** (future respiratory system) will take origin from the caudal part of the pharyngeal tube and will be dealt with later.

The **oesophagus** arises caudal to the pharynx. The epithelium of the tube will proliferate rapidly and fill the lumen of the tube. Later in development, the tube will canalise so that a lumen is again formed. At first the oesophagus is short, but with continued growth of the gut tube it will elongate. This causes the **stomach**, a fusiform dilatation lying distal to it, to descend from its original cervical level to the abdominal cavity.

> **POSSIBLE COMMON ABNORMALITIES**
>
> - If the oesophagus does not elongate, the stomach will remain partly in the thoracic cavity. This is one of the forms of hiatus hernia.
> - If the oesophageal lumen does not re-canalise or only partially re-canalises, this will result in **atresia** (blockage) or **stenosis** (narrowing) of the oesophagus. This may also occur in other parts of the gastro-intestinal tube eg. duodenal atresia or stenosis.

Development of the Stomach and Duodenum

The entire gut tube is suspended from its posterior aspect by a **dorsal mesentery** (Fig. 3.77A,E) which, having two layers, with intervening mesenchyme allows blood vessels, nerves and lymphatics to enter and leave the gut. As the mesentery passes from either side of the gut tube, it encloses the **septum transversum** ventrally (Fig. 3.77A). That part of the gut tube which is opposite the septum will become the stomach (Fig. 3.77B) and is positioned in the cervical region of the embryo at this stage. The developing stomach undergoes a series of changes:

(a) The growth of its dorsal wall is more rapid than that of its ventral wall. This will result in the dorsal curvature of the stomach being longer (greater) than the ventral curvature. The ventral curvature is called the **lesser curvature** and the dorsal elongation the **greater curvature** of the stomach.

(b) At about the 8th week pf, the stomach rotates longitudinally through 90° in a clockwise direction when viewed ventrally. As a result, the lesser curvature will rotate to the right and the greater curvature to the left (Fig. 3.77C).

(c) The stomach elongates while the rotation is occurring. The rotation of the stomach causes the next part of the gut tube, the duodenum, to 'flop over' to the right.

(d) A dilatation, the future **fundus**, occurs at the cranial pole of the fusiform stomach. The simultaneous development of the right lobe of the liver will cause the cranial part of the stomach to lie to the left side.

(e) The stomach descends as the oesophagus elongates. Thus, the stomach arrives at its abdominal position. The descent of the stomach causes the duodenum to form a ventrally directed loop (Fig. 3.77C), which will now come to lie on the right as a result of the rotation of the stomach.

POSSIBLE COMMON ABNORMALITIES

- Abnormalities of the stomach are rare except for **hypertrophic pyloric stenosis.**
- **Duplication** of the lumen of the duodenum may occur. This abnormality may also occur in other parts of the tubular intestine.

There are a number of consequences as a result of the movements of the stomach (consult also the section on the development of the coelom and mesenteries):

(a) The stomach and duodenum move away from the septum transversum to make room for the rotation of the stomach to take place and this creates a mesentery between the gut tube and the septum (Fig. 3.77D). This mesentery is called the **gastrohepatic omentum** (**ventral mesogastrium**) and will form the **lesser omentum** of the adult (Fig. 3.77F).

(b) By this time, the proximity of the gut tube to the septum transversum has allowed the **hepatic diverticulum** from the foregut to penetrate the septum to form the liver cords (Fig. 3.78A). The hepatic diverticulum (part of which will become the gall bladder and the bile duct of the adult) lies between the layers of the ventral mesogastrium at its inferior end. This explains why - after positional adjustments of the organs in the adult - the **bile duct, hepatic artery** and **portal vein** lie in the right free edge of the lesser omentum (Fig. 3.78B).

(c) Once the hepatic diverticulum has entered the septum transversum from the foregut, it follows that the artery supplying the foregut will provide a branch to supply the liver diverticulum. This is the **hepatic branch of the coeliac trunk**. It therefore follows that the hepatic diverticulum and the hepatic artery will lie between the layers of the ventral mesogastrium (lesser omentum of the adult).

(d) The rotation of the stomach to the right creates a posterior 'pocket' to the left of the stomach. The opening of this pocket must therefore be on the right side (Fig. 3.78C). The pocket is called the **lesser peritoneal sac** (vestibule) and the opening, the **omental (epiploic) foramen**. The lesser sac will now develop additional excavations which are termed **recesses**. These are the **retrohepatic, splenic** (Fig. 3.78D) and **omental recesses,** which together constitute the omental bursa.

(e) The dorsal mesentery of the stomach (dorsal mesogastrium) forms the left boundary and posterior wall of the lesser sac. Due to the formation of the omental recess in the lesser sac, the dorsal mesogastrium becomes excavated to hang inferiorly over the transverse colon. This is known as the **greater omentum**. The posterior part of the greater omentum will fuse with the **transverse mesocolon** and the superior part of the transverse colon. The greater omentum in its development is a four-layered structure (Fig. 3.79A,B). With further development, the two interior layers will fuse with one another and partially obliterate the internal cavity (Fig. 3.79C) to a varying extent.

(f) As a result of the formation of the lesser sac, the peritoneal cavity has become divided into two. The smaller **lesser sac** lies posterior to the stomach, while the greater part of the peritoneal cavity lies anterior to the stomach, small intestine and colon and is referred to as the **greater sac**. The greater omentum therefore lies within the greater sac.

(g) The former right side (now dorsal) of the peritoneal surface of the duodenum comes into contact with the coelomic lining (future peritoneum) over the right **mesonephros**. The two coelomic linings (that over the duodenum and that over the mesonephros) fuse so that in the adult, the duodenum is firmly attached to the anterior surface of the kidney (**Fig. 3.80**). After positional adjustment of the organs in the adult, it is evident that the omental (epiploic) foramen lies superior to the adherence between the duodenum and the kidney (**Figs. 3.78D** and **3.80**).

The foregut and its derivatives are supplied with arterial blood by the coeliac trunk, a branch of the abdominal aorta.

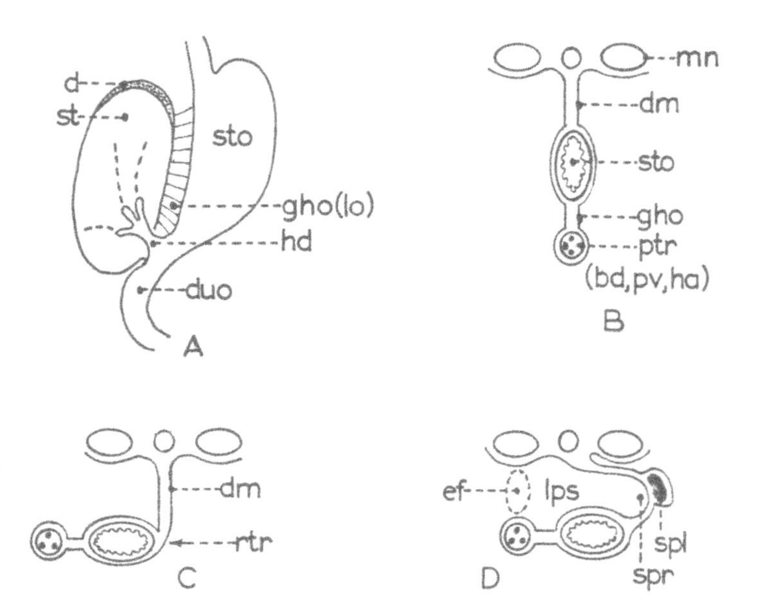

Figure 3.78: Formation of the lesser peritoneal sac. A: Lateral view. **B-D:** Transverse sections. bd = bile duct; d = diaphragm; dm = dorsal mesentery; duo = duodenum; ef = epiploic foramen; ha = hepatic artery; hd = hepatic diverticulum; lo/gho = lesser omentum/gastrohepatic omentum; lps = lesser peritoneal sac; mn = mesonephros; ptr = portal triad; pv = portal vein; rtr = rotates to right; spl = spleen; spr = splenic recess; st = septum transversum; sto = stomach.

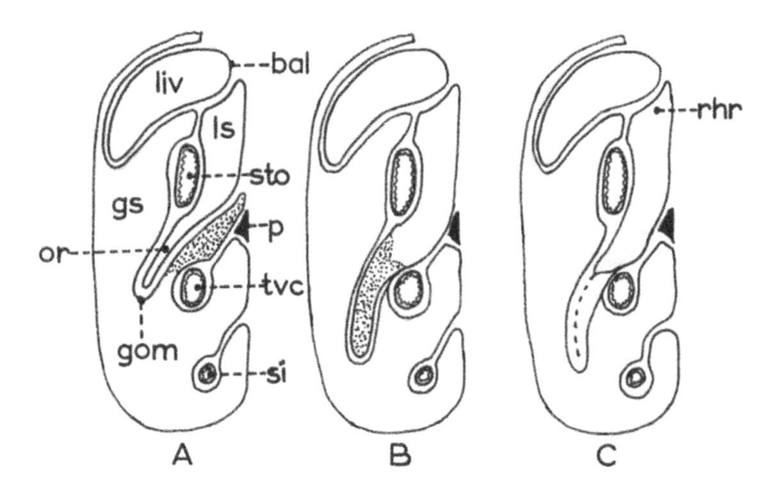

Figure 3.79: Sagittal sections to show the development of the greater omentum. bal = bare area of liver; gom = greater omentum; gs = greater peritoneal sac; liv = liver; ls = lesser peritoneal sac; or = omental recess; p = pancreas; rhr = retrohepatic recess; si = small intestine; sto = stomach; tvc = transverse colon.

Development of the Midgut

The development of the **midgut** is rapid, especially after the vitello-intestinal duct has become elongated and attenuated. It seems possible that with the rapid elongation of the midgut, the abdominal cavity is too small to accommodate the midgut adequately, due to the large mesonephros and liver. As a result, the midgut herniates into the connecting stalk (which later will form part of the umbilical cord) at approximately the 4th week pf. The rapid growth and the herniation of the midgut result in the formation of a U-shaped tube or 'loop' having **proximal** and **distal** limbs (**Fig. 3.81A**).

Note the terms proximal and distal refer to the 'closeness' of a structure to its point of origin. Thus the proximal limb of the midgut is closer to the stomach than the distal limb.

The loop is 'based' upon an artery coming from the dorsal aorta and which later becomes the **superior mesenteric artery** (**Fig. 3.81A,B**). Between the 7th and 10th weeks pf, the presence of an **'umbilical' hernia** is regarded as a normal situation.

Note. A **hernia** is the protrusion of the contents of one cavity into another cavity or to the exterior.

While the midgut is in the umbilical cord, the following changes occur:

(a) When the embryo is viewed from its ventral

aspect, the midgut loop rotates around the superior mesenteric artery through 90° (in an anti-clockwise direction), so that the proximal limb comes to lie on the right and the distal limb on the left of the embryo.

(b) A bulge, the primitive **caecum**, will appear at the transition between the proximal and distal limbs of the midgut loop. An extension of the caecum, the **vermiform appendix**, will also appear. The proximal limb (now on the right) will undergo much greater elongation than the distal limb (now on the left). This elongation leads to the formation of the loops of the future **small intestine** from the proximal limb.

(c) At approximately the 10th week pf, when the abdominal cavity has increased in size (due to the degeneration and disappearance of the mesonephros and the slowing down in the growth of the liver), the midgut loop begins to return to the abdominal cavity. This it does by rotating through 180° (in an anti-clockwise direction) (**Fig. 3.81B**). This rotation continues around the superior mesenteric artery as the axis, with the proximal limb returning first. The proximal limb (future small intestine) will return to a dorsal and inferior position. The distal limb becomes draped around the proximal limb in a more superior and ventral position as a picture frame would surround a picture. The caecum comes to lie inferior to the right lobe of the liver (in a **subhepatic** position) (**Fig. 3.81C**). With further growth of the distal limb 'cranial to' the caecum, the caecum will descend into the right iliac fossa. This elongation of the tube will give rise to the ascending colon (**Fig. 3.81D**).

The part of the tube which originally formed the proximal part of the midgut loop forms the **jejunum** and upper two-thirds of the **ileum** and comes to lie more or less in the centre of the abdomen. The distal limb surrounds the centrally placed jejunum and ileum (**Fig. 3.81D**) and gives rise to the **rest of the ileum, ascending and transverse colon.** The **descending colon** is not part of the midgut loop. (The **descending colon, sigmoid colon** and **rectum** are not derived from the midgut loop and will thus not have herniated into the umbilical cord.) By the end of the 10th week pf, all of the midgut tube has returned to the abdominal cavity.

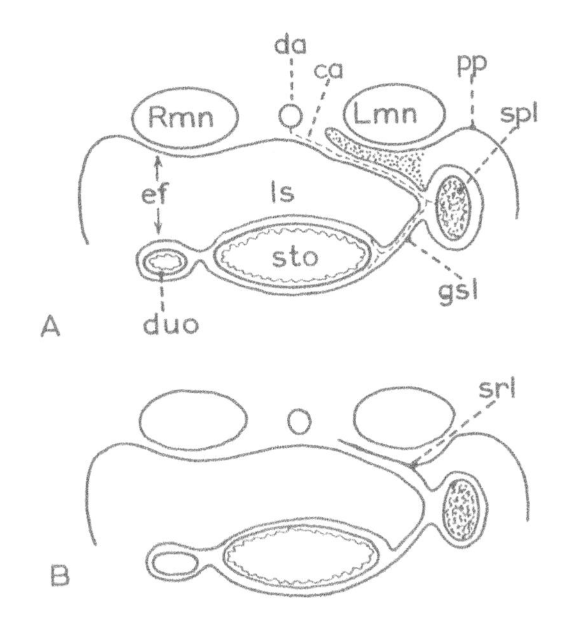

Figure 3.80: Transverse sections of the surroundings of the lesser peritoneal sac.
B indicates the closure of the stippled area in **A**, resulting in the formation of the splenorenal ligament (srl). ca = coeliac artery; da = dorsal aorta; duo = duodenum; ef = epiploic foramen gsl = gastro-splenic ligament; Lmn = left mesonephros; ls = lesser sac; pp = parietal peritoneum; Rmn = right mesonephros; spl = spleen; sto = stomach.

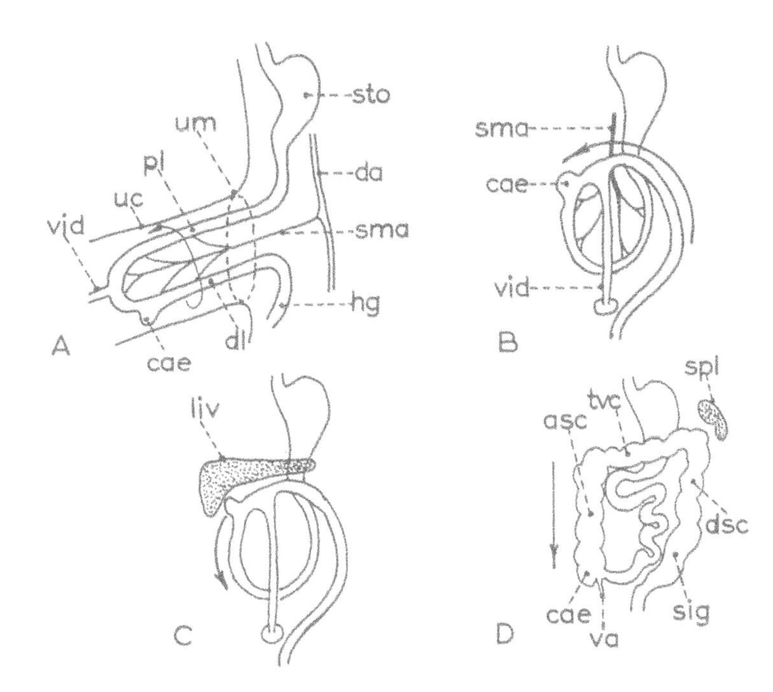

Figure 3.81: Rotation of the midgut loop. A: Lateral view. **B-D:** Ventral views. Arrows indicate direction of rotation. asc = ascending colon; cae = caecum; da = dorsal aorta; dl = distal limb; dsc = descending colon; hg = hindgut; liv = liver; pl = proximal limb; sig = sigmoid colon; sma = superior mesenteric artery; sto = stomach; spl = spleen; tvc = transverse colon; uc = umbilical cord; um = umbilicus; va = vermiform appendix; vid = vitello-intestinal duct.

POSSIBLE COMMON ABNORMALITIES

- If the midgut loop fails to return to the interior, a congenital abnormality known as **exomphalos** results. This herniation will be covered by the membrane of the umbilical cord. If the loop returns to the interior and then herniates out again, it will be covered by skin (as the skin will have formed by this time) and is known as a **secondary umbilical hernia.**
- Normally the return of the intestine to the abdomen takes place in an anti-clockwise direction (when viewing the embryo from its ventral aspect). If this rotation takes place in a clockwise (malrotation) direction, the parts of the midgut loop will be abnormally placed, eg. the appendix may come to lie in the left iliac fossa.
- If the ceacum fails to descend to the right iliac fossa, the caecum and vermiform appendix will remain in a sub-hepatic position. This may lead to an abnormal site of pain in acute appendicitis.

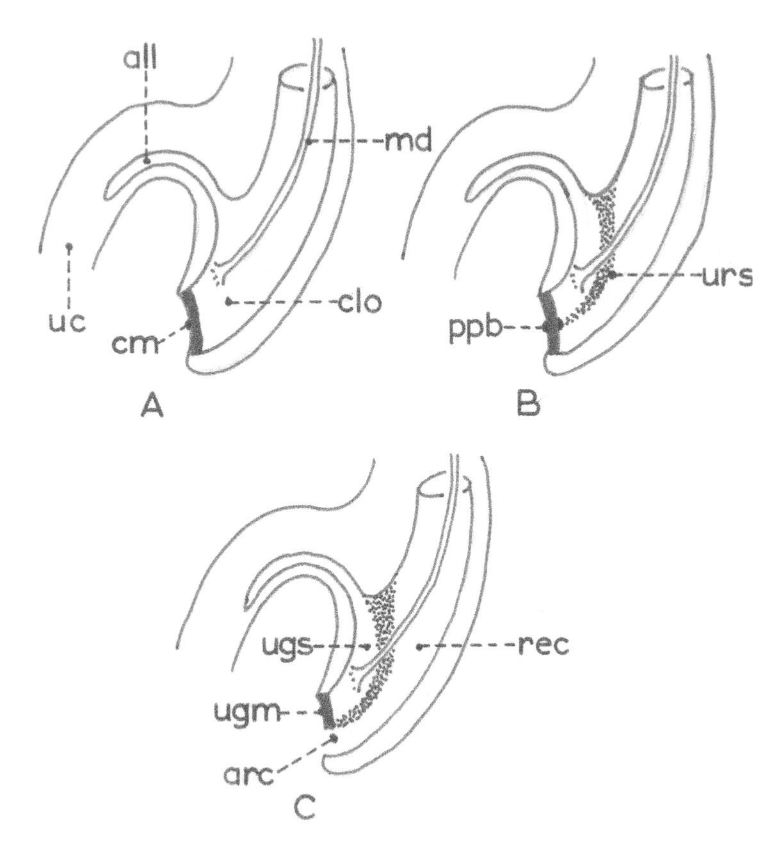

Figure 3.82: Development of the hindgut in lateral view.
all = allantois; arc = anorectal canal; clo = cloaca; cm = cloacal membrane; md = mesonephric duct; ppb = position of perineal body; rec = rectum; uc = umbilical cord; ugs = urogenital sinus; ugm = urogenital membrane; urs = urorectal septum.

Development of the Hindgut

The **hindgut** is a continuation of the midgut and is separated from the exterior, the proctodeum, by the cloacal membrane. The inferior part of the hindgut is slightly more dilated than the rest of the gut and has an extension into the body stalk called the **allantoic diverticulum** (or **allantois**) (Fig. 3.82A). When the urinary system develops, the mesonephric duct enters the hindgut and thus, at this time, the hindgut may be regarded functionally as a **cloaca** (a cavity containing urine and faecal material). The cloaca becomes separated into a ventral **urogenital sinus** and a dorsal **rectum** by the downgrowth of the **urorectal septum** which reaches the perineum at the position of the **perineal body** (Fig. 3.82B). This septum consists of mesenchyme which, together with the mesenchyme surrounding the rectums will form the anal musculature. The rectum is initially separated from the proctodeum by the cloacal membrane so that ultimately, with the dissolution of the membrane at the 7th week pf, the hindgut becomes continuous with the ectodermally-lined proctodeum, forming the anal canal (Fig. 3.82C).

The terminal part of the epithelium of the gastro-intestinal tract (descending colon, sigmoid colon, rectum and superior part of anal canal) is formed from that part of the hindgut lined by endoderm. The inferior part of the anal canal is lined by proctodeal ectoderm.

The arterial supply to the hindgut and its derivatives is via the inferior mesenteric artery, a branch of the abdominal aorta.

Clinical significance of the junction between the hindgut and the proctodeum in relation to its embryological development is detalied below.

The junction between the hindgut and the proctodeum is said to be marked by the pectinate line of the adult and is regarded clinically as a watershed region for the following reasons:

- The hindgut is supplied with arterial blood from the inferior mesenteric artery while the proctodeal derivative (inferior part of anal canal) is supplied by the pudendal artery.
- The venous drainage of the hindgut is into the portal vein, whereas the venous drainage of the proctodeal region is into the pudendal vein, which drains into the inferior vena cava.
- The lymphatic drainage of the hindgut is into the paraortic nodes, while the drainage of the proctodeal region is into the inguinal nodes.

- The nerve supply of the hindgut is by the autonomic system, whereas the nerve supply of the proctodeal region is somatic in nature.

POSSIBLE COMMON ABNORMALITIES

- Congenital **aganglionic megacolon** (Hirshsprung's disease): In this condition part of the colon lacks ganglion cells which are of neural crest origin. The reason for the failure of neural crest migration to this region is unknown. Lack of nerve cells results in the inability of the hindgut to propel its contents. This will result in dilatation of the segment of the gut proximal to the denervated part.
- If the cloacal membrane fails to disintegrate, **imperforate anus** will result.
- Failure of the urorectal septum to completely separate the ventral urogenital sinus from the dorsal rectum may result in a **urorectal fistula.**

Development of Associated Gastro-intestinal Organs and the Spleen

There are three organs closely associated with the development of the gastro-intestinal tract, one (**spleen**) which develops independently in the dorsal mesogastrium, and two which are derivatives of the tract itself (**liver**, **pancreas**).

Spleen

The **spleen** begins its development at about the 5th week pf as a condensation of mesoderm between the layers of the posterior mesogastrium (Fig. 3.78D). As the **lesser peritoneal sac** is formed by the rotation of the stomach, the developing spleen lies within the dorsal mesogastrium at the left extremity of the sac. The posterior layer of the dorsal mesogastrium comes into contact with, and fuses with, the coelomic membrane covering the left mesonephros (Fig. 3.80A,B) (cf. duodenum and right mesonephros). This fusion becomes the **splenorenal ligament**. The part of the former dorsal mesogastrium joining the spleen to the stomach becomes the **gastrosplenic ligament**. Since the foregut is supplied by the coeliac trunk, it follows that the stomach will be supplied by this artery. The spleen likewise will take a branch from the coeliac trunk, which is passing close to it (Fig. 3.80A,B). This artery becomes the **splenic artery**.

The mesoderm forming the spleen becomes transformed into a trabecular network in which red blood cells, white blood cells, platelets and lymphocytes are formed, while branches of the splenic artery form blood sinuses within the network. Thus, the spleen is a haemopoietic organ in the embryo until such time as the bone marrow takes over this function.

POSSIBLE COMMON ABNORMALITY

Accessory spleens may occur. These are often found in relation to the hilus of the normal spleen or the tail of the pancreas.

Liver

The **liver** begins to develop at about the 3rd week pf from a thickening of the endodermal gut tube at the anterior intestinal portal (Fig. 3.83A,B). A diverticulum, known as the **hepatic diverticulum**, protrudes from the thickened endodermal tube. This region of the foregut is separated from the adjacent septum transversum by a short ventral mesentery. The hepatic diverticulum grows anteriorly into this mesentery (Fig. 3.83B,C) as well as into the substance of the septum transversum. The hepatic diverticulum elongates and dilates distally to give rise to the **gall bladder** and its duct, the **cystic duct** (Fig. 3.83D). On the cranial aspect of the hepatic diverticulum, cords of liver cells, the **hepatocytes**, proliferate and grow into the ventral mesentery and the caudal part of the septum transversum. The cords of hepatocytes are linked to the cystic duct via **hepatic ducts**. Together the cystic duct and the hepatic ducts form the **bile duct**, which at this stage is opening into the ventral surface of the duodenum (Fig. 3.83E). The hepatocytes are of endodermal origin and are secretory in nature.

As the hepatic cords proliferate they encroach upon two pairs of veins, the **vitelline** and **umbilical veins** which lie lateral to them. First, they encroach on the paired vitelline veins breaking them up into sinusoids (vessels with a discontinuous endothelium) between the hepatic cords (Fig. 3.84A). As proliferation of the liver continues, it encroaches on the paired umbilical veins, engulfs the vessels and once again breaks them up into sinusoids. The liver reaches a reasonable size and therefore presents a considerable vascular resistance. The vascular resistance is overcome by the formation of a **ductus venosus** which is placed between the distal part of the left umbilical vein and the proximal part of the persistant right vitelline vein (hepatic outflow) (Fig. 3.84B). The oxygenated blood from the left umbilical vein is therefore directed across the developing liver in the ductus venosus. Thus there is a minimal amount of admixture of oxygenated

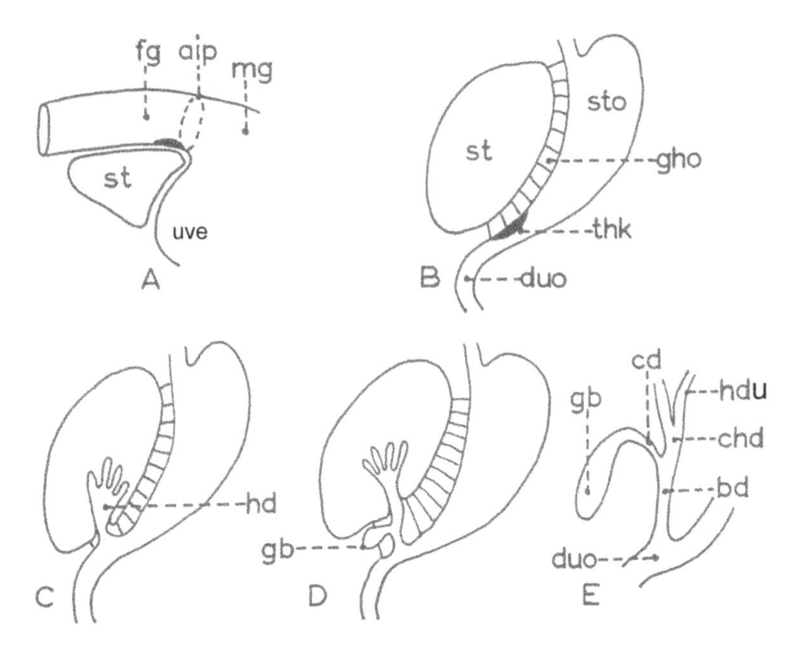

Figure 3.83: Lateral views of the development of the liver and ducts (A-D). E: composite diagram of ductal system. aip = anterior intestinal portal; bd = bile duct; cd = cystic duct; chd = common hepatic duct; duo = duodenum; fg = foregut; gb = gall bladder; gho = gastro-hepatic omentum (ligament); hd = hepatic diverticulum; hdu = hepatic duct; mg = midgut; st = septum transversum; sto = stomach; thk = thickening from which hepatic diverticulum will arise; uve = umbilical vesicle.

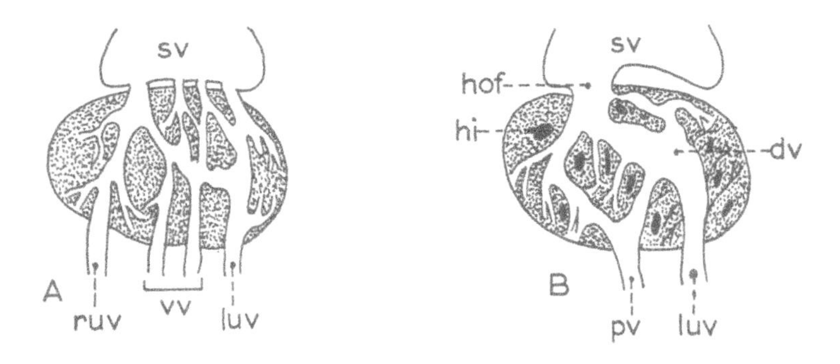

Figure 3.84: Ventral views (A and B) illustrating venous development within the liver. dv = ductus venosus; hi = haemopoietic islands; hof = hepatic outflow; luv = left umbilical veins; pv = portal vein; ruv = right umbilical vein; sv = sinus venosus; vv = vitelline veins.

blood from the left umbilical vein and the de-oxygenated sinusoidal blood. The sinusoids of the liver are derived from the angioblastic tissue of these veins. Islands of **haemopoietic tissue** form between the sinusoids and the hepatocytes (Fig. 3.84B). These islands will form red and white blood cells which enter the blood vessels. The liver acts as a haemopoietic organ between the 3rd and 5th months pf until the definitive bone marrow takes over this function.

Since the foregut is supplied by the coeliac trunk, the liver being a derivative of the foregut will be supplied by the **hepatic artery**, which is a branch of the coeliac trunk.

The ventral mesentery surrounds the developing liver, giving rise to its peritoneal covering. Ventrally (anteriorly) the ventral mesentery extends on to the anterior body wall and persists in the adult as the **falciform ligament** which contains the **ligamentum teres** (obliterated left umbilical vein). Cranially, where the peritoneal reflections of the liver continue on to the inferior surface of the diaphragm, the **coronary ligaments** will form (Fig. 3.85). The lateral extensions of the coronary ligaments are called the **triangular ligaments**. These ligaments surround the **bare area** of the liver and form an attachment of the liver to the diaphragm.

The liver continues to grow rapidly until approximately the 10th week pf after which the growth diminishes. The decrease in growth of the liver will affect the left lobe more than the right lobe, hence the difference in size of the two lobes in the adult. The **quadrate** and **caudate** lobes are said to arise from the right lobe of the liver.

POSSIBLE ABNORMALITIES

- Intra- and extra-hepatic biliary atresia are relatively rare and would result in disturbed liver function.
- Additional biliary ducts are relatively common.

Pancreas

The **pancreas** arises from two diverticula which originate from the caudal part of the foregut close to the origin of the hepatic diverticulum. (Fig. 3.86A,B). The first or dorsal diverticulum arises at approximately the same time as the hepatic diverticulum (4th week pf) and slightly cranial to it but from the dorsal aspect of the

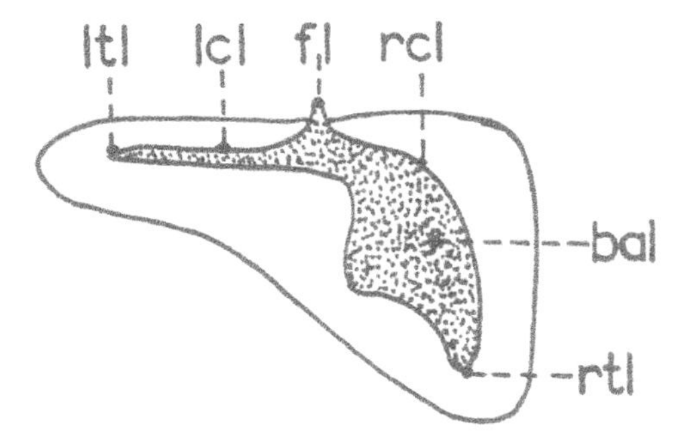

Figure 3.85: Ligaments and bare area of the adult liver from the posterior aspect.
bal = bare area of liver; fl = falciform ligament; lcl = left coronary ligament; ltl = left triangular ligament; rcl = right coronary ligament; rtl = right triangular ligament.

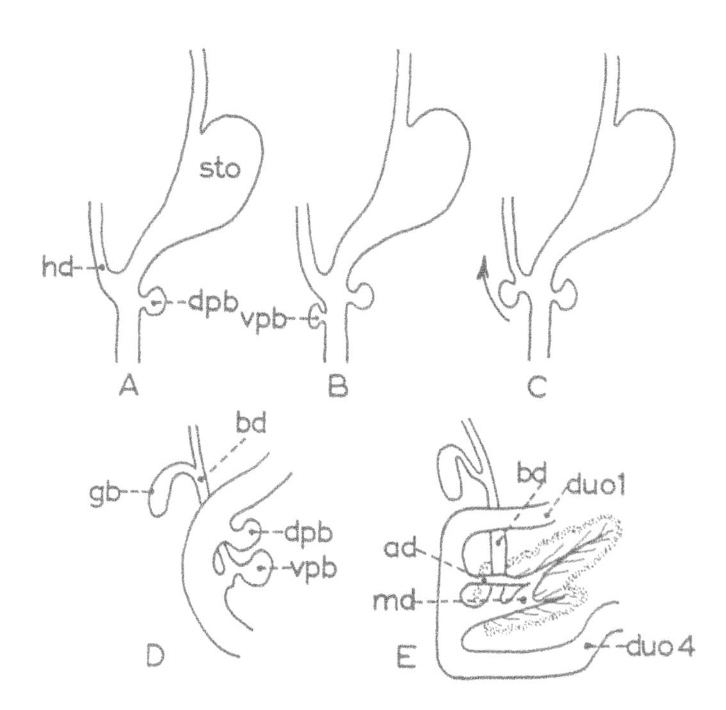

Figure 3.86: Lateral views of the development of the pancreas.
ad = accessory duct; bd = bile duct; dpb = dorsal pancreatic bud; duo1 = 1st part of duodenum; duo4 = 4th part of duodenum; gb = gall bladder; hd = hepatic diverticulum; md = main duct; sto = stomach; vpb = ventral pancreatic bud.

duodenal tube. This diverticulum is known as the **dorsal pancreatic bud**. It grows into the dorsal mesentery. The second diverticulum originates slightly later than the dorsal pancreatic bud and from the ventral surface of the duodenum, just caudal to the hepatic diverticulum, and is known as the **ventral pancreatic bud** (Fig 3.86B). As the hepatic diverticulum elongates, the ventral pancreatic bud is pulled away from the duodenum and appears to come directly off the bile duct (Fig. 3.86C). (There is a view that the ventral pancreatic bud develops directly from the duodenal end of the bile duct.) The ventral pancreatic bud now begins to migrate around the *right* side of the duodenum and as it does so, it pulls the bile duct around with it (Fig. 3.86C). The ventral pancreatic bud (and the bile duct) now come to lie on the dorsal surface of the duodenum, a short distance below the position of the dorsal pancreatic bud (Fig. 3.86D), and between the layers of the dorsal mesentery of the duodenum (**mesoduodenum**). Thus, the ventral pancreatic bud and bile duct have migrated through 180° from ventral to dorsal. The ventral and dorsal buds now fuse (Fig 3.86E). At this time, the stomach has begun its rotation through 90° (along its longitudinal axis) and, as a result, the duodenum 'flops' over to the right. Thus, the pancreatic buds and the bile duct now come to open on to the *left* side of the duodenum.

As a result of the fusion of the two pancreatic buds, the duct of the ventral pancreatic bud joins the duct of the dorsal pancreatic bud. The **main pancreatic duct** (of Wirsung) will form from the distal part of the duct of the dorsal pancreatic bud plus the duct of the ventral pancreatic bud. The proximal part of the duct of the dorsal pancreatic bud will form the **accessory pancreatic duct** (of Santorini).

The glandular parenchyma as well as the duct systems of the two parts of the pancreas fuse with one another to form the completed pancreas. The endoderm gives rise to both the exocrine and endocrine components of the pancreas as well as the duct system.

POSSIBLE ABNORMALITIES

- Ectopic pancreatic tissue may be found in the stomach, duodenum and the diverticulum of ileum.
- The ventral pancreatic bud occasionally splits into two and one part passes with the hepatic diverticulum around the right side of the duodenum, while the other passes around the left side of the duodenum. In this way, the duodenum becomes surrounded by a ring of pancreatic tissue. This is called an anular pancreas and may constrict the duodenum causing obstruction.

Development of the Diaphragm

In the adult, the diaphragm forms the physical separation between the thoracic and abdominal cavities. Anatomically, it consists of a **central tendon** surrounded by radially disposed striated muscle fibres. The muscle fibres are innervated by the **phrenic nerve** (C3, 4, 5), but mostly from the 4th cervical nerve. This nerve supplies the diaphragm from its inferior surface. The diaphragm is covered on its superior surface by diaphragmatic pleura supplied by intercostal nerves. Its inferior surface is covered with peritoneum which is supplied with sensory fibres of the phrenic nerve. As it is a septum between the thorax and the abdomen, structures such as the aorta, oesophagus, inferior vena cava and nerves will pass through it.

In the embryo, the diaphragm originates in a cranial mass of lateral plate mesoderm which has been repositioned by the longitudinal folding of the embryo to lie ventral to the foregut (that part which will form the oesophagus). This collection of mesoderm contains that part of the intra-embryonic coelom which is destined to become the pericardial sac of the adult. During the folding of the embryo the mass of mesoderm undergoes a process of reversal, so that the part which was originally cranial, now comes to lie in a caudal position (Fig. 3.87A). This repositioning results in the mass being wedged transversely between the foregut and the adjacent part of the umbilical vesicle and is thus called the **septum transversum** (Fig. 3.87B).

The septum transversum is not a complete septum across the embryo, since the paired pleural canals are situated directly dorsal to the septum on either side (Fig. 3.87C).

Note. In amphibia, reptiles and birds the communication between pleural and peritoneal cavities is persistent, whereas in mammals the openings between the channels are obliterated so that the diaphragm ultimately forms a complete separation between the pleural and the peritoneal cavities.

The caudal part of the mesodermal mass (septum transversum) will be penetrated by liver cords from the adjacent gut tube. The cranial part, which abuts on the pericardial coelom, will form the greater part of the diaphragm. After the establishment of the septum transversum, the diaphragm develops in a series of phases:

Phase 1: The cranial part of the septum transversum is infiltrated with myoblasts from the 4th cervical myotome (Fig. 3.87D,E). These radially infiltrating myoblasts stop short of the central part of the septum, which becomes the central tendon for the surrounding muscle attachment. These myoblasts will bring with them their nerve supply, from the 4th cervical nerve segment.

Phase 2: The gut tube has developed a dorsal mesentery (**meso-oesophagus**) which lies in the midline, dorsal to the septum transversum (Fig. 3.87F). The mesentery will give rise to the formation of the **crura** (vertebral attachment) of the diaphragm which surround the oesophagus, producing the oesophageal hiatus.

Phase 3: The **urogenital ridge** develops on the dorsal wall of the embryo and projects into the junction between the pleural canals and the peritoneal cavity (Fig. 3.87F). As the ridge increases in size, it protrudes progressively ventrally and the pleuro-peritoneal junction is progressively closed and replaced by the **pleuro-peritoneal membranes** (Figs. 3.87G and 3.88).

Thus, in summary, the diaphragm may be said to be formed by four structures (Fig. 3.87H):

- Septum transversum (central tendon);
- Body wall muscle;
- Mesentery of oesophagus (crura);
- Pleuro-peritoneal membranes.

The diaphragm arises at cervical levels and then descends to a region between the thorax and the abdomen due to the descent of the heart. It, however, retains its innervation from cervical levels.

POSSIBLE COMMON ABNORMALITIES

- **Agenesis** of the diaphragm is rare.
- **Eventration** of the diaphragm is an upward displacement of one of the sides of the central tendon of the diaphragm and is due to the failure of myoblastic infiltration.
- Failure of various parts of the diaphragm to form will result in the passage of organs from the abdominal cavity into the thoracic cavity. This is known as a **diaphragmatic hernia.** A hiatus hernia may also occur in the region of the crura. A **posterior hiatus** (of Bochdalek) results from a persistant pleuro-peritoneal canal. An **anterior** or **parasternal hiatus** (of Morgagni) may allow the passage of abdominal contents into the anterior mediastinum (Larrey's space).

Figure 3.87: A series of diagrams to illustrate the development of the diaphragm. Figure **A** depicts the position of the septum transversum (opstm) at the cranial end of the embryo in a sagittal section. Figure **B** is a lateral view showing the 'reversal' of the cardiac region and the final position of the septum transversum (fpstm). Figure **C** is a ventral view of a transverse section of the diaphragm with the coelom passing though it. ac = amniotic cavity; cm = cardiogenic mesoderm; fg = foregut; lpm = lateral plate mesoderm; n = notochord; pc = pericardial coelom; plc = pleural coelom; om = oropharyngeal membrane; pp = prechordal plate; st = septum transversum; uve = umbilical vesicle. Figures **D-H** are transverse sections at the level of the developing diaphragm. Arrows (tcf) indicate the track of muscle fibres into the crura of the diaphragm. appm = advancing pleuroperitoneal membrane; bwm = body wall muscle; cns = central nervous system; cppm = completed pleuroperitoneal membrane; ct = central tendon; da = dorsal aorta; der = dermatome; dm = dorsal mesentery; gt = gut tube; myo = myotome; n = notochord; ppc = pleuroperitoneal coelom; ppm = pleuroperitoneal membrane; r = rib; scl = sclerotome; som = somite; st = septum transversum; tpm = track of body wall muscle fibres; ugr = urogenital ridge; vc = vertebral column.

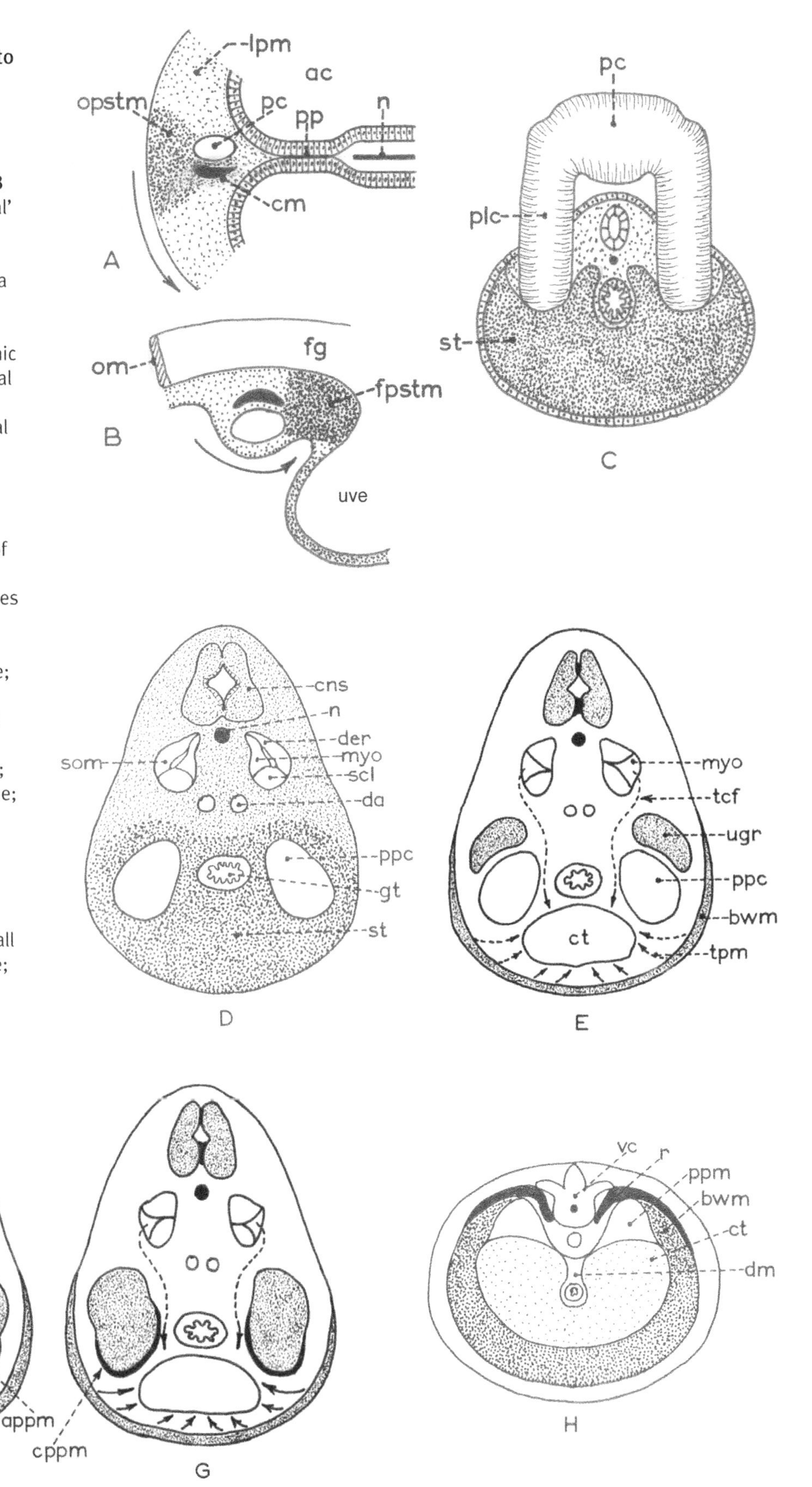

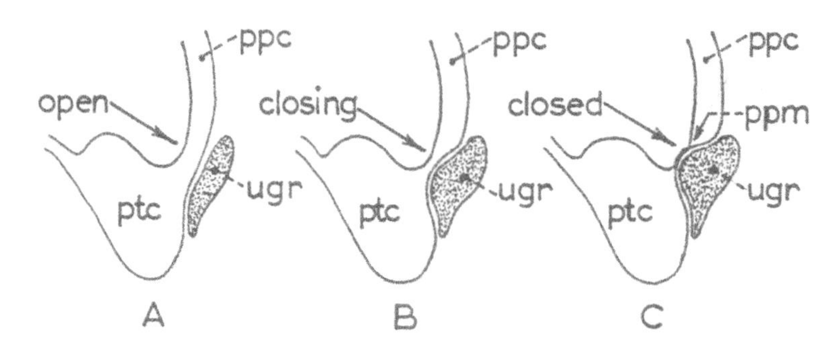

Figure 3.88: Lateral views of the closure of the pleuroperitoneal canal. ppc = pleuroperitoneal canal; ppm = pleuroperitoneal membrane; ptc = peritoneal cavity; ugr = urogenital ridge.

DEVELOPMENT OF THE RESPIRATORY SYSTEM

Before studying this section, revise the section on the development of the pharyngeal arches. Fundamentally the respiratory system is an offshoot of the gastro-intestinal system and begins at the level of the 6th pharyngeal arch.

When the pharyngeal arches are viewed from the interior, there is a transverse groove just caudal to the epiglottis (which arises from the hypopharyngeal eminence) **(Fig. 3.89A,C)**. Extending caudally from the transverse groove for a short distance is a vertical depression, the **laryngotracheal diverticulum** (respiratory groove), in the ventral pharyngeal wall **(Fig. 3.89D)**. Since the larygotracheal diverticulum arises from the pharyngeal wall its mucous membrane is of endodermal origin. The surrounding tissues, eg. cartilages and connective tissue, will be of mesodermal origin.

The vertical depression which will give rise to the **respiratory tract** deepens vertrally forming a hollow evagination **(Fig. 3.90A)**. Soon, two balloon-like dilatations develop at the lower end of the ridge and these are the future **lung buds** **(Fig. 3.90B)**.

The area surrounding the vertical slit is gradually transformed from mesenchyme to cartilage to form the arytenoid cartilages and the ary-epiglottic folds **(Fig. 3.89B)**. As a result, the original slit-like opening is converted into a T-shaped opening. The epiglottis will lie cranial to this opening and is derived from the caudal part of the 6th pharyngeal arch (caudal part of hypopharyngeal eminence). The thyroid and cricoid cartilages are derived from the 4th and 6th pharyngeal arches, ventral to the slit so that they form the skeleton of the larynx. The cavity of the larynx communicates with the pharynx through the slit (**rima glottidis**) **(Fig. 3.89B)**. This area is innervated by the cranial part of the accessory

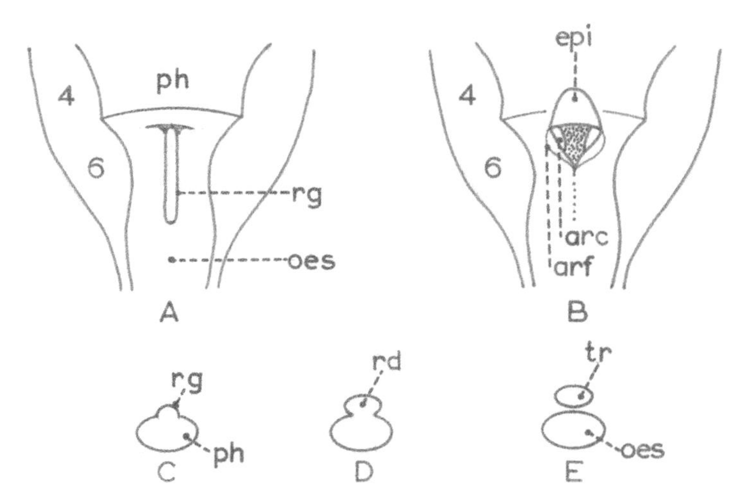

Figure 3.89: Development of the respiratory system. A and **B**: Ventral floor of the pharynx from a posterior view. **C-E**: Transverse sections of pharynx and respiratory diverticulum. 4, 6 = 4th and 6th pharyngeal arches; arc = arytenoid cartilage; arf = aryepiglottic fold; epi = epiglottis; oes = oesophagus; ph = pharynx; rd = respiratory diverticulum; rg = respiratory groove (laryngotracheal diverticulum); tr = trachea.

nerve which runs within the vagal sheath. The **superior laryngeal nerve** of the cranial branch of the accessory nerve is both motor (**to the cricothyroid muscle**) and sensory to the **mucous membrane** of the larynx cranial to the vocal folds. The terminal branch of the cranial part of the accessory nerve is the **recurrent laryngeal nerve** which is both motor (**to the intrinsic muscles of the larynx**) and sensory to the **mucous membrane** of the larynx caudal to the vocal folds (Fig. 3.91).

The original evagination separates from the foregut to form a separate tube (which will ultimately form the **trachea**) (Fig. 3.89E) but will retain its attachment to the pharyngeal wall at the region of the original depression. The primitive trachea has the lung buds at its lower end. The attachment of the primitive trachea to the pharyngeal wall is the region where the larynx will develop.

The lung buds, which form lobes and expand laterally, encounter the pleural canals of the intra-embryonic coelom. The lobes begin to 'push into' the medial sides of the pleural canals (Fig. 3.90C,D). Ultimately, the lung buds will be completely surrounded by coelomic membrane, forming the **visceral pleura**. While the penetration of the pleural coelomata is taking place the tracheal tubular system is undergoing successive divisions. These will form the bronchial tubes, the major ones being three on the right side and two on the left (Fig. 3.90D) which represent the lobar arrangements of the two lungs. Since the developing lungs are originally situated at the level of the 6th pharyngeal arches, the 6th pharyngeal arch arteries are utilised to form the right and left pulmonary arteries.

The lung buds enlarge progressively and gradually expand into the pleural cavities. As a result they are covered with **visceral pleura.** The **parietal pleura** is formed from the outer layer of the coelomic cavity in such a way that a narrow space, the **pleural cavity**, persists between the two layers (Fig. 3.90D). The visceral pleura extends between the lobes of the lungs forming deep fissures which demarcate the lobes.

POSSIBLE COMMON ABNORMALITIES

- An **azygos lobe** may be 'cut off' by the presence of the azygous arch so that the upper lobe of the right lung may be doubled.
- A **tracheo-oesophageal fistula** will result if the separation between the trachea and the gut tube is incomplete.

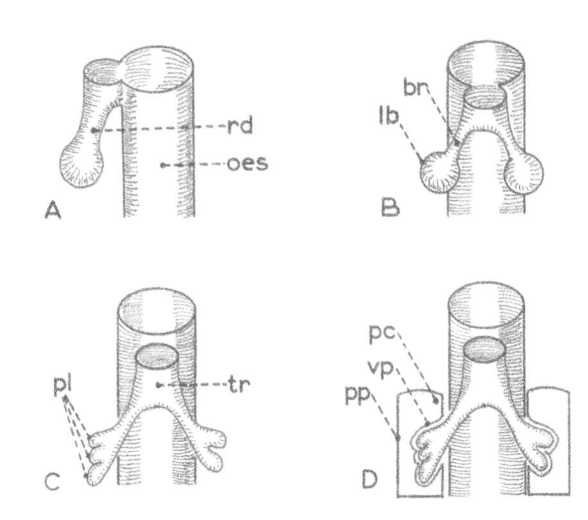

Figure 3.90: Lateral (A) and ventral views (B-D) of the developing lung buds. br = bronchus; lb = lung bud; oes = oesophagus; pc = pleural cavity; pl = pulmonary lobes; pp = parietal pleura; rd = respiratory diverticulum; tr = trachea; vp = visceral pleura.

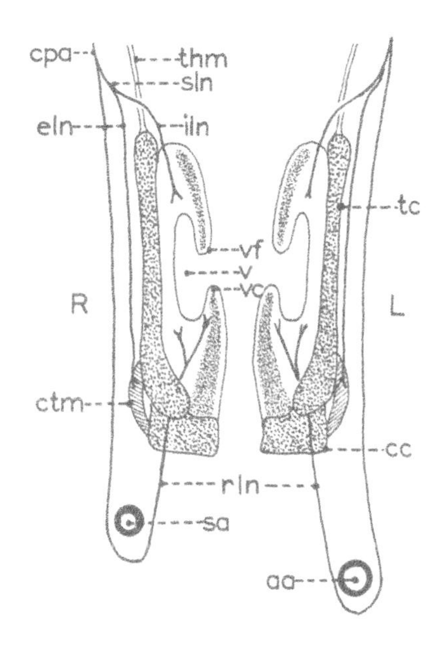

Figure 3.91: Coronal section through the adult laryngeal mechanism. aa = aortic arch; cc = cricoid cartilage; cpa = cranial part of accessory nerve (accompanying vagus nerve, which is the derivative of 4th and 6th pharyngeal arches); ctm = cricothyroid muscle; eln = external laryngeal nerve; iln = internal laryngeal nerve; L = left; R = right; rln = recurrent laryngeal nerve; sa = subclavian artery; sln = superior laryngeal nerve; tc = thyroid cartilage; thm = thyrohyoid membrane; v = laryngeal ventricle; vc = vocal cord; vf = vocal fold.

UROGENITAL SYSTEM

Although in most textbooks the two systems, urinary and genital, are dealt with together, here we shall deal with them separately.

Urinary System

Development of the Kidney

The **kidney** (nephric primordium) develops in the **intermediate mesoderm**. This is also called the **nephrogenic mesoderm** (kidney-maker) and extends from the lower cervical region to the upper sacral segments of the embryo (**Fig. 3.92**). The urinary system develops in the intermediate mesoderm in three distinct stages which take place in temporal sequence, and from cranial to caudal positions:

- formation of the **pronephros** (rudimentary kidney)
- formation of the **mesonephros** (the intermediate kidney)
- formation of the **metanephros** (definitive kidney).

Set out as a table (**Table 3.2**), these are located as follows:

Table 3.2: Disposition of nephric elements

	Somite level	*Region*	*No. of units*	*Age (pf)*
Pronephros	8-14	C + UT	±7	3 wks
Mesonephros	15-26	T + L	±70	4-6 wks
Metanephros	26-28	SS	± 1 million	6-8 wks

C+UT = Cervical and upper thoracic
T+L = Thoracic and lumbar
SS = Superior sacral

Pronephros

In humans, these bilateral structures are rudimentary, temporary and functionless and disappear by the end of the 4th week pf. Each is found as segmentally situated groups of cells in the lower cervical and upper thoracic parts of the intermediate mesoderm (**Fig. 3.92**). The groups of cells form vesicles, the uppermost of which degenerate early and the lower form tubules which join to form the **pronephric ducts** (**Fig. 3.93A**). While all the pronephric vesicles degenerate, the pronephric ducts grow caudally (*see mesonephros below*) (**Fig. 3.93B**) to ultimately open into the cloaca on either side. No glomeruli or renal corpuscles are formed by the pronephros.

Mesonephros

These bilateral structures develop in a craniocaudal direction in the intermediate mesoderm and overlap the lower end of the degenerating pronephroi. Because of this, it is thought that the proximity of the pronephros induces the formation of the mesonephros. The caudal ends of the mesonephroi reach to the level of the

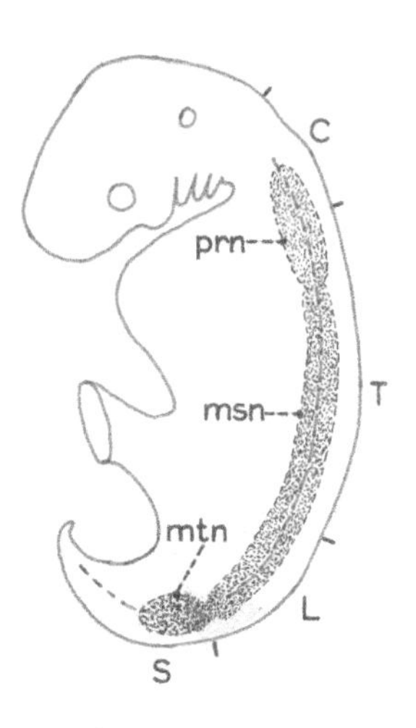

Figure 3.92: Levels at which the pronephros (prn), mesonephros (msn) and metanephros (mtn) develop within the intermediate mesoderm. C = cervical; T = thoracic; L = lumbar; S = sacral.

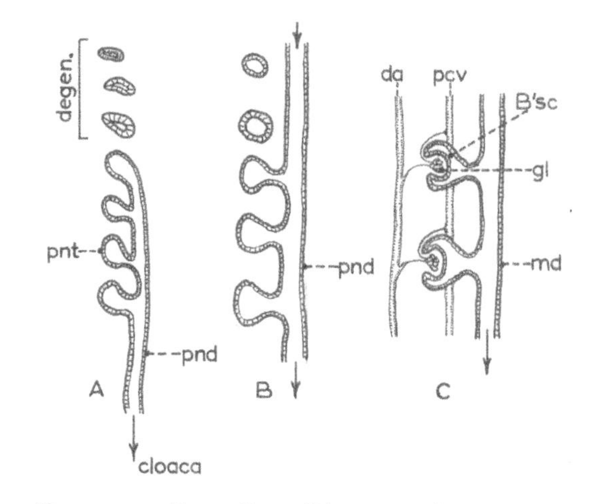

Figure 3.93: Formation of the pronephros. B'sc = glomerular (Bowman's) capsule; da = dorsal aorta; gl = glomerulus; md = mesonephric duct; pnd = pronephric duct; pnt = pronephric tubules; pcv = posterior cardinal vein.

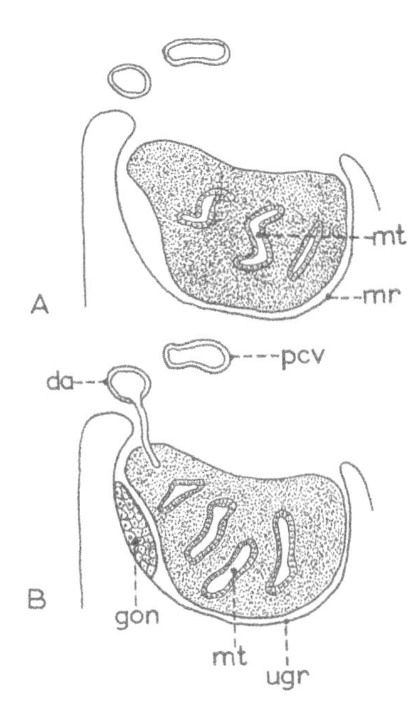

Figure 3.94: Transverse section through the urogenital ridge. da = dorsal aorta; gon = gonad; mr = metanephric ridge; mt = mesonephric tubule; pcv = posterior cardinal vein; ugr = urogenital ridge.

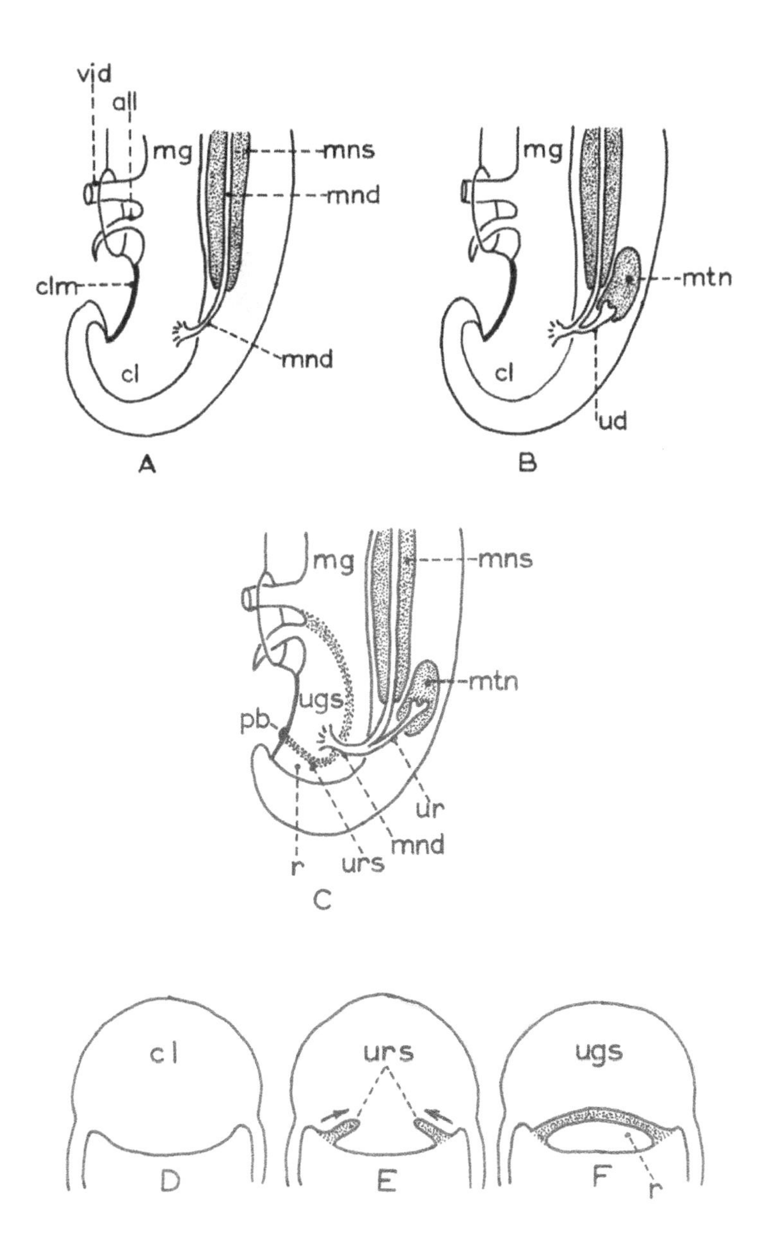

Figure 3.95: A-C: Longitudinal views of the caudal aspect of embryo. **D-F:** Tranverse sections through the cloaca to show division of the cloaca into the urogenital sinus (ugs) and the rectum (r). all = allantois; cl = cloaca; clm = cloacal membrane; mg = midgut; mnd = mesonephric duct; mns = mesonephros; mtn = metanephros; pb = perineal body; ud = ureteric diverticulum; ur = ureter; urs = ororectal septum; vid = vitello-intestinal duct.

glomerular (Bowman's) capsule. Capillary tufts form in the mesoderm of the indentation of the capsule, and these find a vascular attachment to the dorsal aorta of that side. The vessels leaving the tufts drain into the posterior cardinal veins **(Fig. 3.93C)**. The mesonephric tubules undergo elongation to become S-shaped **(Fig. 3.94A)**. These mesonephric tubules continue to multiply until about 70 have formed on each side. As the mesonephros reaches its maximum size, it bulges into the coelomic cavity to form the **mesonephric ridge** **(Fig. 3.94A)** *(see also closure by pleuroperitoneal membrane)*. Later, when the gonad has formed in the mesonephric ridge, it is known as the **urogenital ridge** **(Fig. 3.94B)**.

As progressively more mesonephric tubules join the mesonephric duct, it extends further caudally within the mesonephric ridge until it reaches the lower end of the intermediate mesoderm. The duct then grows caudal to the intermediate mesoderm and under cover of the coelomic epithelium, makes contact with the cloaca into which it opens **(Fig. 3.95A,B)**.

As the mesonephric tubules develop in a cranio-caudal direction, a similar directional process of *degeneration* of these structures takes place. As a result of this degeneration, the mesonephric ridge shortens and extends from about the level of the septum transversum to about the middle lumbar mesonephros is presumed to excrete limited quantities of fluid.

At this stage, the hindgut containing both intestinal and possible urinary excretory products, may be dubbed the **cloaca**.

Some of the tubules of the degenerated part of the mesonephros persist and form appendages in the adult. In the male, the remnants of the uppermost mesonephric tubules become associated with the two ends (head and tail) of the **epididymis.** These are called **aberrant ductules.** Occasionally, a small group of tubules is found in the distal part of the spermatic cord. These form the **paradidymis** and are also thought to be remnants of mesonephric tubules. In the female, when present, they exist in the mesentery of the uterine tube (**mesosalpinx**), in the form of small tubular structures called the **epoophoron**, or as **vesicular appendages** attached to the ovary itself (Fig. 3.97B). The mesonephric duct will persist in the male, but will degenerate in the female to form the epoophorontic duct (duct of Gartner). The caudal mesonephric tubules form the smaller paroophoron which usually disappears.

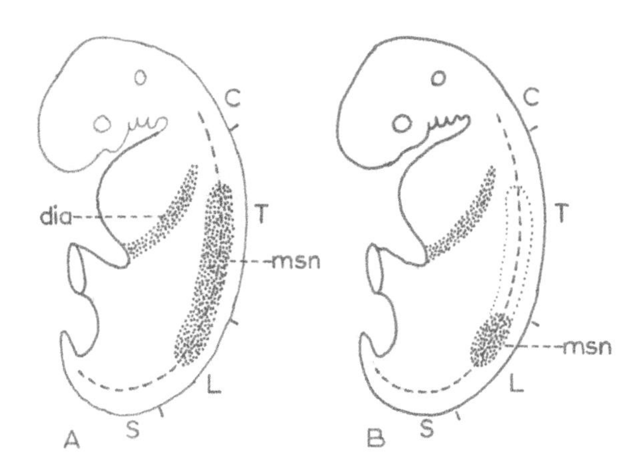

Figure 3.96: Lateral view of the embryo to show degeneration of the mesonephros (msn) in a cranio-caudal direction. C = cervical; T = thoracic; L = lumbar; S = sacral. dia = diaphragm; msn = mesonephros.

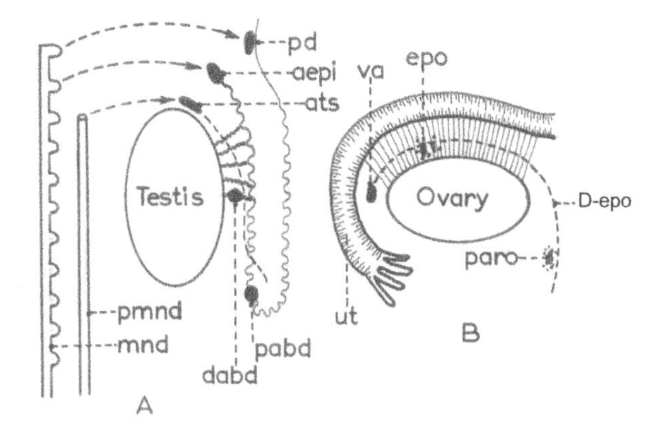

Figure 3.97: Comparative remnants of the ductular system in the male (A) and the female (B). aepi = appendix epidydimis; ats = appendix testis; dabd = distal abberant ductules; D-epo = epoophorontic duct; epo = epoophoron; mnd = mesonephric duct; pabd = proximal abberant ductule; paro = paraoophoron; pd = paradydimis; pmnd = paramesonephric duct; ut = uterine tube; va = vesicular appendix.

Metanephros

The definitive or permanent kidney is derived from two developmental parts, the **metanephrogenic mesoderm** and the **ureteric diverticulum.** The metanephrogenic mesoderm is the inferior part of the intermediate mesoderm. The metanephrogenic mesoderm and the ureteric diverticulum are the forerunners of the excretory part of the kidney. The metanephrogenic mesoderm will give rise to almost the entire **nephron** (the excretory unit of the kidney). The ureteric diverticulum forms the part which is responsible for collecting the excretory products and transferring them to the bladder.

Soon after the mesonephric duct has entered the cloaca, a diverticulum appears from the dorsal aspect of the duct. This is the **ureteric diverticulum** (Fig. 3.98A). The diverticulum elongates in a cranial direction and enters the metanephrogenic mesoderm (Fig. 3.99). As it does so, the cranial end of the ureteric diverticulum forms a funnel-shaped dilatation which divides successively into major calyces, minor calyces and numerous collecting ducts (Fig. 3.98B-D). After approximately five generations of branching, the metanephrogenic mesoderm will form a condensation or cap over the end of each duct

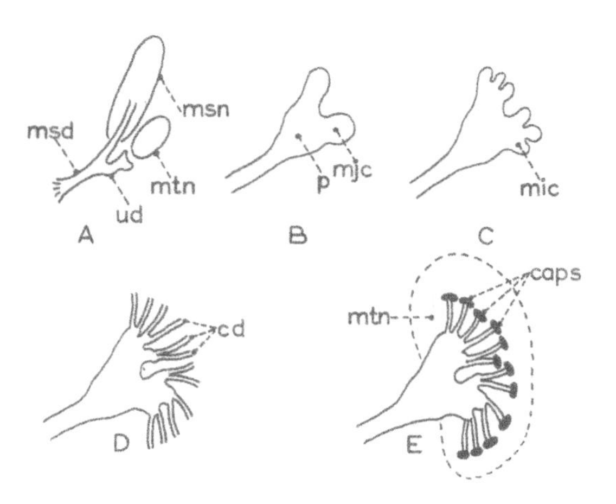

Figure 3.98: Formation of the collecting ducts and the ureter. caps = condensation of metanephrogenic mesoderm; cd = collecting ducts; mic = minor calyx; mjc = major calyx; msd = mesonephric duct; mtn = metanephros; msn = mesonephros; p = pelvis; ud = ureteric diverticulum.

(Fig. 3.98E). The cap is known as the **metanephrogenic blastema**. It is within these small caps that the metanephric tubules develop.

Each metanephric tubule becomes S-shaped. The medial end of each tubule becomes dilated and is closely related to a capillary tuft which emanates from the arteries supplying the metanephros (Fig. 3.100 A-C). The capillary tufts are called **glomeruli** and these invaginate the dilated medial ends of the metanephric tubules to form the **glomerular (Bowman's) capsules** of the definitive kidney (Fig. 3.100C). The combination of the glomerulus and the Bowman's capsule is known as the **renal corpuscle**. As development proceeds (Fig. 3.101 A,B), the S-shaped tubules are drawn out into a **proximal convoluted tubule** (near to the corpuscle), a **loop of Henle** and a **distal convoluted tubule** (far from the corpuscle). Finally, a collecting duct derived from the ureteric diverticulum joins the free lateral end of the distal convoluted tubule (Fig. 3.101B) to form the physiological unit of the kidney called the **nephron.** The collecting tubules enter the minor calyces, which open into the major calyces and then into the pelvis of the kidney. The definitive **ureter** forms from the unbranched part of the ureteric diverticulum.

In a case where the collecting tubule fails to join the distal convoluted tubule, the latter is left with

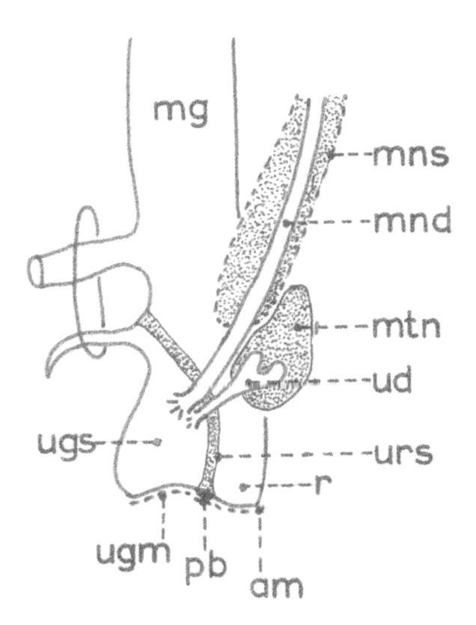

Figure 3.99: Lateral view of the caudal region of the embryo depicting the opening of ducts into the urogenital sinus (ugs). am = anal membrane; mg = midgut; mnd = metanephric duct; mns = mesonephros; mtn = metanephros; pb = perineal body; r = rectum; ud = ureteric diverticulum; ugm = urogenital membrane; urs = urorectal septum.

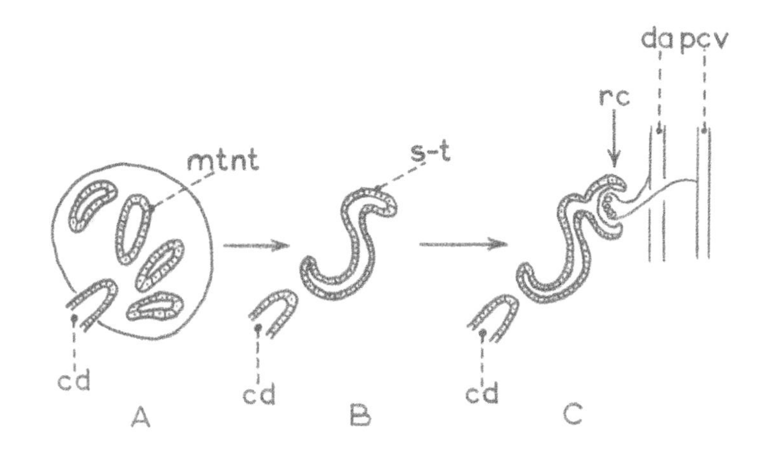

Figure 3.100: Conversion of metanephric tubules into part of the early nephron. cd = collecting duct; da = dorsal aorta; mtnt = metanephric tubules; pcv = posterior cardinal vein; rc = renal corpuscle; s-t = 'S'- shaped tubule.

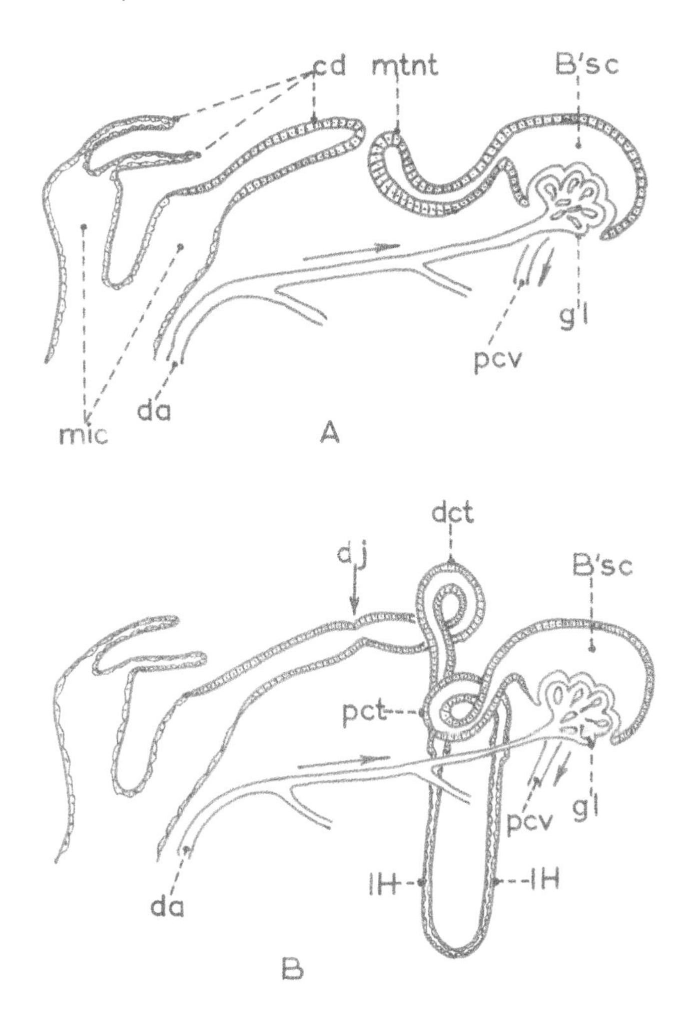

Figure 3.101: Joining of the metanephric tubule with the collecting duct system and formation of the nephron. B'sc = glomerular (Bowman's) capsule; cd = collecting ducts; da = dorsal aorta; dct = distal convoluted tubule; dj = duct junction; gl = glomerulus; lH = loop of Henle; mic = minor calyx; mtnt = metanephric tubule; pct = proximal convoluted tubule; pcv = posterior cardinal vein.

a blind end, and since the metanephric renal corpuscle is an excretory entity, the tubule will fill and expand with urine, forming a cyst in the kidney. Thus, the kidney unit consists of two parts:

- an excretory part from the metanephrogenic mesoderm;
- a collecting part from the ureteric diverticulum.

Ascent of the kidneys

Originally, the kidneys lie in the 'pelvis' of the embryo (**Fig. 3.102A**) and ascend to reach their adult positions in the upper abdomen. It is thought that this takes place partly by a downward extension of the trunk and partly by an upward movement of the kidneys themselves. In the pelvis the kidneys are supplied by vessels from the common iliac trunks. As the kidneys ascend in the abdominal cavity, they are supplied by successively higher branches of the dorsal aorta (**Fig. 3.102B**), until they impinge upon the suprarenal glands where they are arrested by the glands impinging on the diaphragm (**Fig. 3.102C**). Here they are supplied by the definitive renal arteries and frequently by other arteries in the vicinity.

Anatomically, the fetal kidney is markedly lobulated and this lasts for about a year after birth. In addition, the early fetal renal pelves face anteriorly, but as the kidneys ascend into the abdomen, they rotate so that the pelves face medially.

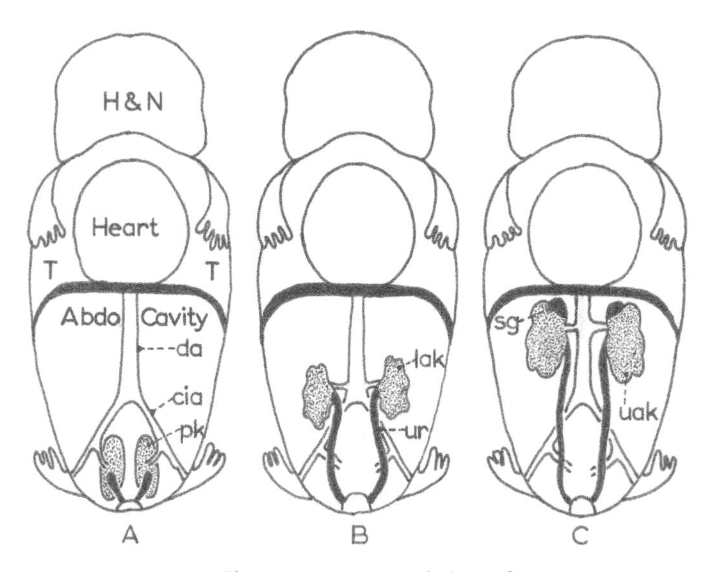

Figure 3.102: Ventral view of the embryo to show the ascent of the kidneys. cia = common iliac artery; da = dorsal aorta; H&N = head and neck; lak = lower abdominal kidney; pk = pelvic kidney; sg = suprarenal gland; uak upper abdominal kidney; ur = ureter.

COMMON ABNORMALITIES

- Renal **agenesis** (non-development) may be unilateral or bilateral.
- **Ectopic kidney**, unilateral or bilateral, results from various degrees of non-ascent of the kidney.
- Occasionally the lower poles of the kidneys will fuse, resulting in a so called **horse-shoe kidney**. This is arrested in its ascent at the level of the inferior mesenteric artery.
- From time to time the ureters are duplicated either unilaterally or bilaterally.

Development of the Urinary Bladder and Urethra

The allantoic diverticulum extends from the cloaca into the umbilical cord (**Fig. 3.95B**). During the 5th week pf, the mesonephric duct has penetrated the dorsolateral wall of the cloaca (**Fig. 3.95B**). The mesoderm surrounding the cloaca now grows inwards forming a septum, beginning just cranial to the intestinal attachment of the allantoic diverticulum. This ingrowth extends caudally, pushing the epithelium of the cloaca inwards. The mesoderm and epithelium form a septum which is known as the **urorectal septum** (**Fig. 3.95C**). This septum separates the cloaca into a larger ventral part, the **urogenital sinus** and a smaller dorsal part, the **rectum** (**Fig. 3.95C**). The septum reaches the position of the future **perineal body** where it divides the cloacal membrane into a ventral **urogenital membrane** and a dorsal **anal membrane** (**Fig. 3.95C**).

Note. If the urorectal septum does not completely separate the rectum from the urogenital sinus, an opening will persist between the rectum and the sinus. This is a **recto-vesical** or **recto-urethral fistula**.

The ventrally positioned urogenital sinus develops a cranial **vesico-urethral dilatation** and a caudal **phallic canal** (**Fig. 3.103A iv**). The superior part of the vesico-urethral dilatation will become expanded to form the bladder. The allantois opens into the vesical part of this dilatation. The caudal part of the vesico-urethral dilatation which is narrow, will form part of the **urethra**. The mesonephric ducts and the ureteric diverticula (ureters) open into the vesical part

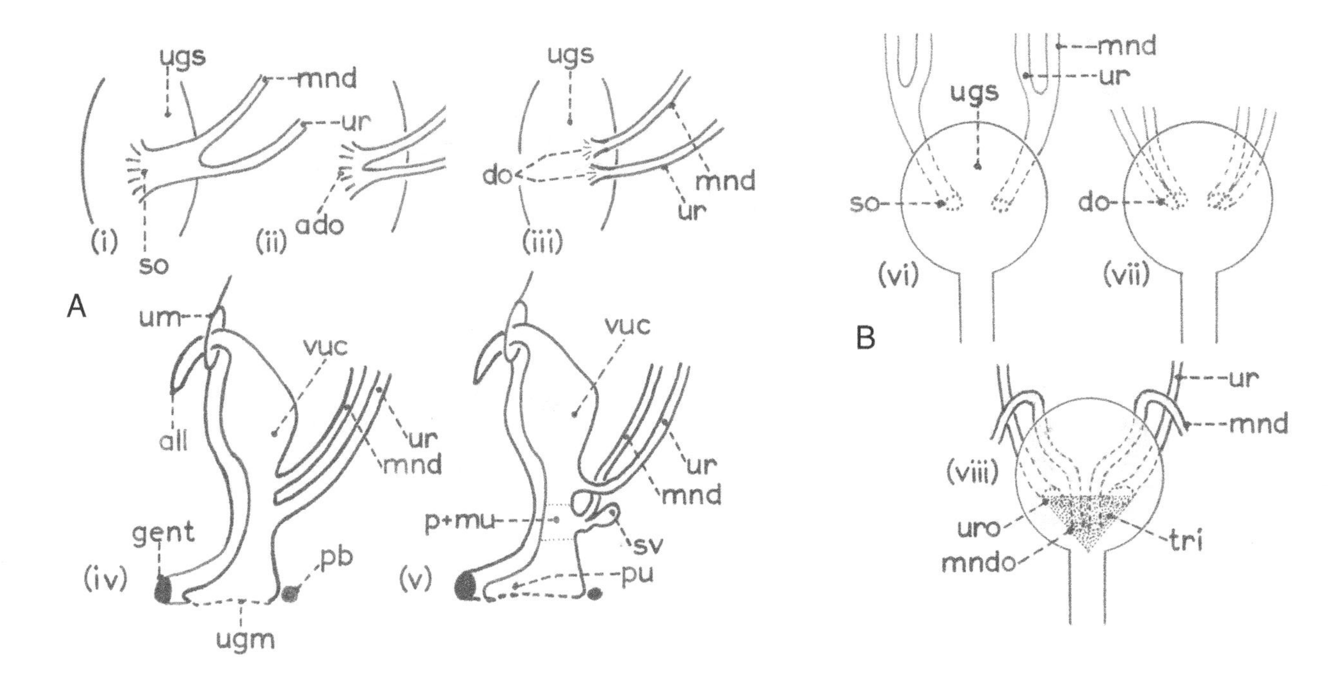

Figure 3.103: Change in position of the openings of the mesonephric and metanephric (ureter) ducts. **A i-v:** Lateral views. **B vi-vii:** Ventral views. ado = almost double orifice; all = allantois; do = double orifice; gent = genital tubercle; mnd = mesonephric duct; mndo = mesonephric duct orifice; pb = perineal body; p+mu = prostatic and membranous urethra; pu = phallic urethra; so = single orifice; sv = seminal vesicle; tri = trigone; ugm = urogenital membrane; ugs = urogenital sinus; um = umbilicus; ur = ureter; uro = ureteric orifice; vuc = vesico-urethral canal.

(future bladder) of the vesico-urethral dilatation (Fig. 3.103A i-iv). The mesonephric ducts are gradually 'absorbed' (taken up) by that part of the vesico-urethral wall which forms the bladder (Fig 3.103A v). In turn, the ureteric diverticula (ureters) will also be absorbed. Thus, four openings into the definitive bladder are found (Fig. 3.103B vi-viii), one for each mesonephric duct and one for each ureter. The openings of the mesonephric ducts are medial to those of the ureters (Fig. 3.103B viii). The entry points of the ureters are at the extremes of an anatomical structure called the **trigone** (triangle) of the bladder (Fig. 3.103B viii). The allantoic diverticulum at the cranial end of the bladder will eventually become obliterated to form the **urachus** of the adult. In the newborn the urachus will become obliterated and will result, in the adult, in the median umbilical ligament.

The caudal narrower part of the vesico-urethral dilatation will give rise to the urethra, i.e. the prostatic and membranous urethra in the male and the urethra proper in the female. The openings of the mesonephric ducts 'slide' down the posterior wall of the urogenital sinus to the approximate position of the **prostatic urethra** where they open as the **ejaculatory ducts** (Fig. 3.103B viii). Prior to the entry of the mesonephric ducts into the prostatic urethra they produce a diverticulum, the **seminal vesicle** or **gland** (Fig. 3.103A v).

POSSIBLE COMMON ABNORMALITIES

- If the urachus remains patent it may form a **urachal fistula** or **urachal sinus** or a **urachal cyst.**
- **Ectopia vesicae** occurs when the posterior wall of the bladder protrudes through a deficiency in the anterior abdominal wall. The anterior wall of the bladder is absent. This abnormality is said to occur very early in embryonic development due to non-invasion of mesoderm from the primitive streak in the region of the lower abdomen and the cloaca. Occasionally the symphysis pubis is also absent. Epispadias (when the urethra opens to the ventral [anterior] surface of the phallus) may be associated with ectopia vesicae (see genital system).

Formation of the Prostate Gland

At about the 12th week of development, solid cords of endodermal cells grow radially and circumferentially from the urethral part of the vesico-urethral canal (Fig. 3.104). These cords later canalise to form the glandular substance of the prostate which, as it enlarges, will surround the urethra and the ejaculatory ducts (Fig. 3.104). The remaining substance of the prostate gland develops from the surrounding mesenchyme.

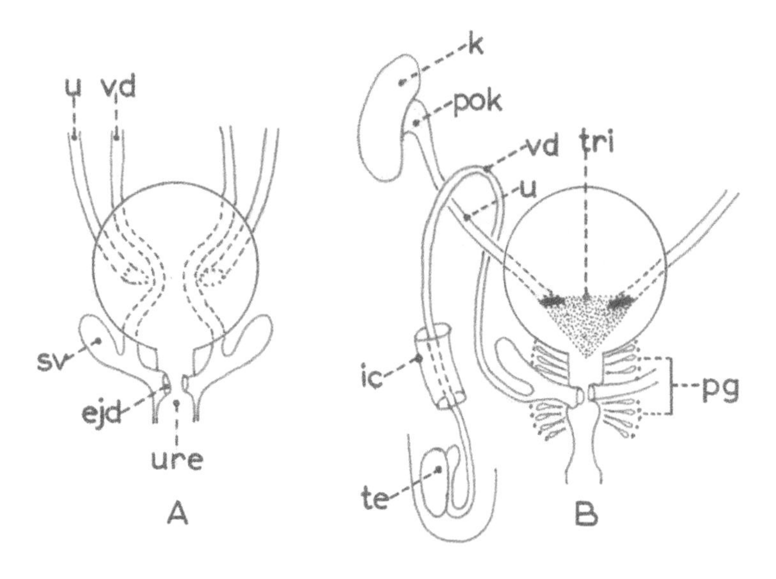

Figure 3.104: Formation of the prostate gland in ventral view.
ejd = ejaculatory duct; ic = inguinal canal; k = kidney; pg = prostate gland; pok = pelvis of kidney; sv = seminal vesicle; te = testis and epidydimis; tri = trigone; u = ureter; ure = urethra; vd = vas deferens (ductus deferens).

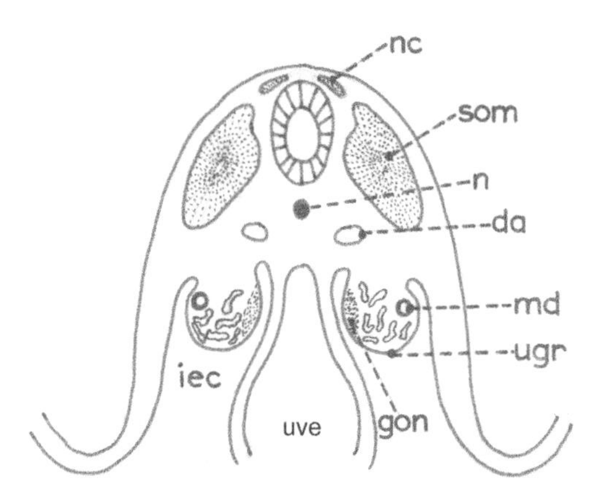

Figure 3.105: Transverse section of the embryo to illustrate the position of the developing testis.
da = dorsal aorta; gon = gonad; iec = intra-embryonic coelom; md = mesonephric duct; n = notochord; nc = neural crest; som = somite; ugr = urogenital ridge; uve = umbilical vesicle.

Genital System

The Indifferent Gonad

As has been mentioned previously, the sex of the embryo is fundamentally determined by the presence or absence of the Y chromosome. However, during early embryonic development, there is a phase during which the gonad and the sexual duct systems of the male and female embryo cannot be distinguished from one another. The time period between the arrival of the sex cells in the gonad and the ability to distinguish true ovary or testis is known as the **indifferent period.** This period lasts until the end of the 6th week pf. As a result of the influence of a gene on the Y-chromosome (testis-determining factor), the indifferent gonad will differentiate into a testis. In the absence of this factor, the indifferent gonad will form an ovary. The formation of definitive gonads will lead to the formation of distinct sexual duct formation. These are the mesonephric and **paramesonephric ducts** (para = next to). The development of the mesonephric ducts has already been described. The paramesonephric ducts will form within the urogenital ridge lateral to each mesonephric duct.

During the 4th week pf, the coelomic epithelium in the middle of the medial surface of the mesonephric ridge proliferates and extends into the underlying mesoderm as **gonadal cords.** In the 6th week these cords are invaded by **primordial germ cells** which have migrated from the endoderm in the region of the allantois (Fig. 1.9).

Development of the Testis and its Relationship to the Mesonephric Duct

The testis develops on the middle half of the medial side of the mesonephros (Fig. 3.105). As the testes enlarge they protrude into the coelomic cavity on a mesentery called the **mesorchium,** which consists of coelomic epithelium and intervening mesoderm (Fig. 3.106A). A layer of mesodermal tissue condenses under the coelomic epithelium to separate the gonadal cords from the overlying epithelium (Fig. 3.106B). This mesoderm forms the **tunica albuginea** of the testis. From the tunica, fibrous septa extend into the root of the mesorchium to form the **mediastinum testis** (Fig. 3.106B). The gonadal cords containing the primordial germ cells will form **seminiferous tubules** (Fig. 3.106B).

The primordial germ cells give rise to the spermatocyte series while the gonadal cords give rise to the **supporting (Sertoli) cells** (Fig. 3.106C). Sperm will not form in the testis until puberty. **Interstitial endocrine (Leydig) cells** are found between the seminiferous tubules and are believed to be derived from mesenchyme (Fig. 3.106D). These cells are responsible for the production of testosterone and this secretion begins at puberty.

The testis obtains its excretory duct by modification of the mesonephric duct. The retained tubules of the mesonephros join the seminiferous tubules forming the **rete testis** and **ductuli efferentes** (Fig. 3.106E). The mesonephric duct gives rise to the **epididymis** and the **vas (ductus) deferens** (Fig. 3.106E). Some non-degenerated mesonephric tubules form the **paradidymis**. A remnant of the mesonephric duct may give rise to an **appendix of the epidydimis** (Fig 3.97A). The rete testis, ductuli efferentes, vas deferens and ejaculatory ducts provide the track along which the sperm reach the prostatic urethra.

In the male, while the mesonephric ducts persist, the paramesonephric ducts degenerate and may give rise to a few remnants, e.g. the **uterus masculinus (prostatic utricle)** found in the posterior wall of the prostatic urethra, which is the approximate position in the male, of the cervix uteri in the female.

Descent of the Testis

As indicated previously, when the kidneys have formed, their ascent seems to coincide with the elongation of the embryonic trunk. In a similar fashion, the testis, being situated in the superior part of the trunk, seems to descend. The testis is a retrocoelomic organ and descends in the retrocoelomic (retroperitoneal) space. By the 12th week pf, the testis has reached to approximately, the **internal inguinal ring** (brim of the pelvis) (Fig. 3.107A). When it reaches this point, a **gubernaculum** (a fibro-muscular band) develops (Fig. 3.107B). This has attachments to the caudal pole of the testis and the labioscrotal fold of the same side (Fig. 3.107B). The gubernaculum lies within the **urogenital (peritoneal) fold.** A protrusion of the coelomic cavity has formed along the length of the gubernaculum, extending from the internal ring to the labioscrotal fold (Fig. 3.107C). The protrusion becomes the **processus vaginalis testis** of the adult. When the processus vaginalis has formed, the testis descends dorsal

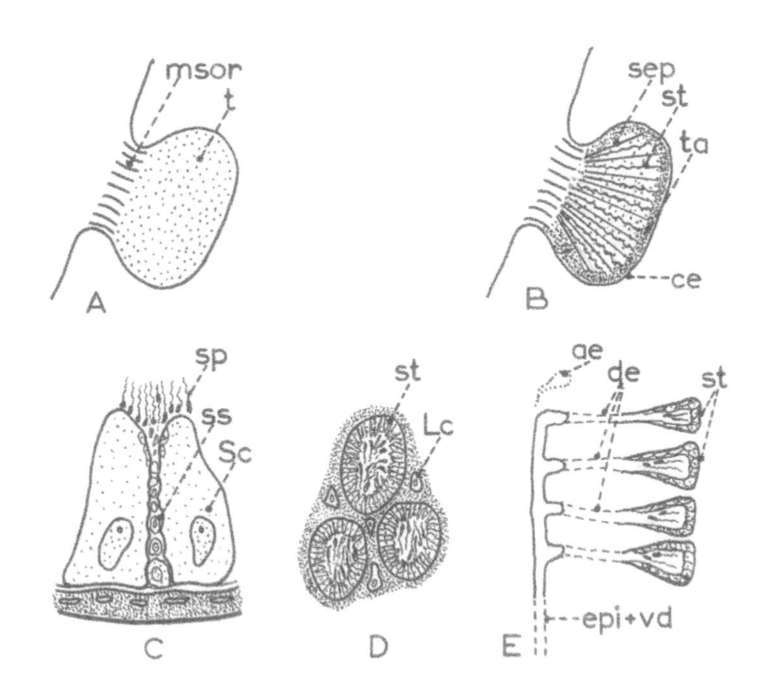

Figure 3.106: Development of the testis. ae = appendix epidydimis; ce = coelomic epithelium; de = ductuli efferentes; epi + vd = epidydimis and vas deferens; Lc = Interstitial endocrine (Leydig) cell; msor = mesorchium; Sc = supporting (Sertoli) cell; sep = septum; sp = spermatozoa; ss = spermatogenic series; st = seminiferous tubules; t = testis; ta = tunica albuginea.

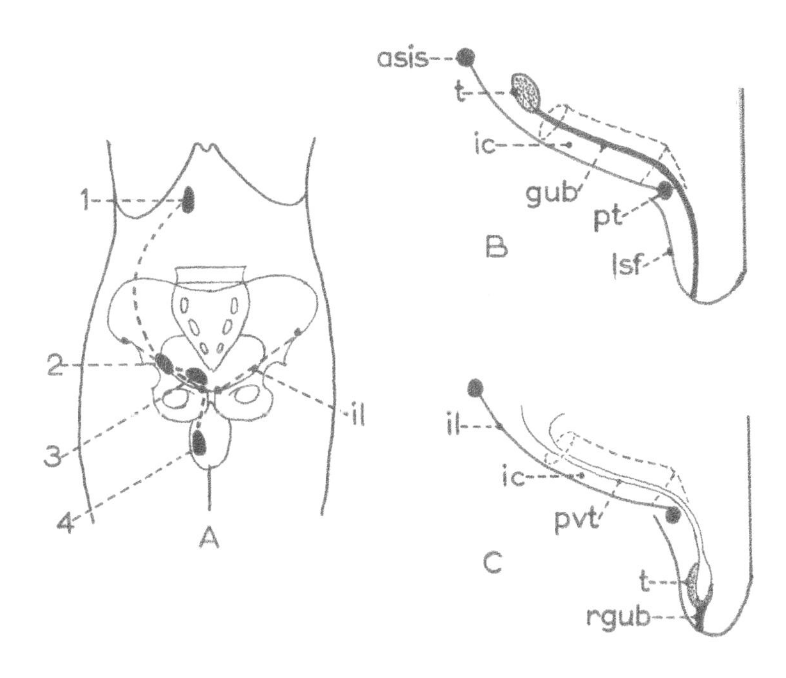

Figure 3.107: Ventral views of the descent of the testis. 1 = urogenital ridge; 2 = internal inguinal ring; 3 = external inguinal ring; 4 = descent into scrotum; asis = anterior superior iliac spine; gub = gubernaculum; ic = inguinal canal; il = inguinal ligament; lsf = labioscrotal fold; pt = pubic tubercle; pvt = processus vaginalis testis; rgub = remnant of gubernaculum; t = testis.

to it, guided by the gubernaculum, through the inguinal canal, to reach the external inguinal ring just before birth. The testes drop into the scrotum a few days after birth (**Fig. 3.107C**).

Note. There are two sources of dispute about the gubernaculum:

(a) It is questionable that the gubernaculum 'pulls' the testis down; and
(b) It is questionable whether the gubernaculum has extensions to the abdominal wall, root of penis, perineum and thigh, which are allegedly the places where **ectopic testes** are found.

POSSIBLE COMMON ABNORMALITIES

- Occasionally the testis fails to descend into the scrotum (**cryptorchidism**).
- Normally the processus vaginalis closes completely, but if it remains patent, the patient is predisposed to an **indirect inguinal hernia** or a **congenital hydrocoele**; if the processus remains partially patent, the patient develops an **encysted hydrocoele** of the spermatic cord.

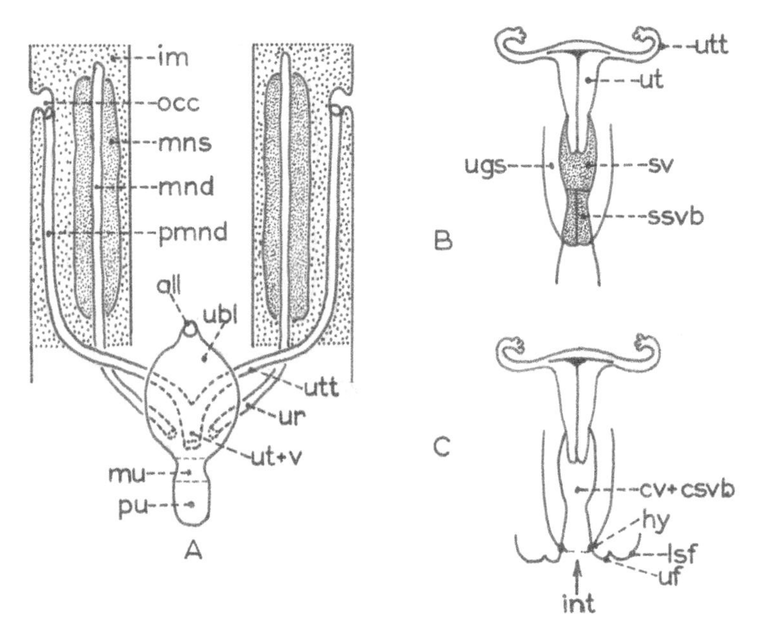

Figure 3.108: Ventral views of the female genital system.
all = allantois; cv + csvb = canalized vagina and canalized sinovaginal bulbs; hy = hymen; im = intermediate mesoderm; int = introitus; lsf = labioscrotal fold; occ = orifice into coelomic cavity; mns = mesonephros; mnd = mesonephric duct; mu = membranous urethra; pmd = paramesonephric duct; pu = phallic urethra; ssvb = solid sinovaginal bulbs; sv = solid vagina; ubl = urinary bladder; uf = urethral fold; ugs = urogenital sinus; ur = ureter; utt = urerine tube; ut + v = uterus and upper part of vagina.

Development of the Ovary and the Uterus

If the embryo is to be a female, the gonads become ovaries. These, like the male gonad, develop on the medial aspect of the urogenital ridges (**Fig. 3.105**). The gonadal cords are broken up into groups of cells by an ingrowth of mesoderm. Each group of cells contains one primordial germ cell which is endodermal in origin. The first follicles which form develop in the centre of the gonad (medulla of ovary). These degenerate. A second group of follicles is formed in the cortical region of the ovary from the ingrowth of the gonadal cords. Only a very slight condensation of mesoderm occurs under the coelomic epithelium. Thus, there is no barrier to the formation of new follicles from the coelomic epithelium.

The ovaries bulge into the coelomic cavity on a mesentery called the **mesovarium**. The urogenital ridge at this stage contains mesonephric tubules and on their lateral side, the mesonephric duct. On the lateral side of the urogenital ridge a groove now develops which is open to the coelomic cavity (**Fig. 3.108**). The edges of the groove fuse to form a tube, the paramesonephric duct, the cranial end of which remains open into the coelomic cavity. Later, the open ends develop 'finger-like' fringes which surround the opening into the peritoneal cavity. These fringes form the fimbriated ends of the uterine (Fallopian) tubes.

The paramesonephric duct burrows caudally through the intermediate mesoderm, lateral to the mesonephric duct until it reaches the lower pole of the mesonephros. The paramesonephric ducts turn medially, ventral to the mesonephric ducts to meet each other in the midline where they fuse (**Fig. 3.108A**). The cranial parts of the paramesonephric ducts form the uterine tubes of the adult while the lower fused parts form the uterus and upper part of the vagina of the adult (**Fig. 3.108 B,C**). The mesenchyme surrounding the ducts condenses to form the smooth muscle of the uterine tubes and uterus. The paramesonephric ducts, mesonephric ducts and the ovaries lie in the retrocoelomic mesenchyme and are, therefore, retroperitoneal in the adult.

The inferior end of the fused paramesonephric ducts come into contact with the dorsal surface of the urogenital sinus, against which the fused paramesonephric ducts grow down (**Fig. 3.109**). At the point of contact, an upgrowth now occurs from the dorsal aspect of the urogenital sinus.

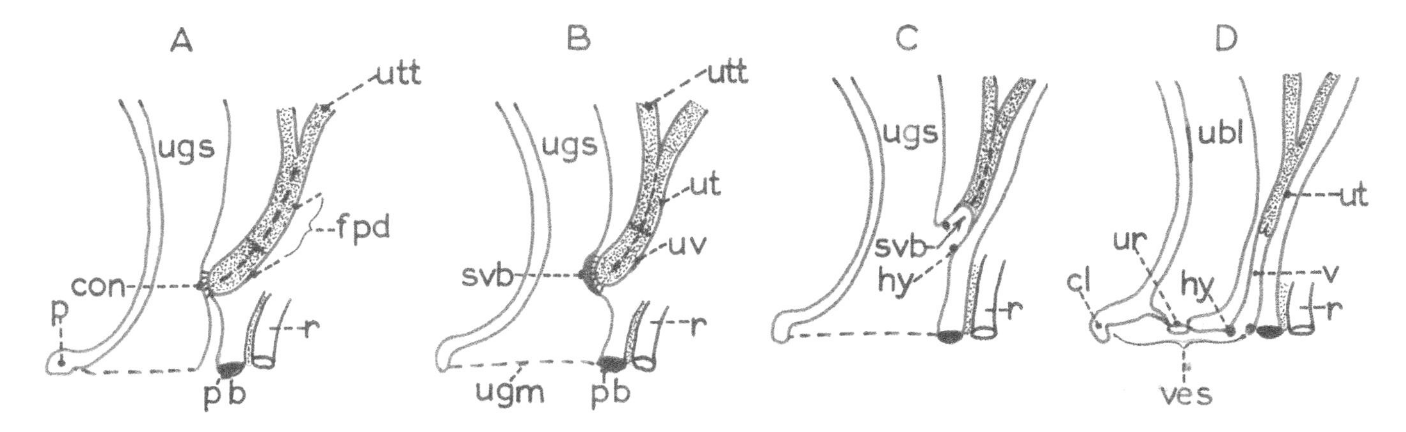

Figure 3.109: Lateral views of the development of the female duct system. cl = clitoris; con = contact; fpd = fused paramesonephric ducts; hy = hymen; p = phallus; pb = perineal body; r = rectum; svb = sinovaginal bulbs; ubl = urinary bladder; ugs = urogenital sinus; ur = urethra; utt = uterine tubes; uv = upper vagina; v = vagina; ves = vestibule.

This forms the **sinovaginal bulbs (uterovaginal plate/mullerian tubercle)**, pushing the fused paramesonephric duct cranially. Up to this point, the fused paramesonephric ducts and the sinovaginal bulbs are solid, but now the two fuse with each other and undergo canalisation (Fig. 3.108B,C and 3.109C,D). The canalised inferior part of the fused paramesonephric ducts become continuous with the fused, canalised sinovaginal bulbs, so that the fused paramesonephric ducts form the uterus and cranial one-third of the vagina, while the canalised sinovaginal bulbs form the caudal two-thirds of the vagina. The canalised caudal end of the fused sinovaginal bulbs breaks through into the urogenital sinus. The 'ragged' edge of the breakthrough forms the **hymen**.

The opening of the vagina into the urogenital sinus 'slides down' the dorsal wall of the sinus until the two tubes (urethra and vagina) are separated from one another at the exterior (Fig. 3.109C,D). Thus, the two openings of the urethra and the vagina come to lie separate from each other in the vestibule between the labioscrotal folds.

Note the formation of the vagina is complex and this creates a variety of viewpoints of its development.

At this time the mesonephric ducts are still present at the lateral sides of the newly formed uterus but are showing definite signs of disintegration. Later, remnants of these ducts may be found close to the lateral sides of the uterus and the vagina, as far as the level of the hymen as the ducts of the epoophoron (ducts of Gartner).

POSSIBLE COMMON ABNORMALITIES

- If the two paramesonephric ducts undergo abnormal fusion, this may result in the formation of a **double uterus** having **two cervices**, a double uterus having one **cervix**, or a **septate uterus** in which the uterine cavities are separated by a septum and which has a single cervix.
- A **vesico-vaginal fistula** may result from non-separation of the openings of the bladder and vagina.

Formation of the Broad Ligament of the Uterus

Each urogenital ridge develops a mesentery, the **urogenital mesentery** which connects it to the dorsal body wall. The ovary is attached to the urogenital ridge by its mesovarium (Fig. 3.110A). With the formation of the urogenital mesentery, the mesonephric and paramesonephric ducts come to lie in the ventral part of the urogenital ridge while the ovaries are attached more medially (Fig. 3.110B). As the urogenital ridges extend ventrally and medially on their mesenteries, carrying with them the ovaries and the contained ducts, the paramesonephric ducts pass ventral to the mesonephric ducts to reach the medial edge of the folds. As the folds reach the pelvis, the two medial edges fuse, bringing together the paramesonephric ducts which also fuse to form the uterus and cranial one-third of the vagina. By now, the ovaries are more or less at the dorsal surface of the fold which may be called the **broad ligament**. The fusion of the two (R&L) folds in the pelvis results in the formation of a shallow ventral 'peritoneal pouch' between the urinary bladder and the uterus (**uterovesical pouch**), and a deep dorsal 'peritoneal cavity' between the uterus, ovaries and the rectum (**recto-uterine pouch**) (Fig. 3.110C). The uterine tubes lie in the

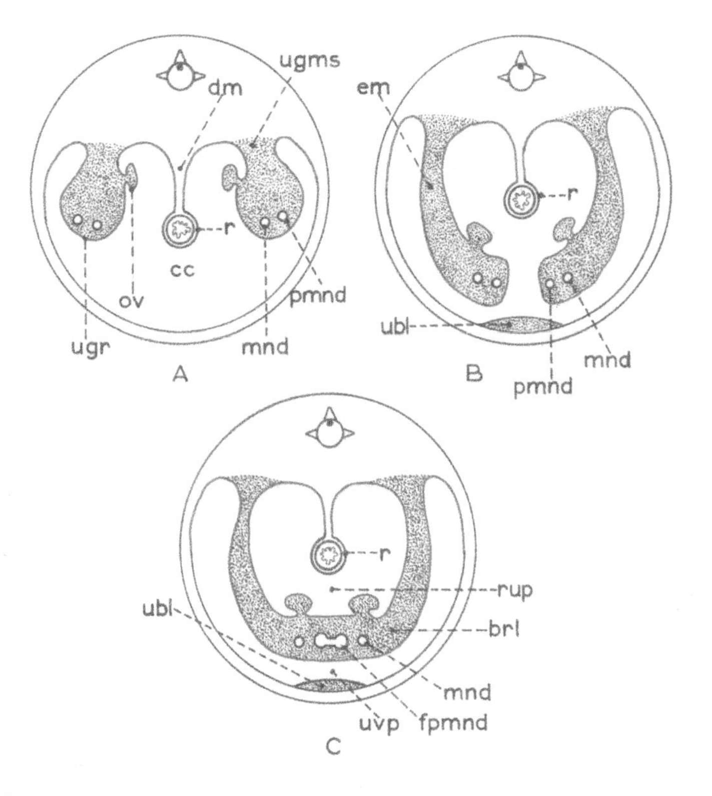

Figure 3.110: Transverse sections to show the development of the broad ligament. brl = broad ligament; cc = coelomic cavity; dm = dorsal mesentery; em = extending mesentery; fpmnd = fused paramesonephric ducts; mnd = mesonephric duct; ov = ovary; pmnd = paramesonephric duct; r = rectum; rup = recto-uterine pouch; ubl = urinary bladder; ugms = urogenital mesentery; ugr = urogenital ridge; uvp = uterovesical pouch.

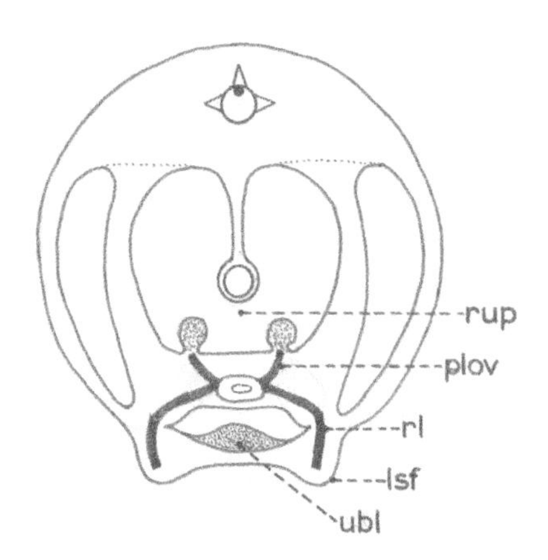

Figure 3.111: Transverse section illustrating the gubernaculum in the female. lsf = labioscrotal fold; plov = proper ligament of ovary; rl = round ligament; rup = recto-uterine pouch; ubl = urinary bladder.

cranial edges of the broad ligament. Since all the organs referred to are retrocoelomic (retroperitoneal in the adult) it follows that all arteries to, and veins and lymphatics from, these organs will necessarily be retrocoelomic.

Descent of the Ovaries

The ovaries, like the testes, descend certainly as far as the pelvic cavity and only very rarely do they reach the labioscrotal folds (labia majora). The possible reason for this is as follows: when the ovary reaches the brim of the pelvis it lies within the urogenital fold, which will become the broad ligament of the uterus, and where the development of a gubernaculum takes place. As in the male, the gubernaculum extends from the caudal pole of the ovary to the labioscrotal fold, after passing through the inguinal canal. This structure is said to 'draw' the ovary downwards but it would seem that the uterine tube obstructs the descent of the ovary, as the gubernaculum becomes 'entangled' with it. The part of the gubernaculum between the ovary and the cornu of the uterus is known, in the adult, as the **ligament of the ovary**, while the caudal part, extending from the cornu to the labioscrotal fold, is known as the **round ligament of the uterus** (Fig. 3.111).

In a similar fashion to the male, the female develops a **processus vaginalis peritonei** (canal of Nuck) from the coelomic cavity. This canal is normally completely obliterated but if it remains open, the subject is predisposed to the formation of an **indirect inguinal hernia**.

Development of the External Genitalia

Up to a certain point of development, the external genitalia are so similar in structure that they are indistinguishable as male and female. This stage is known as the **indifferent stage**.

The primitive streak supplies mesoderm which migrates around the cloacal membrane under the surface ectoderm (Fig. 3.112A). With the cranio-caudal folding of the embryo, this mesoderm also comes to lie between the cloacal plate and the connecting stalk to form the mesoderm of the infra-umbilical part of the future anterior abdominal wall (Fig 3.112B). The urorectal septum becomes attached to the cloacal membrane at the position of the future perineal body (Fig. 3.112B). This divides the cloacal membrane into a ventral **urogenital** component and a dorsal **anal** component (Fig. 3.112C).

Proliferation of the mesoderm around these components results in swellings surrounding the membranes (Fig. 3.112C). At the ventral ends of the swellings, surrounding the urogenital membrane, the mesoderm proliferates to form the genital tubercle which is covered by ectoderm (Fig. 3.112E). The membrane appears to be submerged between the swellings. The perineal body separates the swellings into a genital group and an anal group. The cleft between the genital swellings is the **urogenital sulcus**.

The genital swellings on either side of the urogenital sulcus form two parallel swellings, the medial and lateral genital swellings (Fig. 3.112D). These will later form the labia minora and labia majora respectively in the female, and the floor of the penile urethra and scrotum respectively in the male. The general appearance of the external genitalia at this stage in both sexes is not unlike that of the adult female.

In the next phase of development, the urogenital membrane disintegrates. This opens the inferior part of the urogenital sinus to the urogenital sulcus.

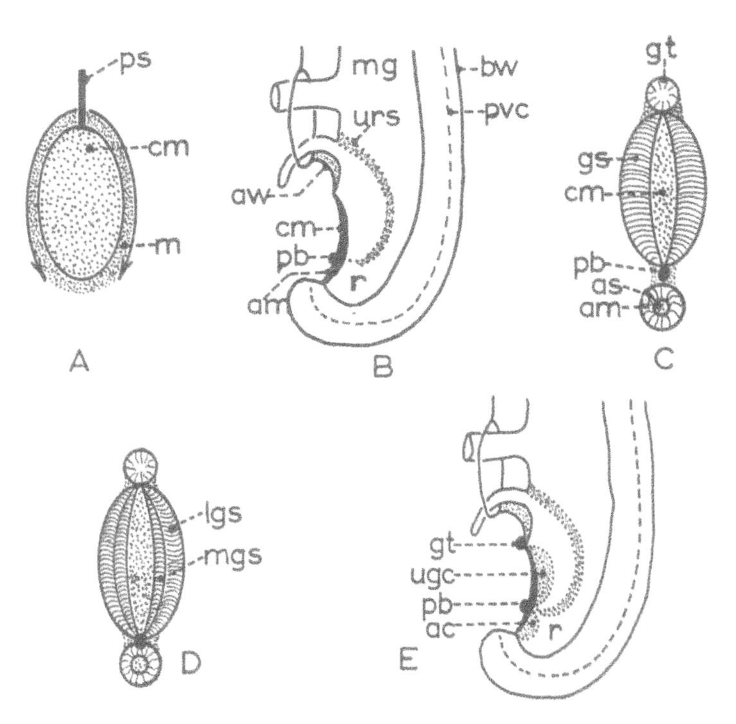

Figurew 3.112: External genitalia. A, C, D: External (ventral) views. **B, E:** Lateral views. Arrows (in A) indicate the direction of growth of the primitive streak mesoderm. ac = anal component; am = anal membrane; as = anal swelling; aw = anterior abdominal wall; bw = body wall; cm = cloacal membrane; gt = genital tubercle; gs = genital swelling; lgs = lateral genital swelling; m = mesoderm; mg = midgut; mgs = medial genital swelling; pb = perineal body; ps = primitive streak; pvc = position of vertebral column; r = rectum; ugc = urogenital component; urs = urorectal septum.

In the Female

Substantially, the genital tubercle and swellings remain unaltered in the female. When the urogenital membrane disintegrates, the opening into the urogenital sinus is exposed. The medial genital swellings, now known as the urethral folds, do not grow significantly and the genital tubercle (**glans clitoris**) collapses between them. The ventral ends of the urethral folds split to enclose the base of the clitoris as the **prepuce of the clitoris** and the **frenulum of the clitoris** (Fig. 3.113A). Because the urethral folds (which will form the labia minora) meet at the clitoris, the ventral extension of the urogenital sinus is obliterated so that the urethra opens into the gap between the labia minora (Fig. 3.113A). As the gap between the labia minora widens to form the **vestibule**, the caudal part of the vagina slides inferiorly until its opening lies in the vestibule (Fig. 3.109). The urethra now has a separate opening into the vestibule. The **introitus** of the vagina and the surrounding **hymen** are level with the roof of the vestibule. The dorsal ends of the urethral folds (**labia minora**) unite dorsal to the vaginal orifice forming the **fourchette** (commissure). The lateral genital swellings, now called the **labioscrotal folds**, persist as separate entities in the female and form the **labia majora** which are the lateral boundaries of the pudendum (Fig. 3.113B).

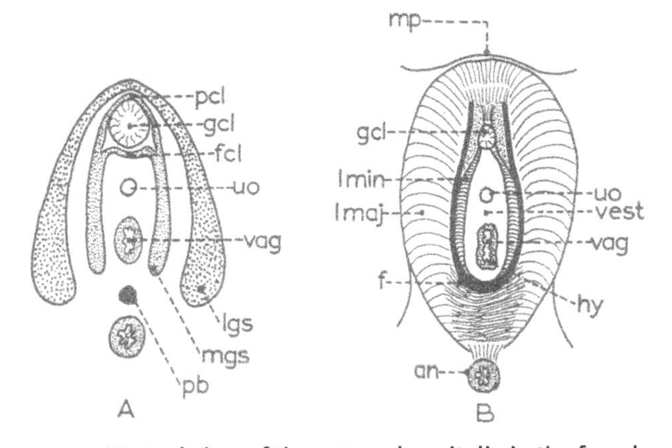

Figure 3.113: **Ventral view of the external genitalia in the female.**
A: Embryo. **B:** Adult. an = anus; f = fourchette; fcl = frenulum of clitoris; gcl = glans of clitoris; hy = hymen; lgs = lateral genital swelling; lmaj = labium majus; lmin = labium minoris; mgs = medial genital swelling; mp = mons pubis; pb = perineal body; pcl = prepuce of clitoris; uo = urethral orifice; vag = vagina; vest = vestibule.

POSSIBLE COMMON ABNORMALITY

Normally the hymen is perforated but from time to time, it is imperforate, with consequent problems at the onset of menstruation.

In the Male

Deep to the urogenital membrane lies the phallic part of the urogenital sinus. The urogenital sinus elongates in a dorso-ventral direction so that its ventral extremity lies just dorsal to the genital tubercle (**Fig. 3.114A,E**). The endodermal lining of the phallic part of the urogenital sinus now proliferates so that the lumen of the sinus is obliterated (**Fig. 3.114 C,G**). This solid cord of endodermal cells forms the urethral plate which carries the genital tubercle in a ventral direction, forming the **phallus.** A groove, the urethral groove, forms on the inferior of the urethral plate (**Fig. 3.114G**). Note that this groove lies between the two medial genital swellings which form the urethral folds. The folds fuse in a dorsal to ventral direction, closing the urethral groove and thus forming the penile (phallic) urethra (**Fig. 3.115 A-D**). This fusion continues until the region of the genital tubercle is reached. The closure of the urogenital folds results in the formation of a **raphe** on the inferior surface of the phallus (**Fig. 3.115D**). The vascular mesenchyme surrounding the newly formed urethra gives rise to the **corpora cavernosa** of the adult.

The genital tubercle forms the **glans penis.** At this stage, the urethra has a 'blind end' (**Fig. 3.116C**). A solid cord of ectodermal cells grows from the glans penis to meet the blind end of the urethra (**Fig 3.116D**). This cord of cells will later canalise and join with the existing phallic part of the urethra, thus completing the penile urethra. The opening in the glans is called the **fossa navicularis** in the adult (**Fig. 3.116E**).

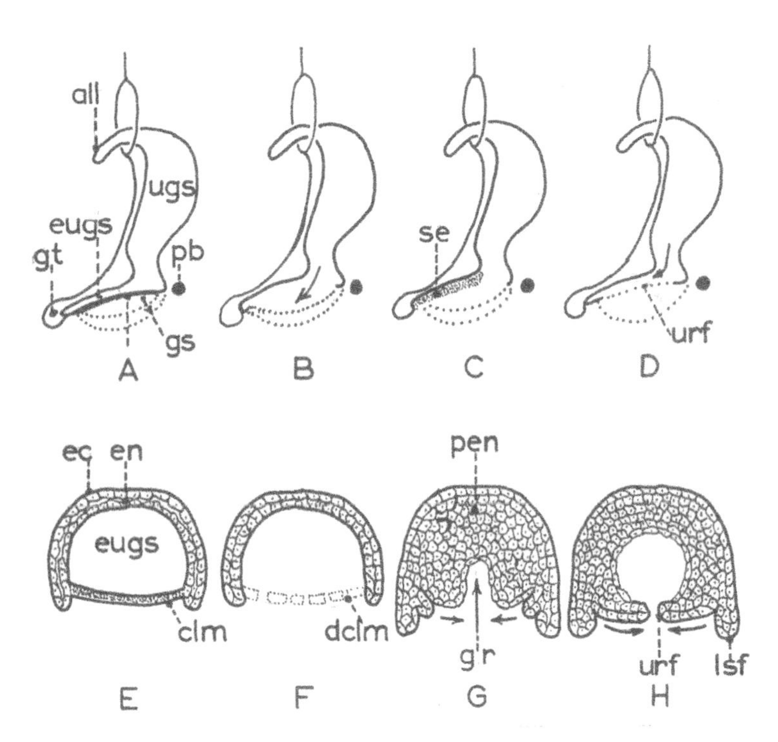

Figure 3.114: Formation of the penile urethra. A-D: Lateral views. E-H: Transverse sections. Arrow in **B** = forward extension of ugs; arrows in **G** and **H** indicate closure of the groove. all = allantois; clm = cloacal membrane; dclm = disintegration of cloacal membrane; ec = ectoderm; en = endoderm; eugs = extension of urogenital sinus; gr = groove; gs = genital swellings; gt = genital tubercle; lsf = labioscrotal fold; pb = perineal body; pen = proliferating endoderm; se = solidification of endoderm; ugs = urogenital sinus; urf = urethral fold.

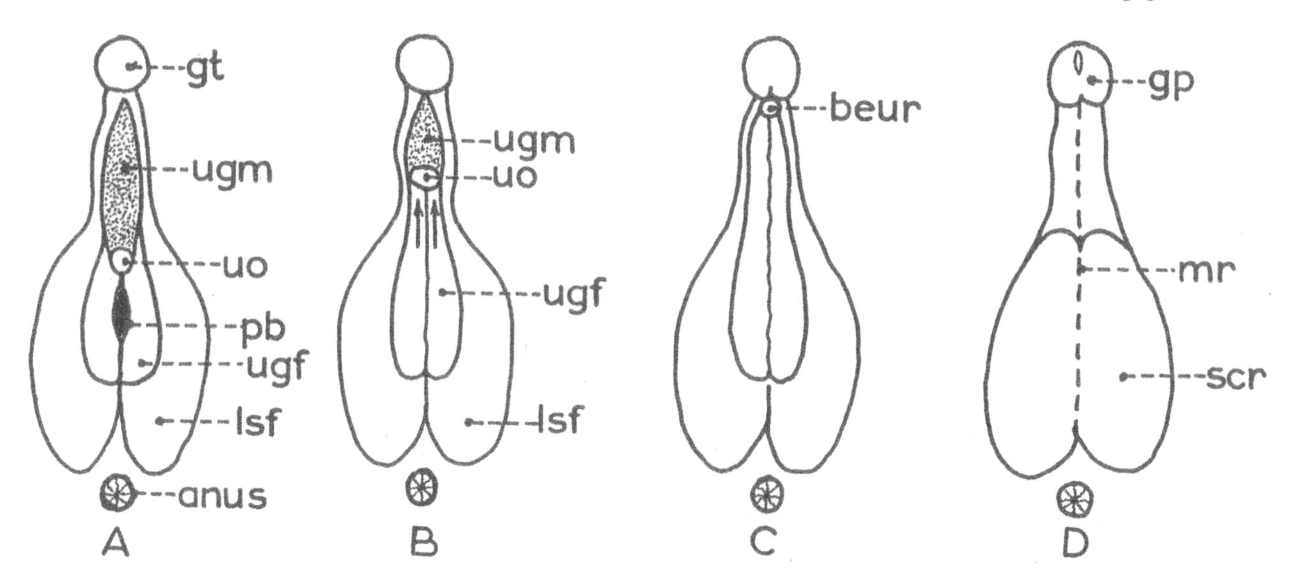

Figure 3.115: External genitalia in the male viewed inferiorly (ventral). beur = blind end of urethra; gp = glans penis; gt = genital tubercle; lsf = labioscrotal fold; mr = median raphe; pb = perineal body; scr = scrotum; ugf = urogenital fold; ugm = urogenital membrane; uo = urethral orifice.

Note on terminology: in this text, the flaccidity of the penis is regarded as its normal state in which it 'hangs downwards'. This means that it has a ventral (anterior) surface (a continuation of the abdominal wall surface), and a dorsal (posterior) surface (related to the scrotum).

A circular cleft develops around the glans penis and deepens ventrally and laterally but not dorsally. Dorsally, the connection between skin and glans is retained, producing the **frenulum preputii** (Fig. 3.116A,B). The cavity produced by the deepening of the cleft is the **preputial sac** with the overhanging skin forming the **prepuce** (Fig. 3.116B). It should be noted that complete retraction of the prepuce is not possible at birth and may be delayed for six or even twelve months after birth due to the presence of epithelial adhesions between skin and glans.

The two lateral genital folds are now called the labioscrotal folds, which enlarge and join in the midline to form the scrotum which also has a central raphe (Fig. 3.115D). This is continuous with the raphe of the penis.

POSSIBLE COMMON ABNORMALITIES

- **Epispadias** results from non-closure of the inferior part of the abdominal wall. The urethra opens on to the ventral (anterior) surface of the penis.
- **Hypospadias** is the failure of the urethral folds to close over and form the corpus spongiosum and urethra. The urethral opening may exist at any position between the perineum and the glans penis on the dorsal (posterior or scrotal) surface.

A tabular summary (Table 3.3) of the homologous structures in the male and female reproductive systems is shown below:

Table 3.3: Comparisons of male and female genital systems

A. Origin of gonadal tissues		
Anlage	**Male**	**Female**
Germ cells: extragonadal in origin, endoderm of umbilical vesicle.	Sperm series	Ovum series
Sustentacular cells: from the gonadal cords which arise from the coelomic epithelium	Supporting (Sertoli) cells	Granulosa cells
Interstitial cells: mesoderm	Interstitial endocrine(Leydig) cells and connective tissue	Stromal cells/ Thecal cells
B. Comparison of the ducts and tubules in the male and female		
Anlage	**Male**	**Female**
Mesonephric tubules	Ductuli efferentes, rete testis, appendix of epididymis, ductuli aberrantes, paradydimis.	Epoophoron, Paraoophoron, Appendix of ovary
Mesonephric duct	Epididymis, vas deferens, seminal gland, ejaculatory ducts	Remnants may persist as duct of the epoophoron
Paramesonephric duct	Appendix of testis, prostatic, utricle	Uterine tube, uterus, part of vagina
C. Comparison of external genitalia in male and female		
Anlage	**Male**	**Female**
Genital tubercle	Penis	Clitoris
Lateral genital swellings	Scrotum	Labia Majora
Urethral folds (medial genital swellings)	Floor of penile urethra	Labia minora

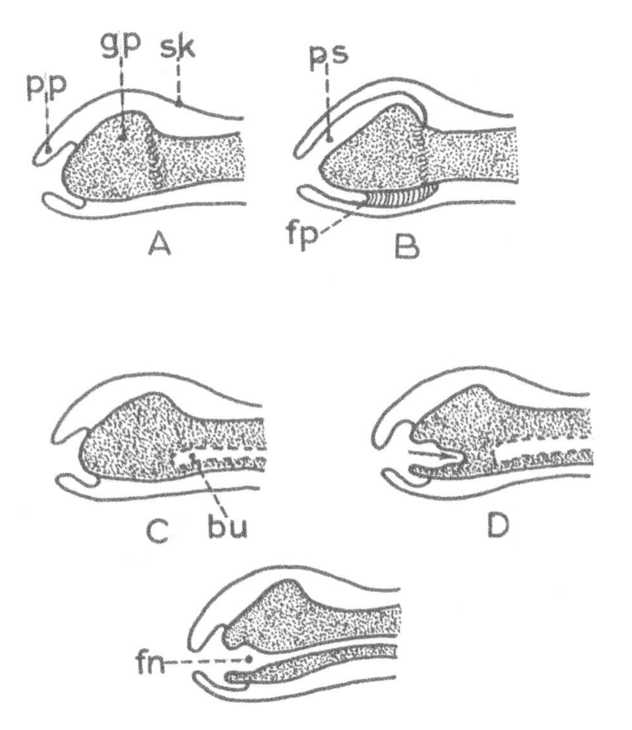

Figure 3.116: Development of the penis in longitudinal section.
Arrow = continuity being established between fn and bu; bu = blind urethra; fn = fossa navicularis; fp = frenulum of penis; gp = glans penis; pp = prepuce; ps = preputial sac; sk = skin.

DEVELOPMENT OF THE NERVOUS SYSTEM

The fully developed nervous system consists of a number of distinct parts:

- **Central Nervous System**: Brain and spinal cord
- **Peripheral Nervous System**: Segmental (spinal) nerves and plexuses
- **Autonomic Nervous System**: Sympathetic and Parasympathetic Nervous Components

Central Nervous System

Early Development

The primitive node gives rise to mesoderm (chorda-mesoderm) which, in turn, gives rise to the notochord (**Fig. 3.117A**). This mesoderm is laid down between the epiblast (later to form ectoderm) and the hypoblast (later to form endoderm). The notochord is responsible for the induction of the nervous system from the overlying ectoderm (epiblast) and this is borne out by the positional relations of the two.

At the beginning of the 3rd week pf, when the embryo is about 1.5mm in length, the notochord induces a strip of columnar ectodermal cells along the central axis of the embryonic disc (**Fig. 3.117B**). This strip is called the **neural plate** and it consists of **neurectodermal cells**. These cells undergo rapid mitotic proliferation and this is said to be one of the causes of the longitudinal and transverse 'bending' of the early embryo. At the edges of the plate, the columnar cells meet the somewhat flattened cells of the 'ordinary' ectoderm which will become the **surface ectoderm**. The surface ectodermal cells will later form the lining of the **amniotic cavity** (**Fig. 3.117C,D**) and the **epidermis** of the entire body.

The rapidly multiplying cells of the neural plate become 'heaped-up' at the junction with the surface ectoderm to form **neural folds** and this

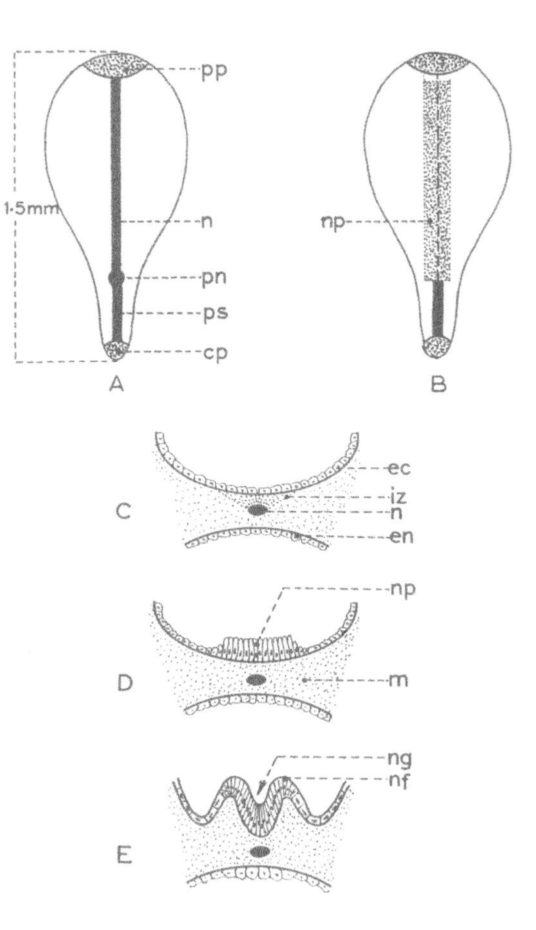

Figure 3.117: A and B: Dorsal view of the early embryo showing induction of the neural plate.
C-E: Transverse sections. cp = cloacal plate; ec = ectoderm; en = endoderm; iz = induction zone; m = mesoderm; n = notochord; nf = neural fold; ng = neural groove; np = neural plate; pp = prechordal plate; pn = primitive node; ps = primitive streak.

results in the formation of a **neural groove** between the folds (Fig. 3.117E). As the folds enlarge, the groove deepens and the folds, bending towards one another, meet and fuse in the middle of the neural plate (Fig. 3.118). This is the beginning of the **neural tube**. The fusion of the folds begins between somite levels 4 and 6 and then proceeds in a cranial and caudal direction until the tube is complete, except for openings at its cranial and caudal ends (Fig. 3.119). These openings are the **rostral** and **caudal neuropores** which remain open temporarily, into the ectodermal (amniotic) cavity. The rostral neuropore will close at the 20 somite stage and the caudal neuropore will close at the 25 somite stage resulting in a completely closed neural tube. The cavity of the tube remains as the **central canal** of the central nervous system.

While the neural folds are developing, a special group of cells differentiates in the crests of the folds. These are the **neural crest cells** (Fig. 3.120A,B). They are special in the sense that they give rise to a multiplicity of adult structures in various parts of the adult body, which they reach by the process of migration in the embryo. When the neural folds fuse, the neural crest cells come to lie on the dorsolateral sides of the tube (Fig. 3.120C). In that position, some will form spinal and sympathetic ganglia, while others will migrate to distant regions of the embryonic body.

By the end of the 4th week pf, the neural folds have fused completely, the neuropores have closed, and the surface ectoderm has separated from the neural tube. This surface ectoderm fuses and becomes continuous dorsal to the neural

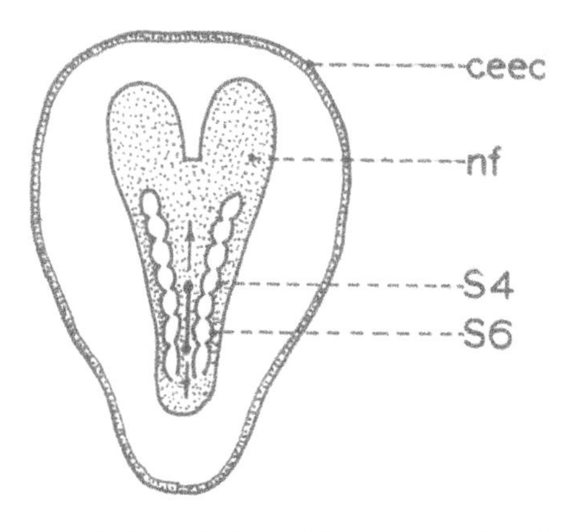

Figure 3.118: Dorsal view of the embryo showing the level of fusion of the neural folds. Arrows = direction of closure of the folds; ceec = cut edge of ectoderm; nf = neural fold; s4, s6 = somites.

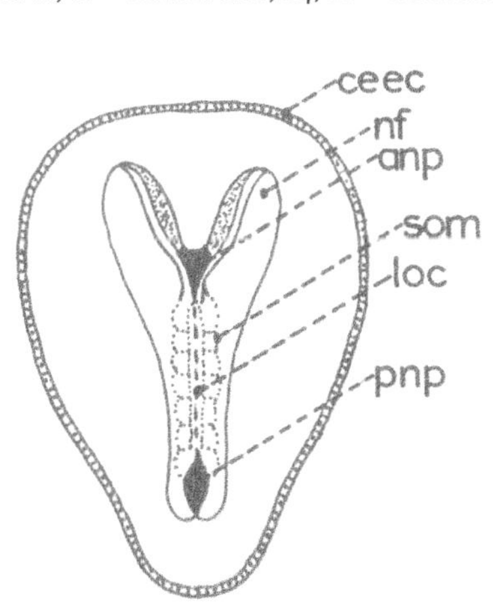

Figure 3.119: Dorsal view of the embryo showing the extension of fusion of the neural folds. anp = anterior neuropore; ceec = cut edge of ectoderm; loc = line of closure; nf = neural fold; pnp = posterior neuropore; som = somite.

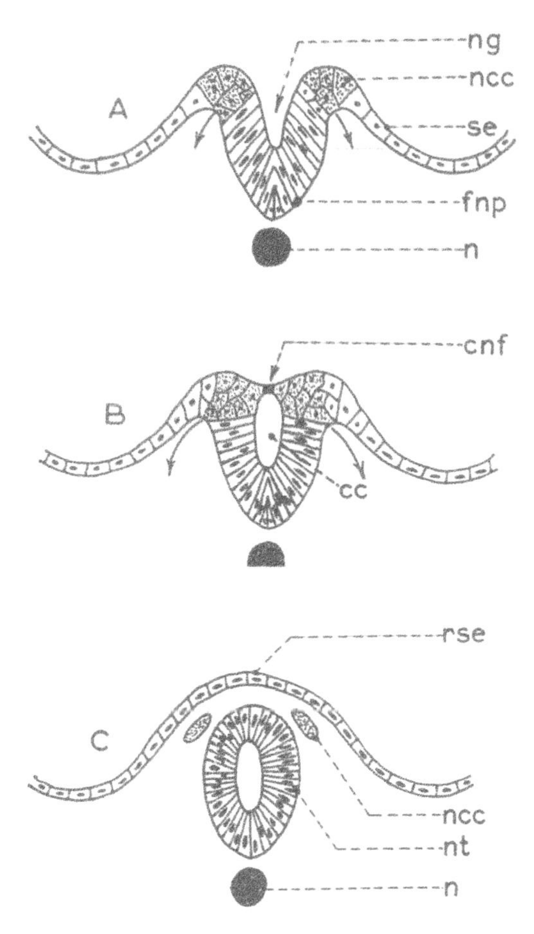

Figure 3.120: Transverse sections showing fusion of the neural folds. Arrows = migration of ncc. cc = central canal; cnf = closure of neural folds; fnp = fold of neural plate; n = notochord; ncc = neural crest cells; ng = neural groove; nt = neural tube; rse = restored surface ectoderm; se = surface ectoderm.

tube, thus preserving the integrity of the ectodermal (amniotic) cavity (**Fig. 3.120C**). In this way the neural tube and neural crest become independent of the ectoderm from which they originally arose.

Development of the Spinal Cord

Since the neural tube is the fundamental progenitor of the nervous system, it is worthwhile understanding something about its anatomy and cytology. Approximately the caudal half of the neural tube will form the spinal cord while the cranial half forms the brain. However, much of the basic structure of the caudal part of the tube will persist in the developing brain.

The columnar cells lining the neural tube may be regarded as **germinal cells** since the majority of cells of the nervous system will arise from them. The columnar germinal cells undergo cycles of mitotic multiplication to produce the appearance of a pseudostratified epithelium. This proliferation of cells results firstly, in thickening of the lateral walls of the tube, while the dorsal (roof plate) and ventral (floor plate) are little affected (**Fig. 3.121A,B,C**). Secondly, the proliferation will result in an alteration of the size and shape of the lumen of the tube. In general, the neural tube consists of three layers of cells and this pattern is maintained throughout development until migration of neuroblasts occurs in the cerebral cortex and cerebellar cortex. The layers may be styled as inner, middle and outer layers, as indicated in Fig 3.121 A–C:

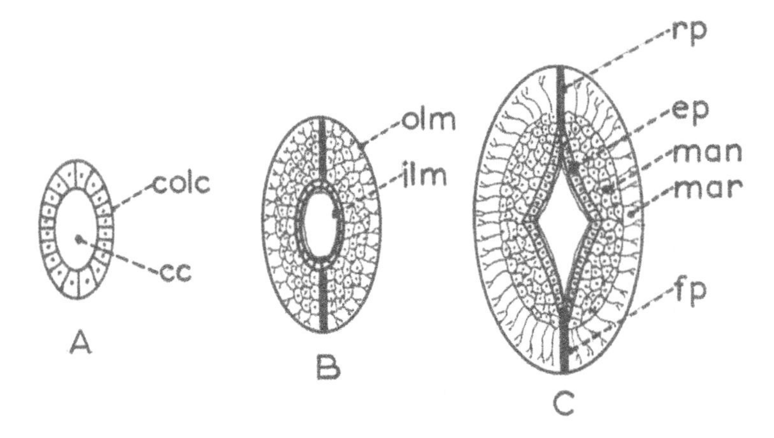

Figure 3.121: Transverse sections to show the development of the neural tube. cc = central canal; colc = columnar cells; ep = ependyma; fp = floor plate; ilm = inner limiting membrane; man = mantle layer; mar = marginal layer; olm = outer limiting membrane; rp = roof plate.

(A) *Inner layer*: The inner lining of the neural tube, known as the **ependymal** cell layer, consists of simple columnar cells, the luminal surfaces of which are attached to an **inner limiting membrane** (**Fig. 3.121B**).

(B) *Middle layer*: The next layer is the **mantle layer** consisting of closely packed cells which differentiate from the germinal cells into **neuroblasts** and **glioblasts**. The neuroblasts will form **neurons**, the electrical functional units of the nervous system. These cells will give rise to processes called axons and dendrites which enable them to make contact with one another at different levels of the nervous system (**Fig. 3.121C**). The glioblasts will give rise to certain of the **neuroglial cells**, such as astrocytes, which are the supporting cells of the nervous system.

(C) *Outer layer*: The outer layer of the tube is the **marginal layer**, which is at first relatively sparsely filled with cells or fibres, but which later becomes packed with axons derived from the developing neurons (**Fig. 3.121C**).

Neuroglial cells are of three types:

1. **Astrocytes**, which may be classified according to the nature of the processes emanating from the cell body:
   - protoplasmic astrocytes, having numerous furry or mossy processes;
   - fibrous astrocytes, which have few radially disposed processes.
2. **Oligodendrocytes**, which have very few surrounding processes. These cells are of neural crest origin. They are responsible for the myelination *within* the central nervous system.
3. There is another group of cells which is found in the nervous system – the **microglia**. These cells are not of neuroblast or glioblast origin but are derived from the mesenchyme of the surrounding blood vessels and are regarded as 'scavenger' cells of the nervous system. They are classified as part of the mononuclear-phagocytic system.

In the adult spinal cord and brain, the inner layer becomes the **ependymal** lining of the central canal of the spinal cord and of the ventricular cavities. This layer is ciliated in parts. The mantle layer becomes the **'grey' matter** of the cord and the marginal layer forms the **'white' matter**.

The outer surface of the neural tube is covered with vascular mesenchyme from which the nervous system will obtain its superficial blood supply. This layer of mesenchyme, which also contains neural crest cells, will ultimately become the **pia** and **arachnoid mater** of the brain and spinal cord. The dura mater develops from mesenchyme of the superfical membrane (see skull development).

In the neural tube, the thickening of the mantle layer of the wall forms dorsal and ventral bulges which protrude into the lumen, reducing its size and also creating in each side a deep cleft, the **sulcus limitans** (Fig. 3.122A). These sulci, which extend along the most of the length of the original neural tube, have considerable significance in that they divide the neural tube into anatomical and functional parts. The dorsal bulges of the mantle layer are called the **alar plates** and the ventral bulges are the **basal plates** (Fig. 3.122B). From a functional point of view, the alar plates contain cells, which receive sensory impulses from the body, and the basal plates contain cells which send motor impulses to the muscles of the body. It is noteworthy that the regions of the mantle layer close to the sulci limitantes are associated with the autonomic nervous system both in structure and function (the **intermedio-lateral region**) (Fig. 3.122B). The alar plates of each side are joined by the **roof plate**, while the basal plates are joined by the **floor plate.** The roof plate and floor plate appear to have no function other than as joining factors.

While in the very early neural tube, the whole lumen (central canal) is lined by germinal cells, some of them will form the ependymal lining of the walls of the canal in the adult. The cells of the roof and floor plates are soon replaced by ependymal and neuroglial cells and, therefore, do not contribute neurons to the formation of the tube.

From the earliest stage of the formation of the neural tube until the age of about 8 weeks pf (30mm CR), the neural tube extends to the level of coccygeal vertebra 4 (Fig. 3.123A). When in the late 8th week of development, the trunk begins to elongate, the growth of the newly formed spinal cord does not keep pace with the elongation of the trunk. By full term, the end of the spinal cord lies at about the level of the 3rd lumbar vertebra. With further growth of the vertebral column, the lower end of the spinal cord (**conus medullaris**) lies at about the level of vertebrae L1 or L2, in the adult (Fig. 3.123B,C).

As a result of the discrepancy in growth between the spinal cord and the vertebral column, the directions of the spinal nerves are progressively angulated caudally until at the lowermost end, the

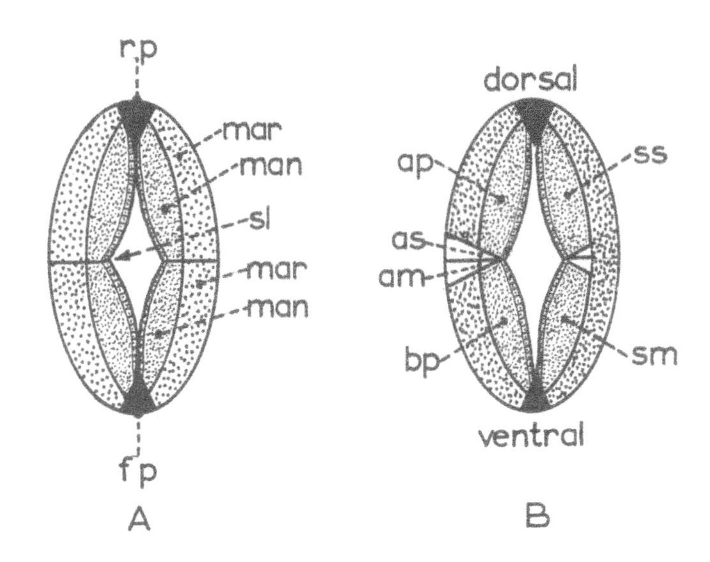

Figure 3.122: Transverse sections of the neural tube. ap = alar plate; am = autonomic motor; as = autonomic sensory; bp = basal plate; fp = floor plate; mar = marginal layer; man = mantle layer; rp = roof plate; sl = sulcus limitans; sm = somatic motor; ss = somatic sensory.

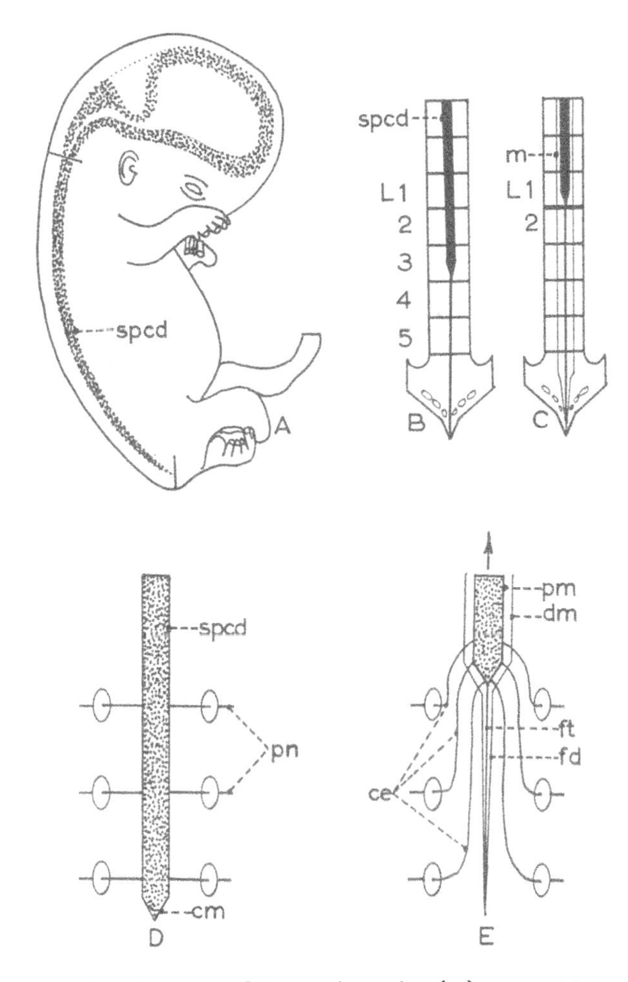

Fgure 3.123: Development of the cauda equina (ce). A: Lateral view. **B-E:** Ventral views. cm = conus medullaris; dm = dura mater; fd = filum of dura; ft = filum terminale; L1, 2, 3, 4, 5 = lumbar vertebrae; m = membranes; pn = peripheral nerves; pm = pia mater; spcd = spinal cord.

nerve roots lie in a vertical direction within the vertebral canal forming the **cauda equina** (Fig. 3.123 D,E), i.e. as a result of the 'retreat' of the spinal cord up the vertebral column, the spinal nerves must descend to exit from the original vertebral foramina. With the elongation of the vertebral column, the membranes surrounding the spinal cord maintain their peripheral attachments so that withdrawal of the cord causes the membranes to collapse, forming the 'thread-like' **filum terminale** (pia mater) surrounded by the arachnoid and the **filum of the dura** (Fig. 3.123E).

POSSIBLE COMMON ABNORMALITIES

- Abnormalities of the spinal cord are associated with a cleft of the vertebral column (**Spina bifida**). These may or may not be associated with herniation of the contents of the vertebral canal, eg. **meningocoele** (meninges only), **meningomyelocoele** (meninges and nervous tissue), and no herniation (**spina bifida occulta**).
- If there is absence of neural tissue in the spinal cord region this is known as **amyelus,** while a **rachischisis** is a closure defect of the spinal cord.

General Development of the Brain

After closure of the neuropores, the central nervous system is a completely closed tube with a slightly dilated cranial end. At this time, the tube develops two constrictions towards its cranial end. These cause the formation of three dilatations, the most caudal being in continuity with the spinal cord (Fig. 3.124A,B). These dilatations are called the **primary brain vesicles** and are named, from cranial to caudal, as the **prosencephalon** (forebrain), **mesencephalon** (middle brain) and **rhombencephalon** (hindbrain) (Fig. 3.124B).

Note. The term 'rhombencephalon' is derived from the fact that its cavity later becomes rhomboid-shaped.

The cell types and layers of the brain conform to the pattern found in the spinal cord.

While the constrictions are occurring, an extrusion forms on either side of the prosencephalon; these are the **optic stalks** which have a blind and somewhat bulbous end (Fig. 3.125B).

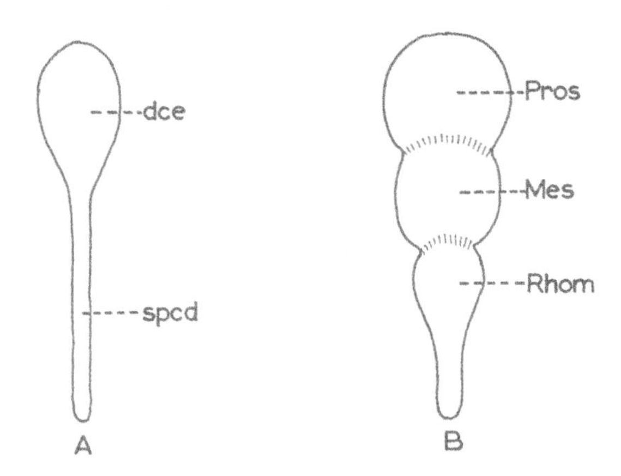

Figure 3.124: Development of the primitive brain.
dce = dilated segment of neural tube;
mes = mesencephalon; pros = prosencephalon;
rhom = rhombencephalon; spcd = spinal cord.

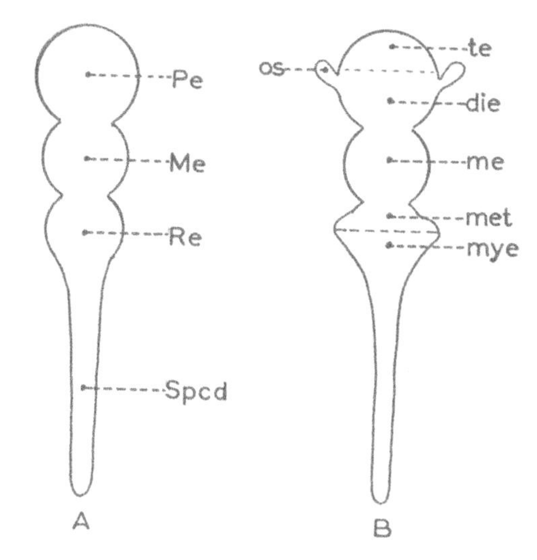

Figure 3.125: Development of the brain vesicles.
die = diencephalon; me = mesencephalon;
met = metencephalon; mye = myelencephalon;
os = optic stalk; pe = prosencephalon; re = rhombencephalon; spcd = spinal cord; te = telencephalon.

Later, the bulbous end becomes invaginated and forms part of the eyeball. The prosencephalon and rhombencephalon will become subdivided to form two further vesicles each. The presence of the optic stalks divides the prosencephalon into two parts. The part rostral to the stalks is the **telencephalon** (end-brain), while that part giving rise to the stalks is the **diencephalon** (the 'in-between' brain) and is that part between the telencephalon and the mesencephalon (Fig. 3.125B). The rhombencephalon now undergoes a dorsal bend (pontine flexure) resulting in the formation of a cranially placed **metencephalon** (later forming pons and cerebellum) and a more caudal part, the **myelencephalon** (later to form the medulla oblongata). Note that the mesencephalon does not divide. The brain now consists of five components: a telencephalon, diencephalon, mesencephalon, metencephalon and myelencephalon, which constitute the **secondary brain vesicles**. The cavities within the secondary brain vesicles extend into the spinal cord as its central canal. The cavities in the brain will expand to give rise to the **ventricular system**.

In its development the brain undergoes three flexion processes. These are the **cephalic, cervical** and **pontine** flexures. As the embryo is undergoing its 'general' bending process, the head end bends ventrally over the cranial end of the notochord to form the **cephalic flexure** (Fig. 3.126A,B). This flexure lies in the region of the mesencephalon. With further bending, the junction between the rhombencephalon and the spinal cord undergoes flexion ventrally to form the **cervical flexure** (Fig. 3.126A,B). Then a third flexure takes place in the opposite direction and this tends to 'dorsiflex' the brain. This is the **pontine flexure** and it divides the brain into a cranially placed metencephalon and a caudally placed myelencephalon (Fig. 3.126A,B) and is largely due to growth changes in the brain itself. The pontine flexure forms an acute forward angle and, as it occurs, the neural tube widens laterally (as does a rubber tube when acutely bent) (Fig. 3.127A,B). The widening takes place at the expense of the roof plate, which is ultimately reduced to a thin membrane of ependyma (Fig. 3.127C,D) covered with vascular mesenchyme.

The widening causes the alar and basal plates to lie side by side, rather than in the dorso-ventral position (Fig. 3.127C,D) and causes the floor of the neural tube to assume a diamond shape (rhomboid), thus forming the floor of the fourth ventricle.

Note. With the expansion of the telencephalon, the vesicles come to overlie the 'flexed' neural tube (Fig. 3.126C,D).

Development of the Telencephalon

The most cranial part of the telencephalon now gives rise to two large diverticula, the telencephalic vesicles which appear at about the 7th week pf and result in the formation of the cerebral hemispheres. These two telencephalic vesicles expand from the original single telencephalon, leaving between them the **telencephalon medium** bounded rostrally by

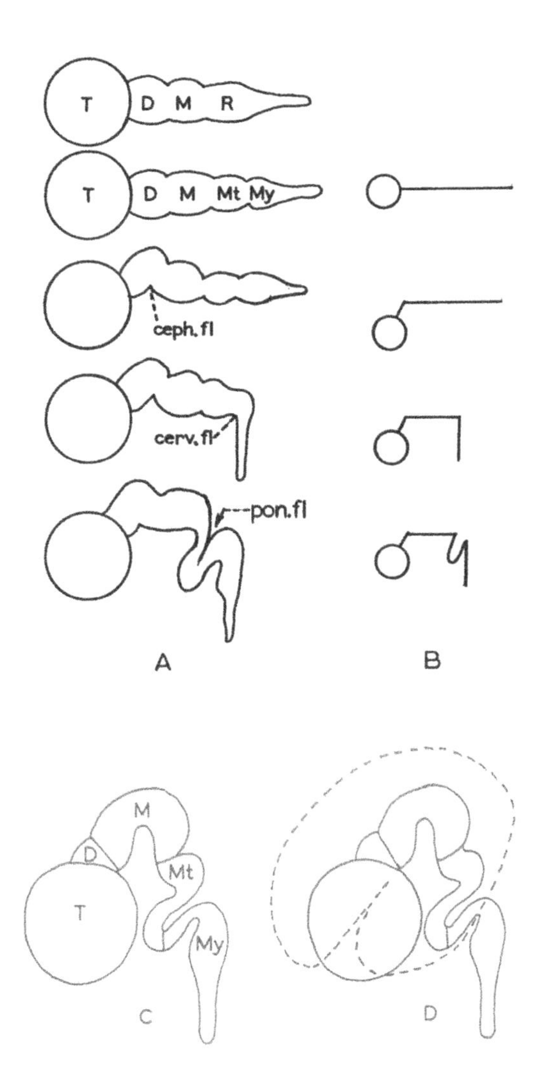

Figure 3.126: Flexures of the brain and expansion of the telencephalon in lateral view.
D = diencephalon; M = mesencephalon; Mt = metencephalon; My = myelencephalon; T = telencephalon; ceph.fl = cephalic flexure; cerv.fl = cervical flexure; pon.fl = pontine flexure.

the **lamina terminalis** (which marks the position of closure of the cranial neuropore). The cavities within the telencephalic vesicles are the **lateral ventricles**, which initially communicate with the telencephalon medium (the unexpanded central part of the telencephalon), through large openings. With further development of the telencephalic vesicles these openings become narrowed to form the **interventricular foramina** (of Monro) (Fig. 3.128). The cavity within the telencephalon medium is continuous with the cavity inside the diencephalon. These two together will form the **3rd ventricle**. The development of the choroid plexuses of the lateral and third ventricles are dealt with later.

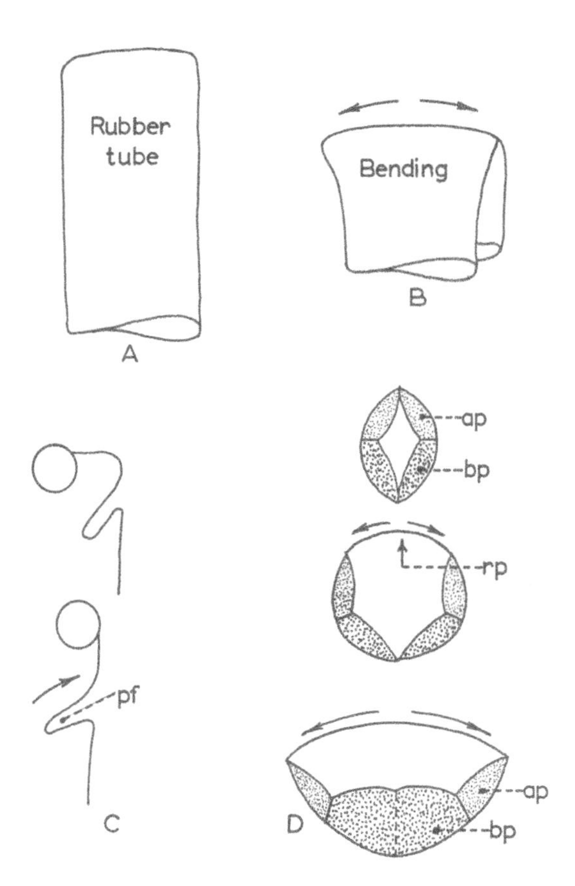

Figure 3.127: A-C: The bending of the neural tube. **D:** Transverse sections of the hindbrain. ap = alar plate; bp = basal plate; pf = pontine flexure; rp = roof plate.

As the telencephalic vesicles grow and expand they undergo a rotational process **(Fig. 3.129A,B,C)**. In this way they cover the rostral and lateral surfaces of the diencephalon and mesencephalon. As this takes place, the dorsal and lateral parts of the vesicles become thinned and are called the **pallium**, while the floor of each vesicle becomes thickened to form the **colliculus ganglionaris (Fig. 3.130A)**. The colliculus is invaded by fibres from the pallium, which divide it into the **caudate** and **lentiform nuclei (Fig. 3.130B,C)**. These fibres are called the **internal capsule (Fig. 3.130F)**. The fibres leave strands of grey matter (mantle layer) between the caudate and lentiform nuclei giving the arrangement a striped appearance. This formation is called the **corpus striatum** (the striate body) **(Fig. 3.130D)**. The caudate and lentiform nuclei are collectively known (amongst others) as the **basal nuclei (Fig. 3.130E,G)**.

Other fibres from the frontal and parietal regions of the pallium form a separation between the lentiform nucleus and an overlying layer of grey matter called the **claustrum (Fig. 3.130G)**.

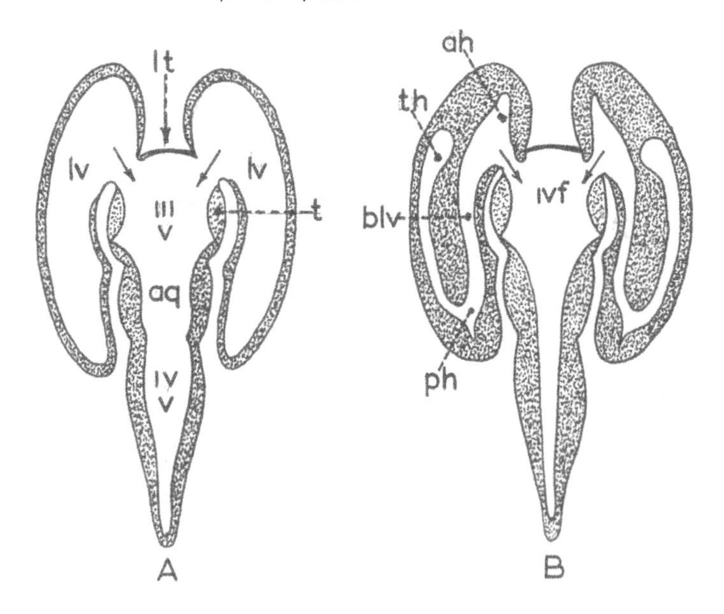

Figure 3.128: Horizontal section depicting the expansion of the telencephalon and the position of the interventricular foramen. IIIv = third ventricle; aq = aqueduct; ivv = fourth ventricle; ah = anterior horn; blv = body of lateral ventricle; ivf = interventricular foramen; lt = lamina terminalis; lv = lateral ventricle; ph = posterior horn; t = thalamus; th = temporal horn.

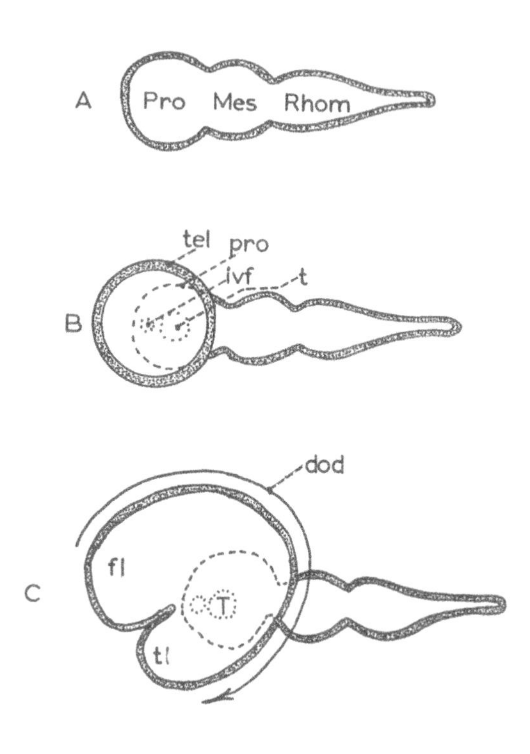

Figure 3.129: Longitudinal section indicating the direction of growth (dod) of the telencephalon (tel). M = mesencephalon; Pro = prosencephalon; Rhom = rhombencephalon; fl = frontal lobe; ivf = interventricular foramen; t = thalamus; tl = temporal lobe.

Note there is dispute about the origin of the claustrum; some embryologists believe that it should be included in the classification of the basal nuclei.

Ultimately the two telencephalic vesicles fuse with the diencephalon where they overlie it.

Note on the mantle layer and the alar and basal plates in the telencephalon

It appears that the sulcus limitans is the major factor in dividing the neural tube into alar and basal laminae. It seems that the sulcus limitans ends at the rostral end of the mesencephalon. Therefore, the dividing factor (the sulcus limitans) does not exist in the forebrain and there is no division of the mantle layer into alar and basal plates. For this reason the telencephalon has been considered to be devoid of separate plates (see also diencephalon) and is said to be composed of alar plate only.

Formation of the Cerebral Cortex

From the mantle layer of each pallium, cells migrate into the existing marginal layer to form a thin cellular layer of grey matter which now constitutes a primitive **cortex** (Fig. 3.131A,B). Over a period of time, this primitive cortex expands enormously. To be accommodated within the confines of the developing skull, numbers of gyri and sulci are formed (Fig. 3.131C). This expansion is particularly noticeable in the parietal and temporal lobes, while the intervening part of the cortex shows a relatively slow degree of growth. This results in the formation of a depressed area of cortex known as the **insula** (of Reil) (Fig. 3.131C). The overlapping parietal and temporal lobes form the **opercula** over the insula and the cleft between the overhanging opercula forms the **lateral cerebral fissure** (of Sylvius). By full term, the cortex contains six layers of cells, some of which arise by proliferation of existing cells and others by additional migration from the mantle layer.

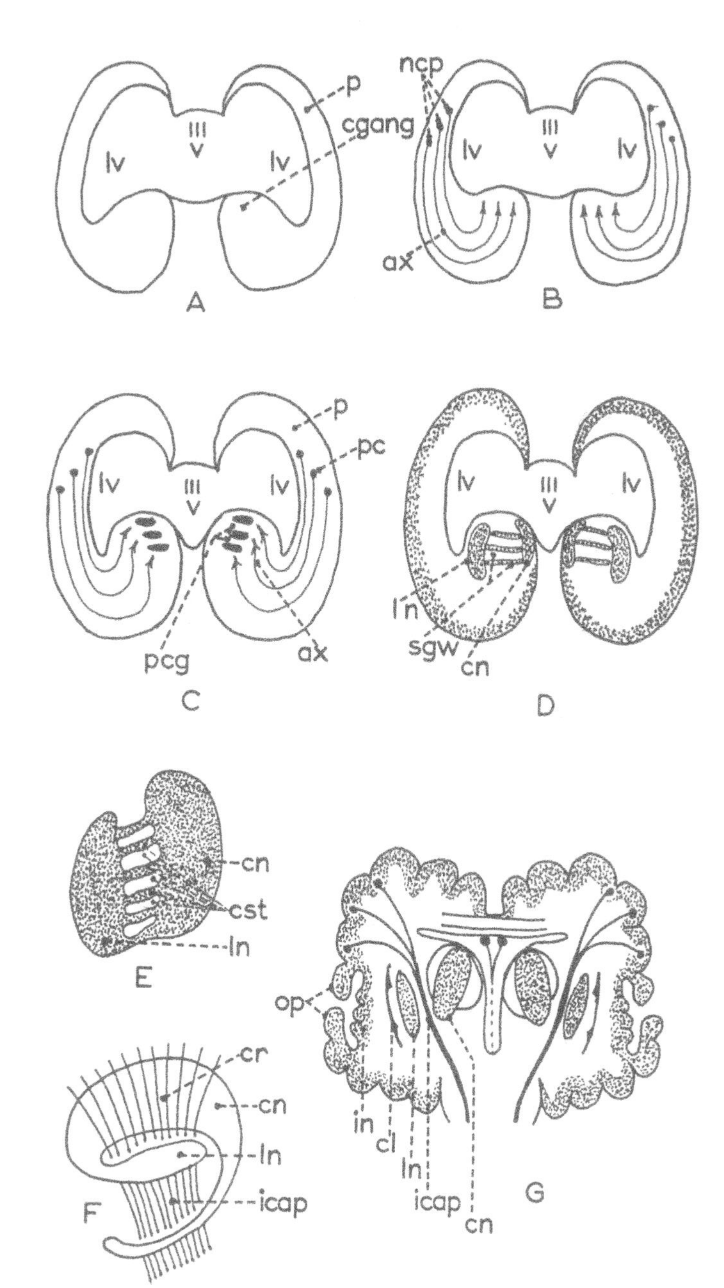

Figure 3.130: Coronal sections to demonstrate the development of the telencephalon and corpus striatum (A-G) and lateral view (E, F).
ax = axons; cgang = colliculus ganglionaris; cl = claustrum; cn = caudate nucleus; cr = corona radiata; cst = corpus striatum; icap = internal capsule; in = insula; ln = lentiform nucleus; op = operculum; p = pallium; pc = pallial cells; pcg = penetration of colliculus ganglionaris by axons (ax); sgw = strands of grey and white matter.

Development of the Diencephalon

While the pattern of the three layers of the neural tube extends throughout the tube, the division of the mantle layer into alar and basal plates runs rostrally from the spinal cord as far as the mesencephalon only. As a result, the sulcus limitans is said to end at the rostral end of the mesencephalon. The **hypothalamic sulcus** replaces the sulcus limitans in the diencephalon, thus separating the thalamus and the hypothalamus (autonomic components of the neural tube) (Fig. 3.132). The hypothalamic sulcus ends at the level of the interventricular foramen.

Immediately caudal to the interventricular foramina, the **thalami** (sing. thalamus) develop in the lateral walls of the diencephalon (Fig. 3.128B). The thalami arise dorsal to the hypothalamic sulci. The thalami are the major sensory stations of the brain. The parts of the lateral wall ventral to the sulci contain the **hypothalamic nuclei** and these extend also into the floor of the **IIIrd ventricle**. In later development the thalami may grow across the IIIrd ventricle to abutt on each other and form the **interthalamic adhesion** or **connexus**.

Note the interthalamic adhesion is *not* a commissure.

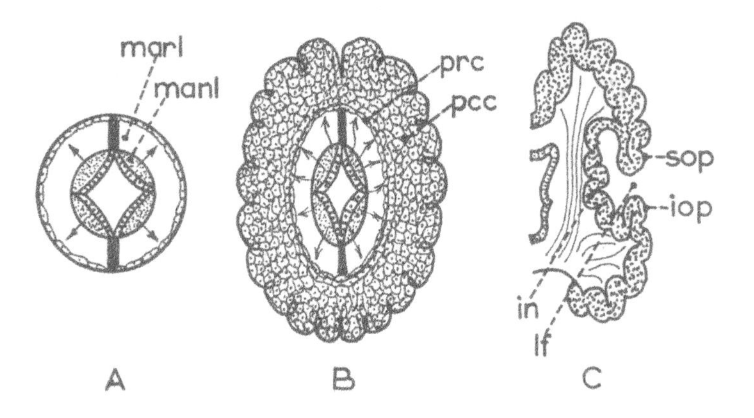

Fig. 3.131: Transverse sections illustrating the development of the primitive cortex. in = insula; iop = inferior operculum; lf = lateral fissure; manl = mantle layer; marl = marginal layer; pcc = primitive cortical cells; prc = primitive cortex; sop = superior operculum.

The hypothalamus is the primary centre for regulation of autonomic and endocrine function. It controls such activities as:

- the response to increased and decreased body temperature,
- hunger and feeding,
- thirst,
- sexual function,
- secretion of anti-diuretic hormone (ADH) and
- emotional activity.

As a result of these functions, many of the neurones in the hypothalamic nuclei send axonic projections to various parts of the brain as well as to the spinal cord.

The rostral part of the roof of the diencephalon remains thin and consists only of a single layer of ependymal cells covered by a vascular mesenchymal layer. The two together give rise to the choroid plexus of the IIIrd ventricle.

Dorsal to the thalamus, there is a shallow lateral indentation called the **epithalamus** bounded dorsally by the thinned roof of the IIIrd ventricle. This is the region where the **pulvinar** (dorsal prominence) of the thalamus lies and where the **lateral** and **medial geniculate bodies** develop in relation to the pulvinar (Fig. 3.133). These bodies are associated with vision and hearing respectively. Caudal to the epithalamus, the thickened caudal roof of the third ventricle gives rise to the **pineal body** (Fig. 3.133), which secretes melatonin and acts in an inhibitory way on the production of pituitary

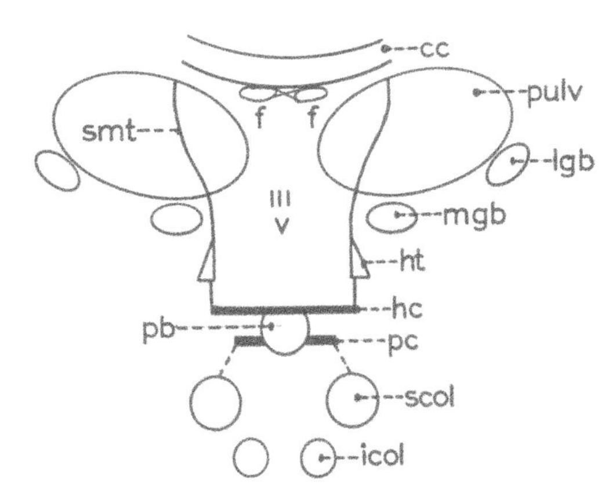

Figure 3.133: Dorsal surface of the diencephalon and the mesencephalon. cc = corpus callosum; f = fornix; hc = habenular commissure; ht = habenular trigone; icol = inferior colliculus; lgb = lateral geniculate body; mgb = medial geniculate body; pb = pineal body; pc = posterior commissure; pulv = pulvinar of thalamus; scol = superior colliculus; smt = stria medullaris thalami.

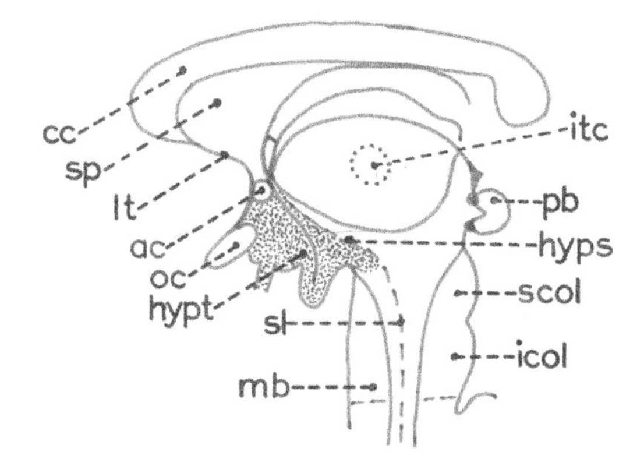

Figure 3.132: Midsagittal section through the brain illustrating the development of the diencephalon. ac = anterior commissure; cc = corpus callosum; hypt = hypothalamus; hyps = hopothalamic sulcus; icol = inferior colliculus; itc = interthalamic connection; lt = lamina terminalis; mb = midbrain; oc = optic chiasma; pb = pineal body; scol = superior colliculus; sl = sulcus limitans; sp = septum pellucidum.

hormones. The **habenular** and **posterior commissures** will develop in the roof of the diencephalon in relation to the pineal body. By this time, the two optic stalks have moved medially to join at the **optic chiasma** and with the lamina terminalis.

From its position in the dissected adult brain, the hypothalamus occupies the position of the autonomic motor component of the embryonic neural tube (Fig. 3.134A,B).

Combined Development of the Telencephalon and Diencephalon

Each expanding telencephalic vesicle comes to overlie the dorso-lateral surfaces of the developing diencephalon (Fig. 3.135A) and the three will fuse (Fig.3.135B). The inferior wall of each telencephalic vesicle consists of a thin layer of ependymal cells which lies between the choroid plexus of the lateral ventricle and the thalamus. This thin layer is known as the **lamina affixa** (Fig. 3.135B). Just medial to the lamina affixa there is a thickening of the medial wall of each telencephalic vesicle which forms the **hippocampus** (Fig 3.135B). This also gives rise to the **induseum griseum** of the adult. Figure 3.135C illustrates the anatomy of the adult.

As the roof and lateral wall of the telencephalic vesicles (pallium) grow and expand, they do so in a 'circular' fashion, covering the sides of the diencephalon and mesencephalon (Fig. 3.129C). Because of the circular type of growth, structures which develop in the medial wall of the telencephalon such as the hippocampus, caudate nucleus, **stria terminalis**, **thalamostriate vein**, **fornix** and choroid plexus of the lateral ventricle are pulled

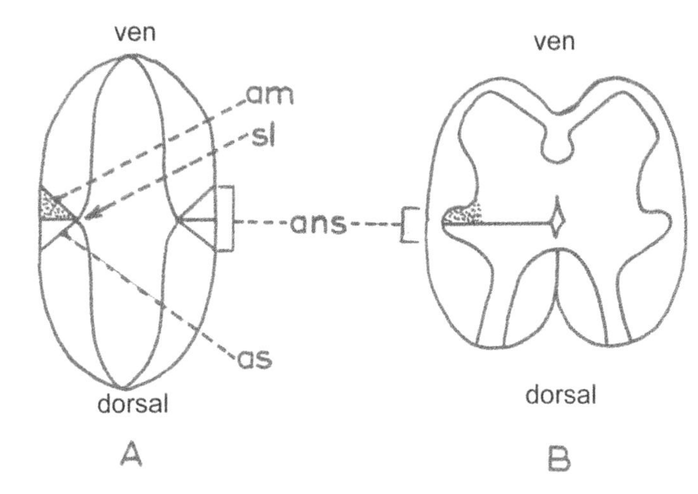

Figure 3.134: Transverse sections of the neural tube (A) and spinal cord (B). am = autonomic motor; ans = autonomic nervous system; as = autonomic sensory; sl = sulcus limitans.

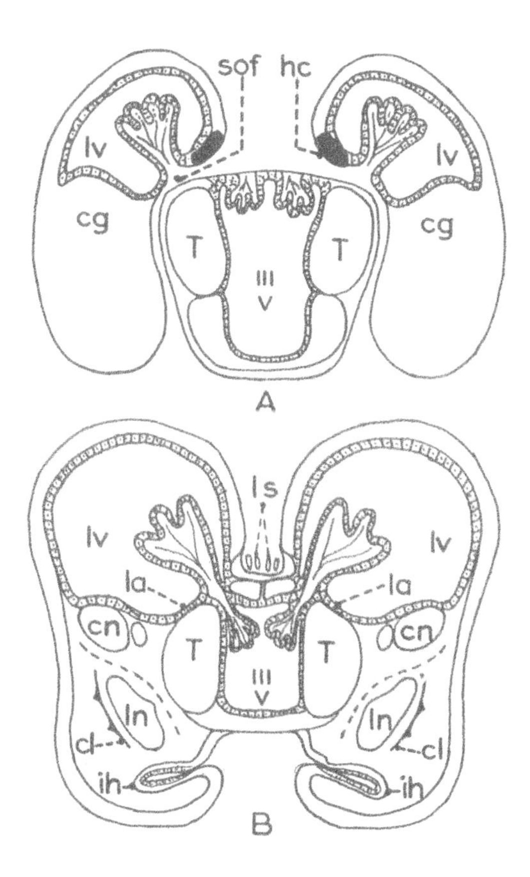

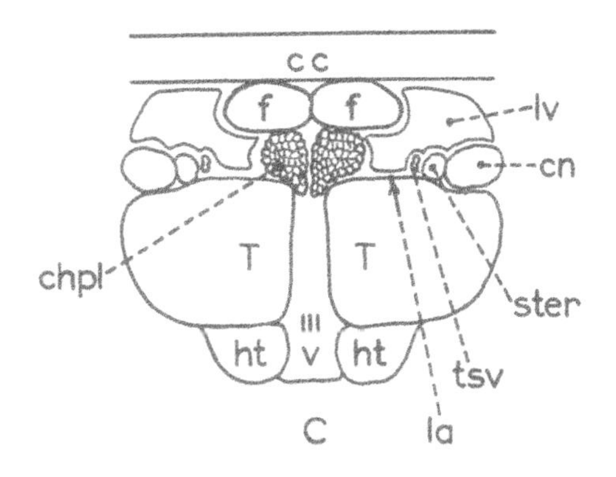

Figure 3.135: Coronal sections to show the fusion of the telencephalon with the diencephalon. IIIv = third ventricle; chpl = choroid plexus; cc = corpus callosum; cg = colliculus ganglionaris; cl = claustrum; cn = caudate nucleus; f = fornix; ht = hypothalamus; ih = inferior horn; la = lamina affixa; ln = lentiform nucleus; ls = longitudinal striae; lv = lateral ventricle; sof = site of fusion; hc = hippocampus; ster = stria terminalis; T = thalamus; tsv = thalamostriate vein.

around the thalamus in a C-shaped direction (Fig. 3.136A-F). It stands to reason that the cavities (lateral ventricles) of the two telencephalic vesicles will also conform to the general C-shape (Fig. 3.136A). With growth, the ventricular cavity will have a body superior to the thalamus and a temporal lobe extension inferior and lateral to the thalamus. The extensions of the ventricles into the frontal and occipital lobes are called the **anterior** and **posterior horns** in the adult (Fig. 3.136A). The contained structures will come to lie, respectively, cranial to the thalamus, dorsal to it, caudal to it, and ventral to it. With the downward growth of the telencephalic vesicles, part of the lateral ventricle is pulled down into the temporal lobe forming the **inferior horn** of the lateral ventricle. Thus, the caudate nucleus and stria terminalis, which lie in the floor of the anterior horn, come to lie in the roof of the inferior horn.

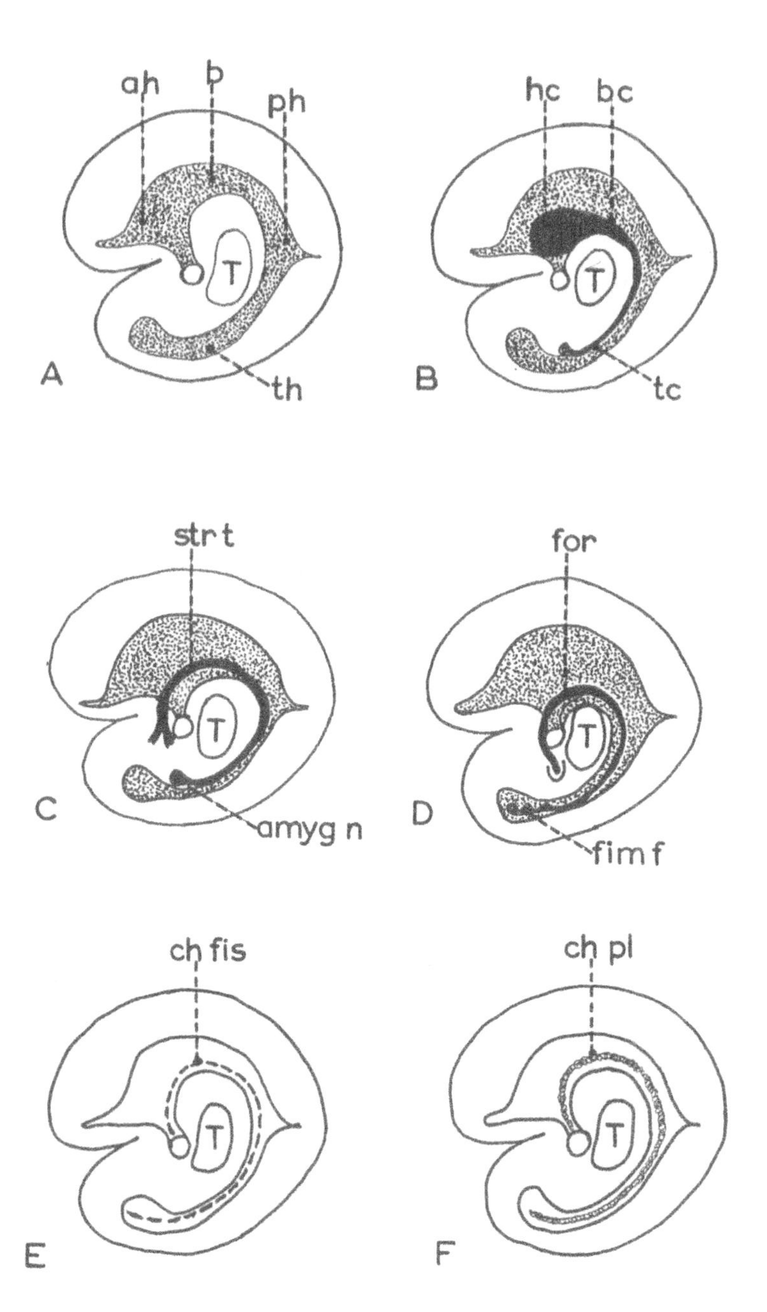

Figure 3.136: Lateral views of the telencephalon illustrating the development of structures in a 'C' shaped pattern in relation to the lateral ventricle. ah = anterior horn; amygn = amygdaloid nucleus; b = body; bc = body of caudate nucleus; chfis = choroid fissure; chpl = choroid plexus; fimf = fimbria of fornix; for = fornix; hc = head of caudate nucleus; ph = posterior horn; strt = stria terminalis; tc = tail of caudate nucleus; th = temporal horn.

Development of the Midbrain

The mesencephalon (later to be the midbrain) consists of the three layers of the neural tube surrounding a relatively large cavity, the mesencephalic ventricle (Fig. 3.137A). The interior of the mesencephalic tube is divided into dorsal and ventral compartments by the **sulci limitantes** (Fig. 3.137A,B) and thus has alar and basal plates. The interior of the tube is lined by ependyma and the parts of the tube adjacent to the sulci are destined to become the elements of the autonomic nervous system.

The mesencephalon forms the connection between the diencephalon and the lower reaches of the brainstem (pons, cerebellum, medulla oblongata). As fibres from the cerebral hemispheres (motor) and fibres from the periphery of the embryonic body (sensory) pass to their destinations, they must necessarily pass through the mesencephalon and thus occupy 'space' in it. In order for the mesencephalon to accommodate the increasing number of fibres, its lumen becomes progressively smaller until it exists as a narrow passage. This is the **cerebral aqueduct** (aqueduct of the midbrain; aqueduct of Sylvius) which connects the IIIrd and the IVth ventricles of the brain.

In addition to the passage of fibres through the mesencephalon, the cells of the mantle layer of the basal and alar plates proliferate and also form an encroachment on the cerebral aqueduct. The cells of the alar plates multiply to form the **pretectal nuclei** and the dorsal part of the **peri-aqueductal grey matter** (Fig. 3.137C). The cells of the basal plates proliferate to form the ventral part of the peri-aqueductal grey matter, the **reticular formation**, the **red nuclei** and the **substantia nigra** (Fig. 3.137C).

Note. There is some dispute about the origin of the substantia nigra and the red nuclei being from the basal plates.

The proliferation of the cells of the alar plates results in the formation of two longitudinal ridges, separated by a groove, on the dorsum of the mesencephalon. The longitudinal ridges are called the **corpora bigemina** (Fig. 3.138A,B,C). These ridges are later separated by a transverse groove converting them into four **corpora quadrigemina**, known also as the **superior and inferior colliculi** (Fig. 3.138A-D).

The adult form of the midbrain is best shown in a section through the superior colliculi (Fig. 3.138E).

The midbrain can be subdivided into three regions: the **tectum** (that part dorsal to a line joining the two sulci limitantes) (Fig. 3.138E); the cerebral fibres traversing the midbrain are situated in the ventral component (**basis pedunculi**); and the intervening part is the **tegmentum** (that part ventral to a line joining the two sulci limitantes and the ventral extreme of the substantia nigra). The tectum consists of the roof plate and the alar laminae. The cells of the latter differentiate into the pretectal nuclei concerned with vision and audition. The basis pedunculi contains **cortico-spinal** (to spinal nerves) and **cortico-nuclear** (to nuclei of cranial nerves) fibres together with **fronto-pontine** and **temporo-pontine** fibres. The tegmentum consists of the floor plate and the basal laminae which are thought to give rise to the substantia nigra and the red nuclei. The tegmentum and basis pedunculi (crus cerebri) together form the **cerebral peduncles**.

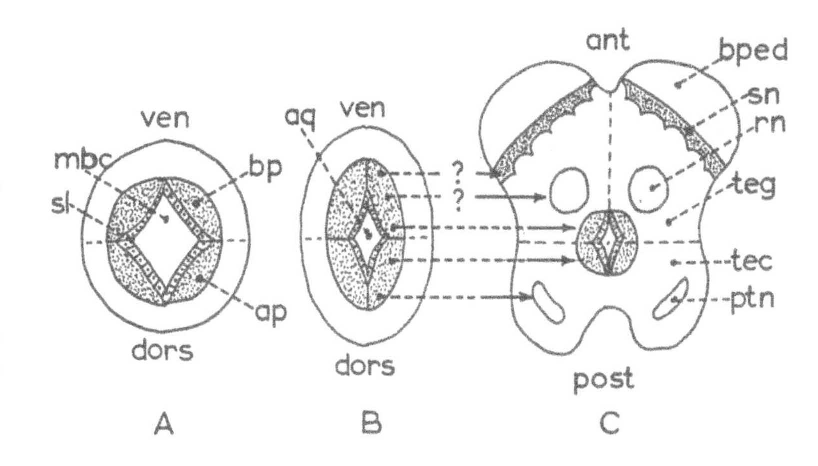

Figure 3.137: Transverse section showing development of the midbrain. Interrupted lines indicate origin of the adult structures. ap = alar plate; aq = aqueduct; bp = basal plate; bped = basis pedunculi; mbc = midbrain canal; ptn = pretectal nucleus; rn = red nucleus; sl = sulcus limitans; sn = substantia nigra; tec = tectum; teg = tegmentum; ?? = uncertainty of origin.

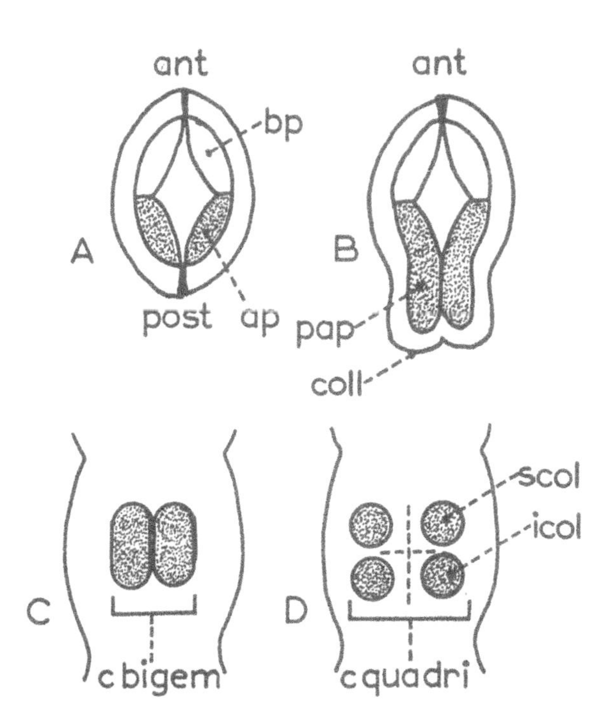

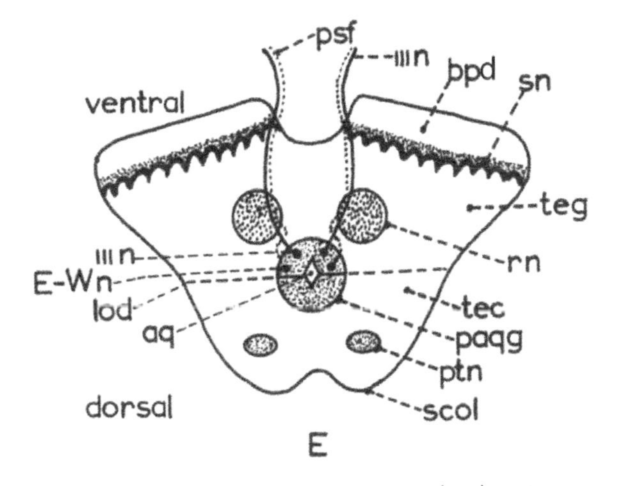

Figure 3.138: Posterior aspect of midbrain (A-D) and transverse section through superior colliculus (E). IIIn = oculomotor nerve; ap = alar plate; aq = aqueduct; bp = basal plate; bpd = basis pedunculi; cbigem = corpora bigemini; coll = colliculus; cquadri = corpora quadrigemini; E-Wn = Edinger- Westphal nucleus; icol = inferior colliculus; lod = line of division across sulci limitantes; paqg = peri-aqueductal gray matter; pap = proliferation of alar plate; psf = parasympathetic fibres; ptn = pretectal nucleus; rn = red nucleus; scol = superior colliculus; sn = substantia nigra; teg = tegmentum; tec = tectum.

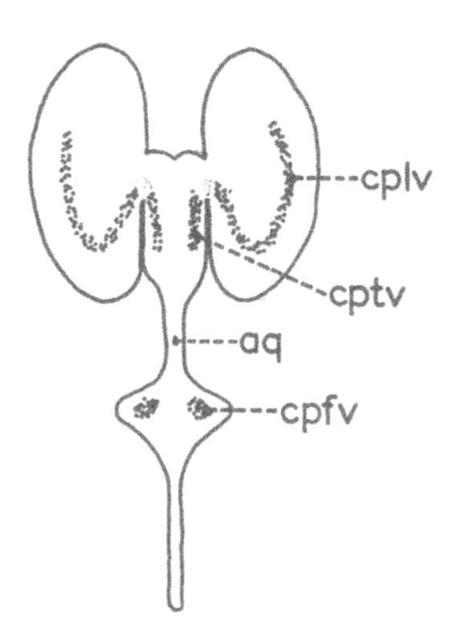

Figure 3.139: Dorsal view of the ventricular system showing positions of the choroid plexuses. aq = aqueduct; cpfv = choroid plexus of fourth ventricle; cplv = choroid plexus of lateral ventricle; cptv = choroid plexus of third ventricle.

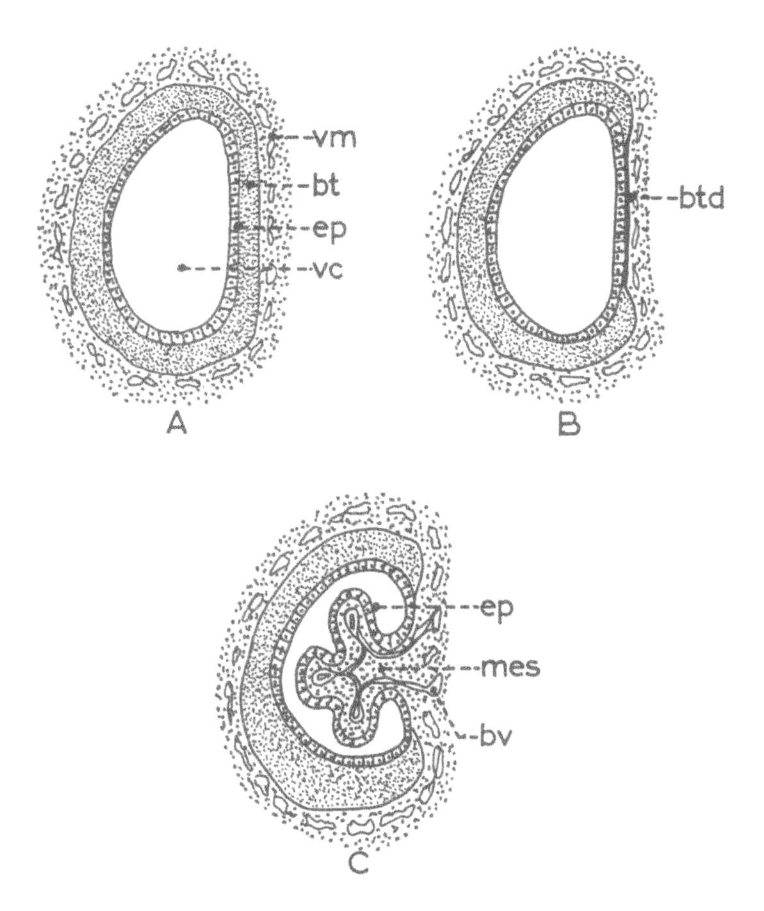

Figure 3.140: Coronal sections through the cerebral hemisphere to show the development of the choroid plexus. bt = brain tissue; btd = brain tissue disappears; bv = blood vessels; ep = ependyma; mes = mesenchyme; vc = ventricular cavity; vm = vascular mesenchyme.

The cranial extension of the gray matter of the spinal cord forms the **peri-aqueductal gray matter** of the midbrain and surrounds the aqueduct. The aqueductal gray matter occupies the same position in the mesencephalon (around the aqueduct), as is occupied by the gray matter of the spinal cord (around the central canal). The peri-aqueductal gray matter will contain the nuclei of the **oculomotor** and **trochlear** nerves as well as the **parasympathetic nuclei** (of Edinger and Westphal) next to the sulci limitantes (**Fig. 3.138E**).

Formation of the Choroidal Fissure and Choroid Plexus

As noted previously, the primitive brain is a 'hollow' organ and is lined with a layer of ependyma. It is covered with a layer of vascular mesenchyme which is heavily populated with neural crest cells (later forming the pia-arachnoid mater). Choroid plexuses are secretory structures which form in four places – in each of the telencephalic vesicles (lateral ventricles), in the roof of the diencephalon (IIIrd ventricle) and in the roof of the IVth ventricle (**Fig. 3.139**). These plexuses are responsible for the formation of the **cerebrospinal fluid** (CSF), which is contained within the ventricular system and the **subarachnoid space** of the brain and spinal cord. CSF also enters the central canal of the spinal cord.

The mode of formation of the choroid plexuses is similar in all situations. The brain tissue becomes thinned so that the ependyma comes into contact with the vascular mesenchyme (**Fig. 3.140A,B,C**). The lack of support of the brain tissue allows the pia mater to 'fall' into the underlying ventricle; thus, the choroid plexus consists of vascular mesenchyme covered on the ventricular aspect by ependyma (**Fig. 3.140C**).

In the case of the lateral ventricles, the thinning of the brain takes place on the medial aspect of the telencephalic vesicles (**Fig. 3.135B**) and since the vesicles grow and enlarge in a 'circular' fashion, the thinning will follow this pattern, beginning at the interventricular foramina and ending in the temporal lobes.

The 'slit' produced by the thinning is called the **choroidal fissure** and may be easily demonstrated in the dissected adult brain by removing the choroid plexus along its route. Also noteworthy is the fact that upon removal of the choroid plexus in the dissected brain, the choroidal fissure can be seen to be continuous with the interventricular foramen. This is because the upper boundary of the foramen is formed by the choroid plexus itself (Fig. 3.141 A,B).

When the fissure reaches the temporal lobe, the formation of the choroid plexus is associated with the development of the **hippocampus.** With the formation of the fissure, the vascular pia mater invaginates the ependyma forming the choroid plexus, but the inferior lip of the fissure folds medially to form the **dentate gyrus** (medial wall of the pallium) (Fig. 3.142 A,B). This gyrus continues laterally into a series of folds having an S-shape (hence the term hippocampus– sea horse). The hippocampus has the following parts: **dentate gyrus, cornu ammonis, subiculum** and **parahippocampal gyrus** (Fig. 3.142A). Axons of cells in the sub-ependymal layer of the ventricular cleft, track towards the dentate gyrus and 'overrun' it so that they form a fringe called the **fimbria** which overhangs the dentate gyrus (Fig. 3.142B). The fibres in the fimbria then become part of the **fornix.** The track of the axons is called the **alveus** (Fig. 3.142B). Thus, the dentate gyrus lies between the fimbria and the parahippocampal gyrus.

In the case of the diencephalic (IIIrd) ventricle, the roof becomes thinned and the same process of a 'falling in' of a choroid plexus takes place in that situation. In the case of the IVth ventricle, the pontine flexure results in the dorsum of the ventricle being covered by membrane only. This membrane also consists of ependyma and overlying vascular mesenchyme. Here, the choroid plexuses form on either side of the midline while the median part of the membrane becomes perforated to form the median foramen (Foramen of Magendie) to allow outflow of CSF from the ventricular system into the subarachnoid space. There are two additional lateral foramina (foramina of Luschka) which also allow outflow of CSF (Fig. 3.143).

Formation of the Commissures of the Brain

A commissure may be defined as a band of white matter connecting **similar** parts of the two opposite sides of the central nervous system.

The commissures of the brain may be divided into two groups:

- A **rostral group** arising in the lamina terminalis;
- A **posterior group** arising in the caudal part of the roof of the diencephalon.

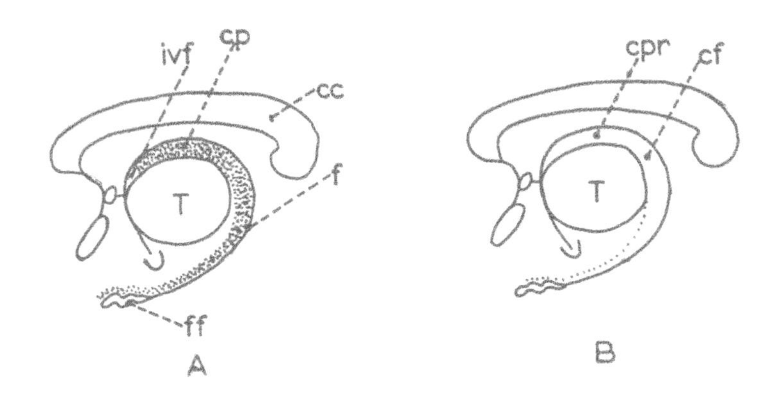

Figure 3.141: Lateral section to depict formation of the choroid fissure. cc = corpus callosum; cf = choroid fissure; cp = choroid plexus; cpr = choroid plexus removed; f = fornix; ff = fimbria of fornix; ivf = interventricular foramen; T = thalamus.

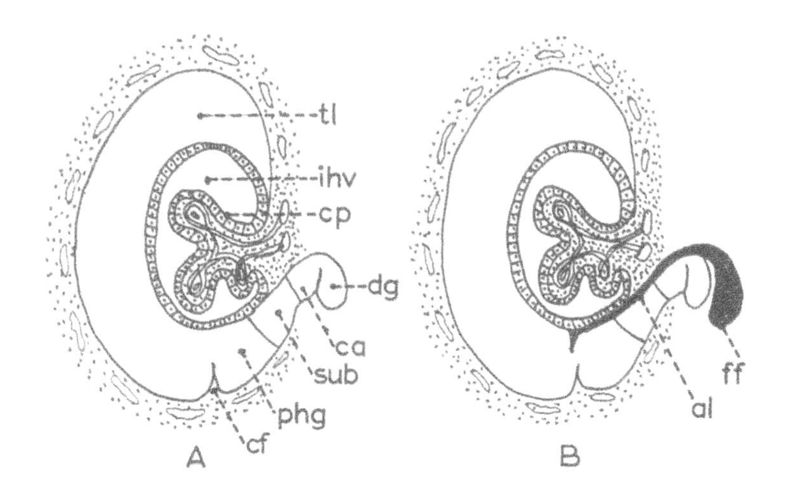

Figure 3.142: Coronal section through the cerebral hemisphere to show the development of the hippocampus. al = alveus; ca = cornu ammonis; cf = collateral fissure; cp = choroid plexus; dg = dentate gyrus; ff = fimbria of fornix; ivh = inferior horn of ventricle; phg = parahippocampal gyrus; sub = subiculum; tl = temporal lobe.

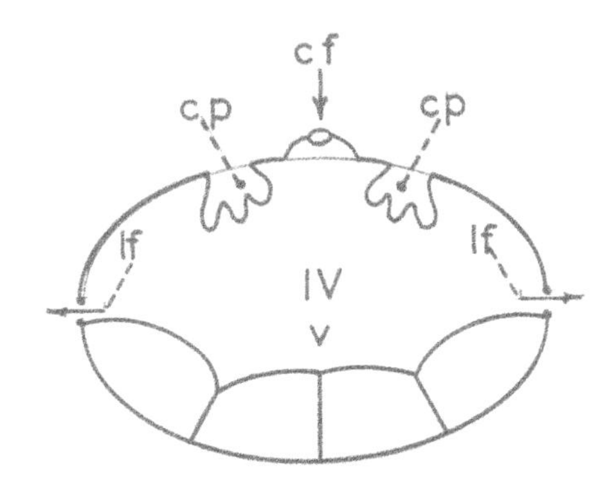

Figure 3.143: Transverse section through the medulla oblongata diplaying the choroid plexuses and formina. IV v = fourth ventricle; cf = central foramen; cp = choroid plexus; lf = lateral foramen.

The Rostral Group

By the time the telencephalic vesicles have formed, the optic stalks have migrated ventrally and are integrated into the lamina terminalis where they form the ventral end of the lamina (Fig. 3.144A). This is the optic chiasma and is not a true commissure, since some of its fibres do not cross.

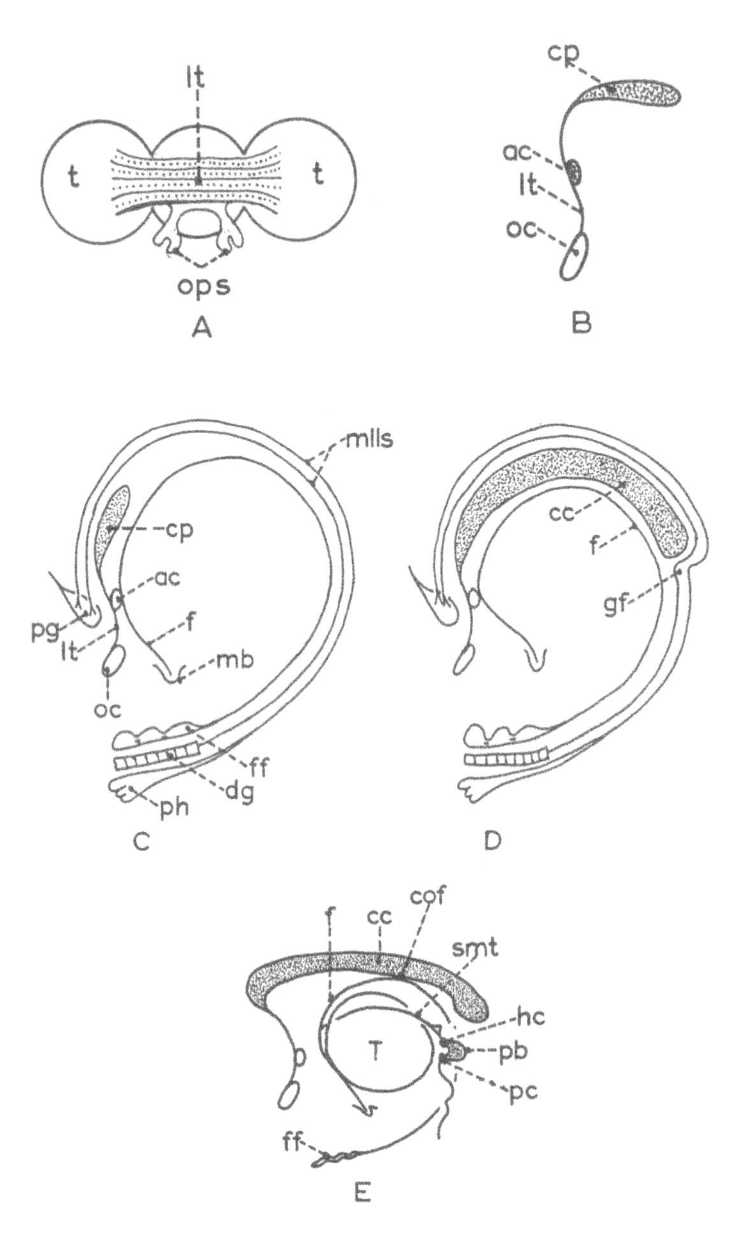

Figure 3.144: Formation of the major commissures of the brain.
A: Frontal view. **B-E:** Sagittal sections. ac = anterior commissure; cc = corpus callosum; cof = commissure of fornix; cp = commissural plate; dg = dentate gyrus; f = fornix; ff = fimbria of fornix; gf = gyrus fasciolaris; hc = habenular commissure; lt = lamina terminalis; mb = mamillary body; mlls = medial and lateral longitudinal striae; oc = optic chiasma; ops = optic stalks; pb = pineal body; pc = posterior commissure; pg = paraterminal gyrus; ph = pes hippocampi; smt = stria medullaris thalami; t = telencephalon.

At first the lamina terminalis or commissural plate, which lies in the rostral part of telencephalon medium, is fairly thick and since the telencephalic vesicles are relatively small at the beginning of development, the plate contains groups of fibres connecting the two telencephalic hemispheres across the laminar 'bridge' (Fig. 3.144A,B). The position of the lamina terminalis forming a connection between the telencephalic vesicles thus lends itself well to the crossing of fibres from one hemisphere to the other.

The first commissure to become apparent is the **anterior commissure** (Fig. 3.144B) which consists of axons coming from the medial olfactory striae to join those parts of the telencephalic vesicles which will become the **piriform cortices.** These are situated in the superior parts of the temporal lobes and have an olfactory function.

As the telencephalic vesicles enlarge and expand caudally, the fibres crossing between the two vesicles increase in number, and the superior part of the commissural plate likewise expands caudally to form the **corpus callosum** (Fig. 3.144B,C). The corpus callosum thus extends to the roof of the diencephalon where the ventral fibres form the **commissure of the fornices** (Fig. 3.144E). As the growth of the commissural plate proceeds caudally it meets and attaches to the two fornices, thus forming a more or less triangular area on either side between the fornix, the anterior commissure and the curved part of the commissural plate. This area becomes filled with a membrane on either side which is called the bilateral **septum pellucidum** (Fig. 3.145A). These membranes usually fuse completely to form a single septum but occasionally a space persists between them forming the **cavum septi pellucidi.** This cavity, which is sometimes referred to as the 'fifth ventricle', has no connection with the normal ventricular system and does not contain cerebro-spinal fluid. The septum pellucidum separates the anterior horns of the lateral ventricles from one another (Fig. 3.145B,C).

As the corpus callosum extends caudally, it separates a thin layer of gray matter and a number of fibres from the upper surface of the fornices so that the gray matter and fibres come to lie on the dorsal surface of the corpus callosum. The grey matter forms the **induseum griseum** and the fibres form the **medial** and **lateral longitudinal striae** (Fig. 3.144D). The fibres may be traced posteriorly, through the **gyrus fasciolaris** (gyrus splenius) to the dentate

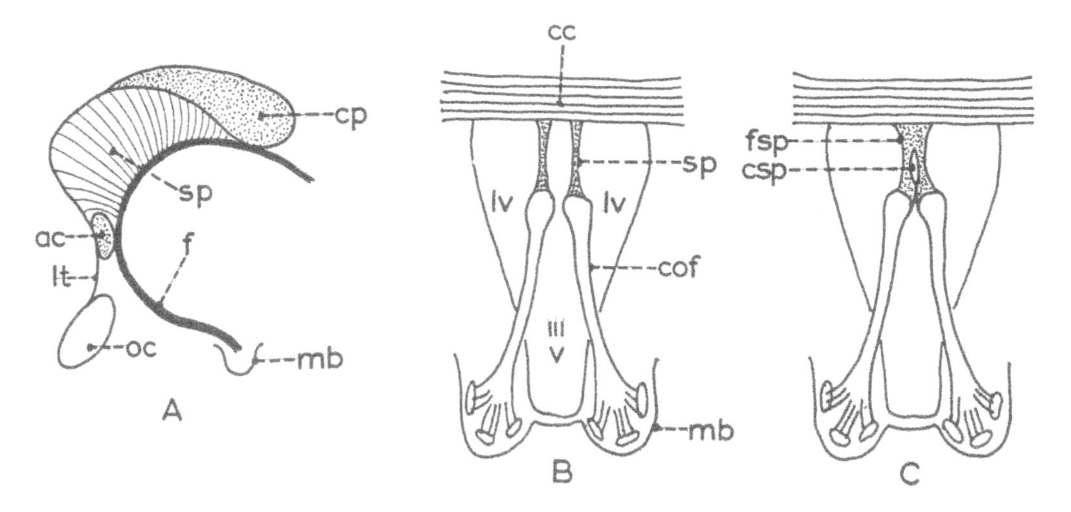

Figure 3.145: Formation of the septum pellucidum in lateral view (A) and coronal views (B,C).
ac = anterior commissure; cc = corpus callosum; cof = column of fornix; cp = commisural plate; csp = cavum septi pellucidi; f = fornix; fsp = fused septi pellucidi; lt = lamina terminalis; lv = lateral ventricle; mb = mamillary body; oc = optic chiasma; sp = septum pellucidum; iiiv = third ventricle.

gyrus and the hippocampus, which lie in the floor of the temporal horn of the lateral ventricle. Ventrally, both induseum griseum and the fibres end in the **paraterminal gyrus** (Fig. 3.144C,D).

The Posterior Group

At the caudal end of the diencephalic roof, two commissures arise; the superior one will become the **habenular commissure** which joins the striae medullares thalami of the two sides (Fig. 3.144E). The superior lamina of the stalk of the pineal gland will become attached to the habenular commissure. The habenular commissure is related to the function of olfaction. Inferior to the habenular commissure is the second commissure, which will become the **posterior commissure** to which is attached the inferior lamina of the stalk of the pineal gland. This commissure carries axons from the **pretectal nuclei** amongst numerous other fibres and is largely related to the visual system (Fig. 3.144E).

Development of the Rhombencephalon

Between the late 4th week and early 5th week of development, a series of segments (**rhombomeres**) appears in the hindbrain. These segments, although transient, are believed to be important in that they provide a basis for the fundamental organisation of the adult brain. The rhombomeres are known to be associated with the expression of particular genes (eg. Hox genes).

With the formation of the acute pontine flexure, the alar plates separate from one another so that the roof plate is drawn out into a thin membrane (Fig. 3.146A). When this process of 'separation' is completed, the alar and basal plates lie next to one another in a horizontal position in the floor of the rhombencephalon, and not as formerly, in a dorso-ventral relationship.

Since the rostral entrance to this part of the neural tube is narrow (cerebral aqueduct) and the caudal end is also narrow (central canal of the spinal cord), the bilateral spreading of this part of the neural tube results in the formation of a diamond-shaped cavity covered dorsally by a thin membrane (Fig. 3.146B). This diamond-shaped cavity is the **IVth ventricle** and the membrane covering it is the **tela choroidea.** The tela choroidea is formed from vascular mesenchyme (later to become pia-arachnoid) on the outer surface and lined by ependyma on the inner surface. These conditions are eminently suitable for the formation of choroid plexuses which occur, in the case of the IVth ventricle, on either side of the midline (Fig. 3.146A). Later, in the midline, the ependyma forms an evagination into the pia-arachnoid membrane. When the roof of the evagination ruptures, an opening, the median foramen, is formed in the ependyma. This allows CSF to escape from the vetricular system into the subarachnoid space.

Towards the rostral end of the roof of the IVth ventricle, a thickening occurs in the alar plates. This extends from side to side and is called the **rhombic lip**; it will give rise to the **cerebellum.**

If we were to remove the tela choroidea at this stage, the **floor** of the IVth ventricle would be exposed. The floor is rhomboid-shaped with a deep longitudinal cleft running from the rostral to the caudal end in the centre of the rhomboid (Fig. 3.146B). In the depths of this cleft resides the original floor-plate of the neural tube. The bulges on either side of the cleft are the **medial eminences** which are formed by the basal plates on either side. Lateral to each eminence are shallow grooves, the sulci limitantes. Lateral to the sulci are the flattened **vestibular areas**, which are formed by the alar plates. The lateral 'corners' of the IVth ventricle are the lateral recesses of the ventricle. (Note that the sulci limitantes have changed position from the sides of the neural tube to the floor of the ventricle. This is due to the widening of the floor resulting from the pontine flexure.)

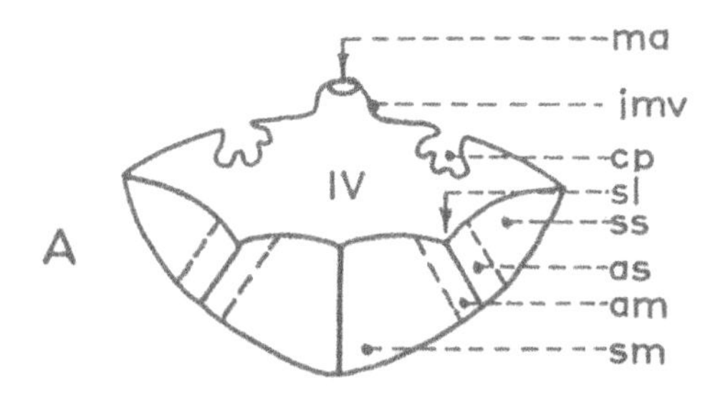

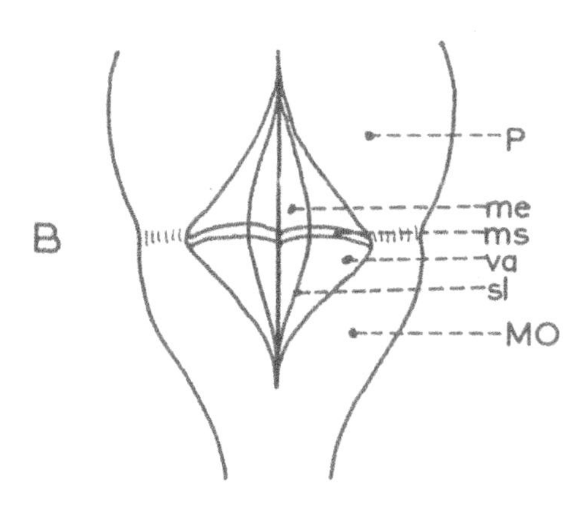

Figure 3.146: Development of the floor of the hindbrain in coronal section (A) and in dorsal view (B). am = autonomic motor; as = autonomic sensory; cp = choroid plexus; imv = inferior medullary velum; ma = median aperture; me = medial eminence; MO = medulla oblongata; ms = medullary striae; P = pons; sl = sulcus limitans; ss = somatic sensory; sm = somatic motor; va = vestibular area.

At approximately the middle of the floor are a number of ridges running transversely over the vestibular areas and the medial eminences to reach the midline cleft. These are the **striae medullares** which represent underlying rhombomeres (neural segments). These striae divide the rhomboid-shaped floor into rostral and caudal triangles or regions. The rostral region is the **metencephalon** which will give rise to the **pons** and **cerebellum,** while the caudal region, the **myelencephalon,** will give rise to the **medulla oblongata** (Fig. 3.146B).

As was stated previously, the alar plates are **sensory** in function and the basal plates are **motor** in function while the regions adjacent to the sulci limitantes are **autonomic** in function (Fig. 3.147). If one knows the position of the various cranial nerve nuclei, the function of the nerve may be easily determined. On the other hand, if one knows the function which the various nerves subserve, then the nucleus must lie in the relevant alar or basal plate. In the case of nerves containing autonomic components, the autonomic nucleus will lie in the region of the relevant sulcus limitans.

Development of the Cerebellum

While the pontine flexure is occurring and the roof membrane is being thinned, the alar plates in the metencephalon develop as the **rhombic**

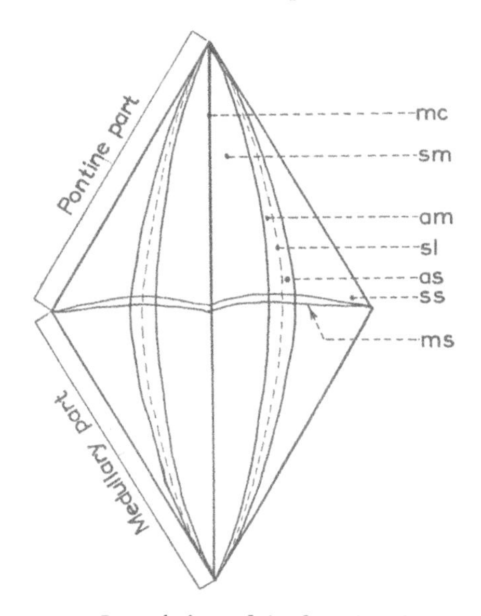

Figure 3.147: Dorsal view of the functional components of the floor of the fourth ventricle. am = autonomic motor component; as = autonomic sensory component; mc = midline cleft; ms = medullary striae; sl = sulcus limitans; sm = somatic motor component.

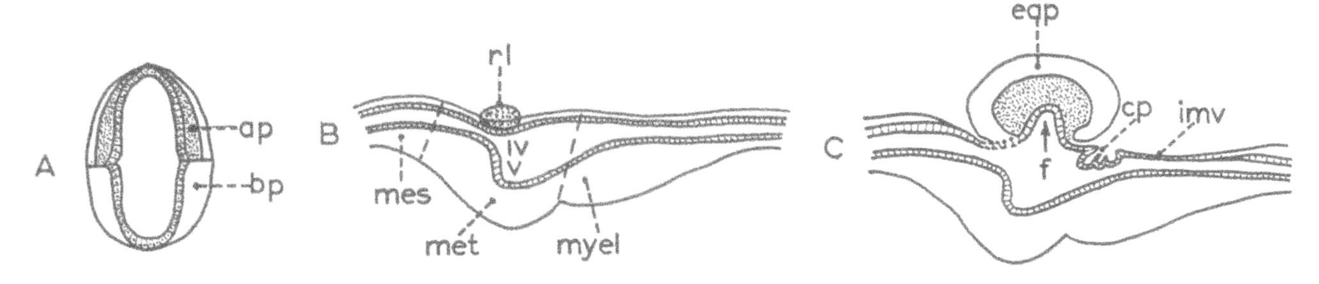

Figure 3.148: Development of the cerebellum in transverse section (A) and sagittal section (B,C). ap = alar plate; bp = basal plate; cp = choroid plexus; eap = expanding alar plates; f = fastigium; imv = inferior medullary vellum; mes = mesencephalon; met = metencephalon; myel = myelencephalon; rl = rhombic lip.

lip (Fig. 3.148A). The part of the membrane between the caudal part of the mesencephalon and the rhombic lip is the **superior medullary velum**, and that between the rhombic lip and the choroid plexuses is the **inferior medullary velum** (Fig. 3.148C). The lateral parts of the rhombic lip enlarge and bulge into the cavity of the IVth ventricle (Fig. 3.149A,B). These fuse in the midline forming a narrow attachment which is called the **superior vermis**. As the lateral parts of the lip increase in size to form the **cerebellar hemispheres**, the bulges are extruded from the ventricle so that the hemispheres come to lie on the surface (Fig. 3.149C). With increasing growth, the **folia** (leaves) of the hemispheres appear, while in the part between the hemispheres a number of folds develop. The most rostral fold is the **lingula** which is part of the superior vermis; the most caudal fold is the **nodule** which becomes part of the inferior vermis (Fig. 3.150A). The nodule has a lateral extension which ends in an enlargement called the **flocculus**. The nodule and the flocculus on each side lie in a fissure which surrounds both hemispheres and is known as the **horizontal fissure** in the adult (Fig. 3.149C). With growth and positional change, the flocculus comes to lie on the anterior surface of the adult cerebellum. However, it retains its connection with the nodule and the combination is called the **flocculonodular lobe** of the cerebellum.

At first the layers of the cerebellum resemble those of the primitive neural tube, with an inner layer of ependyma, a middle mantle layer of neurons and neuroglia, and an outer marginal layer of mainly nerve fibres. In the cerebellum, the cells of the mantle layer give rise to all its cellular components such that cells of the mantle layer migrate; firstly, to form an outer **molecular layer** (Fig. 3.150B); then another group of cells migrates to differentiate into **Purkinje cells** (middle layer) (Fig. 3.150B,C) and finally a third group migrates to form a **granular layer** (Fig. 3.150B,C). These three layers constitute the cortex of the cerebellum.

Simultaneously, the mantle layer produces cells which migrate to form the deep nuclei (**fastigial**, **globose**, **emboliform** and **dentate**) of the cerebellum (Fig. 3.150C).

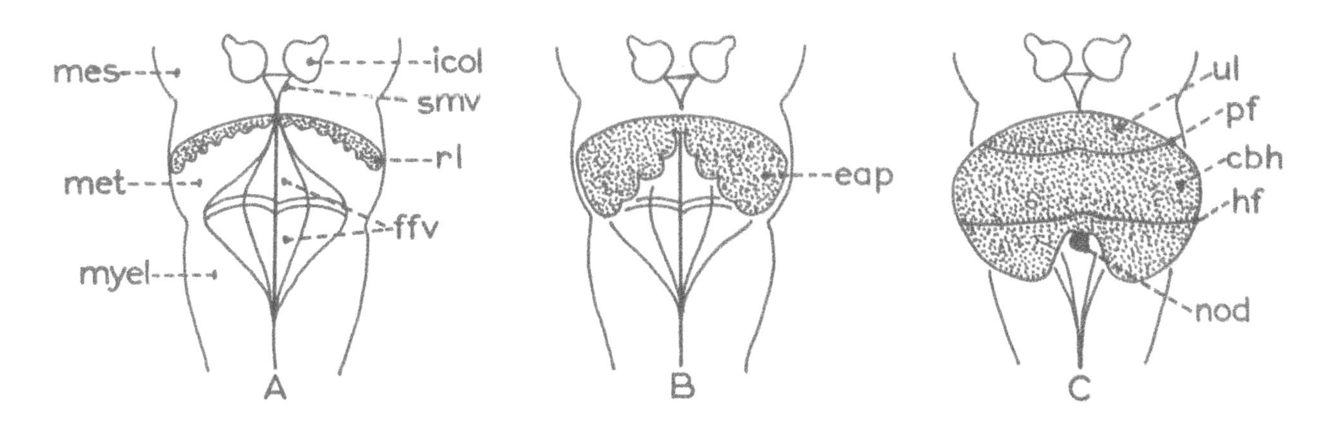

Figure 3.149: Dorsal view of the developing cerebellum. cbh = cerebellar hemisphere; eap = expanding alar plates; ffv = floor of fourth ventricle; hf = horizontal fissure; icol = inferior colliculus; mes = mesencephalon; met = metencephalon; myel = myelencephalon; nod = nodule; pf = primary fissue; rl = rhombic lip; smv = superior medullary vellum; ul = upper lobe of cerebellum.

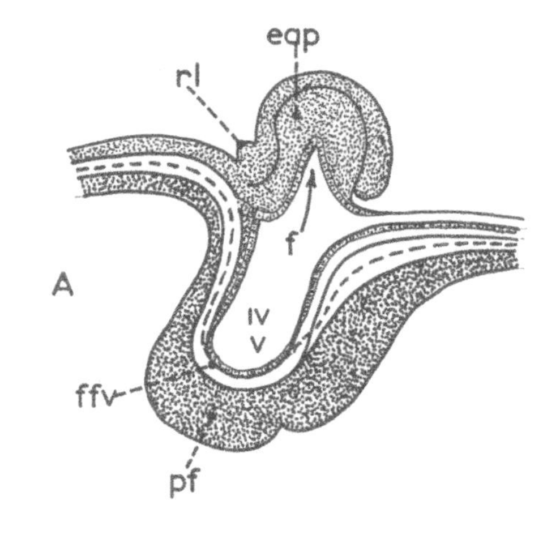

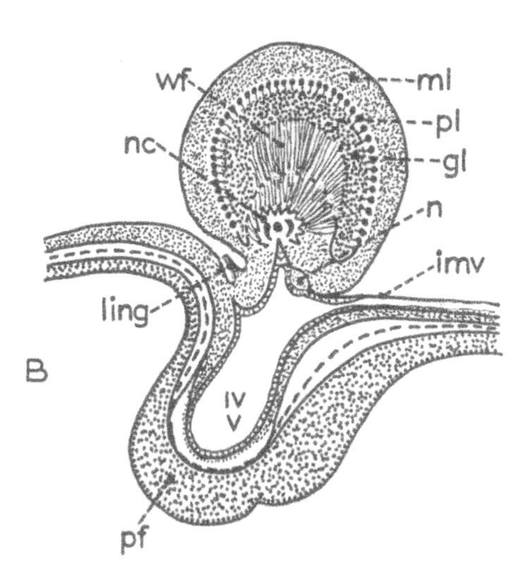

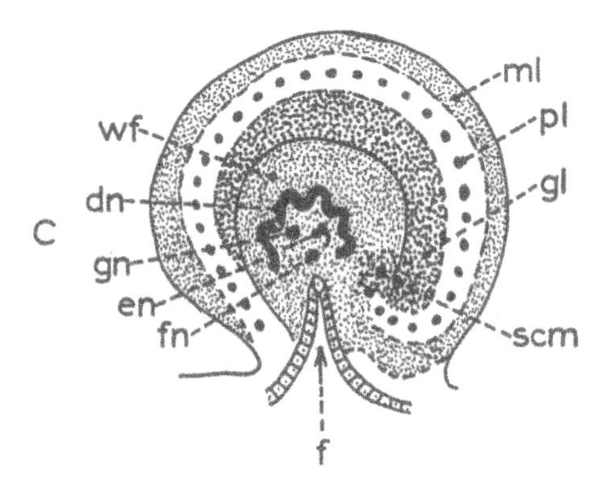

Figure 3.150: Development of the layers of the cerebellum seen in sagittal section. IV v = fourth ventricle; dn = dentate nucleus; eap = expanding alar plates; en = emboliform nucleus; f = fastigium; ffv = floor of fourth ventricle; fn = fastigial nucleus; gl = granular layer; gn = globose nucleus; imv = inferior medullary velum; ling = lingula; ml = molecular layer; n = nodule; nc = nuclear complex; pf = pontine flexure; pl = purkinje layer; rl = rhombic lip; scm = source of cell migration; wf = white fibres.

POSSIBLE COMMON ABNORMALITIES OF THE BRAIN

- These are generally associated with clefts of the skull **(cranioschisis/cranium bifidum).** These will include **cerebral meningocoel** (herniation of meninges) and **meningoencephalocoel** (herniation of meninges and brain tissue).
- **Mero-encephaly (previously anencephaly)** is a condition where there is a reduced amount of nervous tissue. This may effect some areas more than others. This is said to be due to non-closure of the anterior neuropore.
- **Microcephaly**, an underdeveloped or small brain, is associated with a small cranium.
- **Hydrocephalus** occurs when the drainage of CSF is prevented and expansion of the ventricular system of the brain occurs, leading to an enlarged brain and skull.

Development of the Peripheral Nervous System

The peripheral nervous system may be divided into:
(i) Cranial Nerves and
(ii) Spinal Nerves
Some of the cranial nerves are associated with autonomic nerves but all of the spinal nerves are associated with sympathetic nerves.

Cranial Nerves may be classified as:
(i) Motor
(ii) Sensory
(iii) Mixed Motor and Sensory Modalities

All nerve cells have anatomical processes, some of which bring information *to* the cell and others which take information *from* the cell. The main property of these processes is to make contact with other cells or tissues. The mechanisms by which the cells discover where their processes should go, or to which cells or organs they should make contact with, are not clearly known. The term **axonal guidance** has been applied to the selective migration of axons for reaching specific regions of the body or particular cells or tissues. One of the key features of neuron development is the extension of its axon by the development of a **growth cone**. The activity of the growth cone is influenced by contacts which its filopodia make with the extracellular matrix and with other cells.

The Cranial Nerves

The origins of all the cranial nerves except the olfactory and optic nerves are from nuclear masses in the alar and basal plates of the mesencephalon, metencephalon and myelencephalon. These regions, in the adult, constitute the brain-stem (midbrain, pons and medulla oblongata). Some of the nuclei related to some of the cranial nerves (V, VII, VIII, IX and X) are also derived from migrated neural crest cells.

The **olfactory** (CN I) and **optic** (CN II) nerves are associated with *protrusions* from the prosencephalic vesicle and because of this, they are sometimes not regarded as 'true' cranial nerves. In addition, both have anatomical tracts which exist outside the central nervous system whereas all other tractal systems are within the central nervous system (**Fig. 3.151**).

There are functionally two types of nerves, those carrying motor impulses and those conveying sensory impulses. The nerves conveying motor and sensory impulses to and from the viscera are in the autonomic component of the nervous system. The axons conveying motor impulses to the somatic muscles arise from cells in the basal plates of the neural tube while those conveying sensation from the skin or other structures enter the alar plates of the neural tube but have their cell bodies in the posterior (dorsal) root ganglion.

All the cranial nerves contain one or both of these types of fibres and some, in addition, contain fibres emanating from the autonomic parts of the neural tube.

In the brain stem, the grey matter, in the form of the alar and basal plates, are homologous throughout **and** with the grey matter of the spinal cord. In the midbrain, the peri-aqueductal grey matter occupies the same position in relation to the aqueduct as the grey matter of the spinal cord occupies in relation to the central canal, and the alar and basal plates lie in a dorso-ventral relationship to one another. In the metencephalon and myelencephalon, the plates lie in a medio-lateral relationship to one another, due to the effects of the pontine flexure of the neural tube. The motor and sensory autonomic components lie in close relationship to the sulcus limitans on either side of the tube.

If the regional origin of the cranial nerve nucleus is known (**Fig. 3.151**), it is possible to determine the function of the nerve arising from that nucleus. It follows therefore, that if the functional aspects of a cranial nerve are known, it is possible to determine accurately, the position of the brain-stem nucleus. This identification is of value in the clinical diagnosis of brain-stem lesions and holds also for somatic and autonomic nerves.

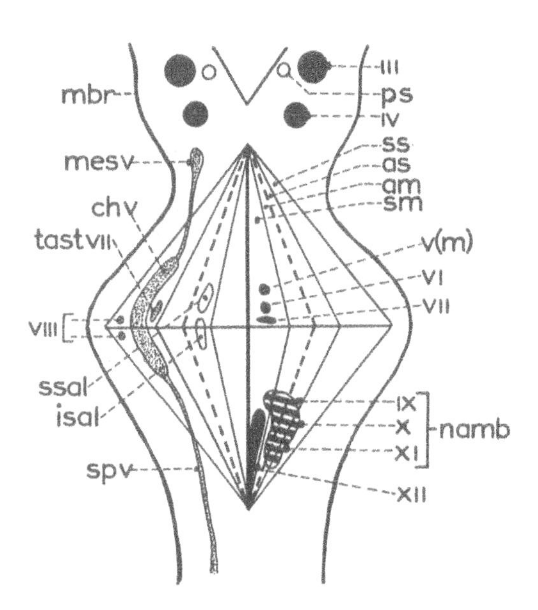

Figure 3.151: Dorsal view of the midbrain and hindbrain depicting positions of the cranial nerve nuclei. III = oculomotor nerve nucleus; IV = trochlear nerve nucleus; V(m) = trigeminal (motor) nerve nucleus; VI = abducent nerve nucleus; VII = facial nerve nucleus; VIII = vestibulocochlear nerve nucleus; (IX,X,XI) namb = nucleus ambiguus; XII = hypoglossal nerve nucleus; am = autonomic motor; as = autonomic sensory; chv = chief nucleus of trigeminal; isal = inferior salivatory nucleus; mbr = midbrain; mesv = mesencephalic nucleus of trigeminal; ps = parasympathetic nucleus; sm = somatic motor; spv = spinal nucleus of trigeminal; ss = somatic sensory; ssal = superior salivatory nucleus; tast VII = taste nucleus of facial.

The cranial nerves supply the structures of the head and neck which arise largely from the pharyngeal arches. The cranial nerves will supply, therefore, structures arising from the pharyngeal arches. These arches contain the relevant nerves namely the trigeminal nerve (CN V) for the 1st arch, the facial nerve (CN VII) for the 2nd arch, the glossopharyngeal nerve (CN IX) for the 3rd arch and the vagus nerve (CN X) for the 4th and 6th arches (which carries fibres of the cranial part of the accessory nerve with it). The facial, glossopharyngeal and cranial accessory nerves also contain autonomic elements.

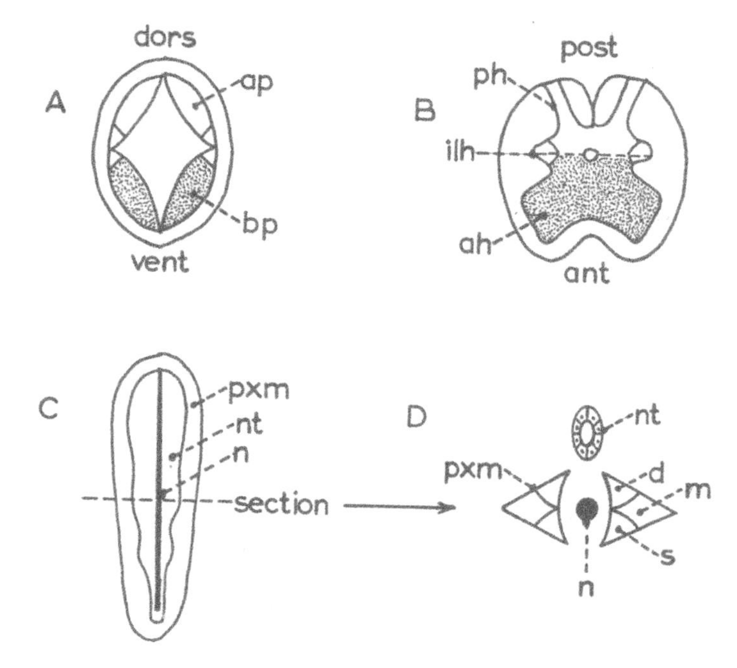

Figure 3.152: Development of the spinal nerves. A, B: Transverse sections of the neural tube (A) and spinal cord (B). **C:** Dorsal view of the embryo to indicate the level of tranverse section **D**. ah = anterior horn; ap = alar plate; bp = basal plate; d = dermatome; ilh = intermedio-lateral horn; m = myotome; n = notochord; nt = neural tube; ph = posterior horn; pxm = paraxial mesoderm; sc = sclerotome.

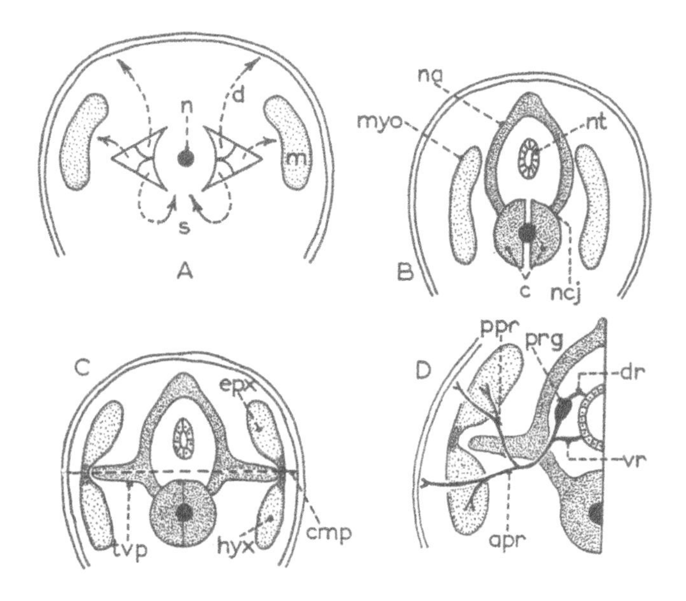

Figure 3.153: Transverse sections of the development of the spinal nerves. Arrows = cellular dispersal. apr = anterior primary ramus; c = centra; cmp = coronal morphological plane; d = dermotome; dr = dorsal root; epx = epaxial muscle; hyx = hypaxial muscle; m and myo = myotome; n = notochord; na = neural arch; ncj = neurocentral junction; nt = neural tube; ppr = posterior primary ramus; prg = posterior root ganglion; s = sclerotome; tvp = transverse process; vr = ventral root.

Spinal Nerves

With the proliferation of cells in the neural tube, aggregations are formed in the dorsal (alar) and ventral (basal) components of the tube, separated by the sulci limitantes (**Fig. 3.152A**). The aggregations ultimately form the grey matter of the anterior and posterior horns of the spinal cord (**Fig. 3.152B**). The regions in relation to the sulci limitantes are the intermedio-lateral horns, which are autonomic in function.

A band of paraxial mesoderm lies on either side of the developing neural tube. The formation of the spinal nerves is closely related to this band of mesoderm. The paraxial mesoderm is segmented and each segment forms three elements of the embryonic body (**Fig. 3.152C,D**). The dorsal element is the **dermatome**, which gives rise to the dermis of the skin; the lateral element is the **myotome**, which gives rise to the skeletal muscle of the trunk, and the ventral element is the **sclerotome** which gives rise to the vertebral column.

With development, the neural arches of the vertebrae surround the neural tube (**Fig. 3.153A,B**), and the dermatomes spread dorsally around the body while the myotomes come to lie lateral to the developing vertebral column. The neural arch produces transverse processes which, passing laterally, divide the myotomes into dorsal (epaxial) and ventral (hypaxial) muscle masses. A line drawn transversely through the transverse processes is called the **coronal morphological plane** which divides the trunk into dorsal and ventral parts (**Fig. 3.153C**).

The cells in the basal plates of the neural tube send axons (motor component of future spinal nerve) to the segmental myotomes (**Fig. 3.153D**). Cells in the dorsal (posterior) root ganglia (of neural crest origin) (ganglion = group of cells outside the central nervous system) receive dendrites (sensory component of future spinal nerve) from the dermatomal segments and send axons to the alar plates of the neural tube (**Fig. 3.153D**). The mixed spinal nerve is thus formed by the approximation of the motor and sensory fibres (**Fig. 3.153D**).

Note:

(a) The nervous elements entering and leaving the primitive spinal cord are the **roots** of the spinal nerves.

(b) Only the posterior root possesses a ganglion, the **dorsal root ganglion** (sensory).

(c) Where the roots come together the motor (anterior root) and sensory (posterior root) fibres become mixed and are called the **mixed spinal nerve**.

It follows therefore that all branches after the mixed spinal nerve will contain both **motor** and **sensory** fibres.
From Figure 3.153D it is thus evident that the structures posterior (dorsal) to the coronal morphological plane are supplied by a nerve called the **posterior primary ramus** (branch) and structures anterior (ventral) to the plane are supplied by a nerve called the **anterior primary ramus** of the mixed spinal nerve. In the case of the position of the limb buds, these develop as antero-lateral structures (Fig. 3.154A,B). Since they are ventral to the coronal morphological plane they will be supplied entirely by anterior primary rami (motor and sensory).

Since the nerves and muscles are segmentally disposed, the embryonic and adult body may be divided into dermal and muscle segments in relation to the segmentation of the paraxial mesoderm.

The Autonomic Nervous System

The term 'autonomic' was coined at a time when it was believed that the actions of this part of the nervous system were outside the control of the cerebrospinal system. It was thought to act as a 'law unto itself'. This idea is now known to be false. The autonomic nervous system is highly dependant upon and largely controlled by the cerebrospinal system, although it would appear that in some respects, it acts 'automatically and involuntarily' to control heart rate, blood pressure, respiration, bowel movement and gland secretion.

The autonomic nervous system may be classified as having **sympathetic** and **parasympathetic** components and their functions may be construed as being complementary. Both components arise from neuroblasts which differentiate from neural crest cells, which are found both outside and in the wall of the neural tube. Those neuroblasts, which are outside the neural tube, come to lie on either side of the tube, while those in the wall of the tube will migrate to the regions of the sulci limitantes and form the **intermedio-lateral horn of grey matter** in the adult spinal cord (Fig. 3.152B). Some of the neuroblasts adjacent to the neural tube will form the cell bodies of both somatic and autonomic sensory neurons of the dorsal (posterior) root ganglia (Fig. 3.155A), while others will migrate to either side of the aorta to form ganglia from the vertebral level of T1 to L2. These ganglia are the **paravertebral ganglia**.

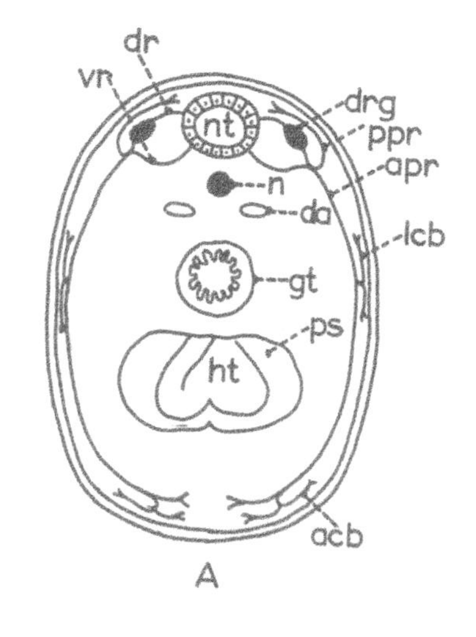

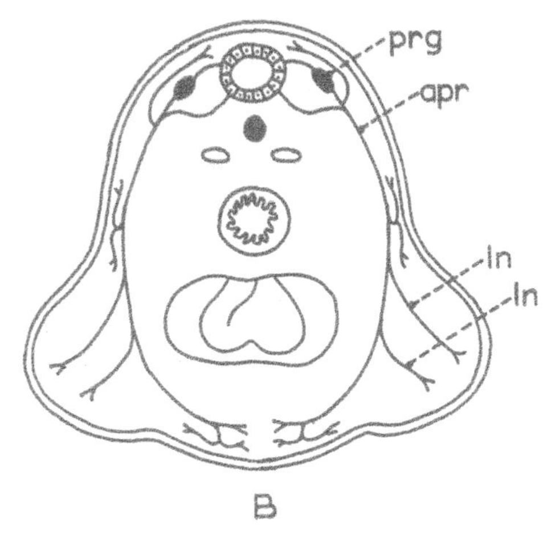

Figure 3.154: Transverse sections to show the development of the innervation to the limbs. acb = anterior cutaneous branch; apr = anterior primary ramus; da = dorsal aorta; dr = dorsal root; drg = dorsal root ganglion; gt = gut tube; ht = heart; lcb = lateral cutaneous branch; ln = limb nerves; n = notochord; nt = neural tube; ppr = posterior primary ramus; prg = posterior root ganglion; ps = pericardial sac; vr = ventral root.

Development of Pre- and Postganglionic Fibres

The paravertebral ganglia receive axons from cells in the autonomic component of the basal plate (autonomic motor part of the intermedio-lateral grey column) of the neural tube. These axons emerge from the neural tube together with the axons of the motor cells (anterior horn cells) of the basal plate (**Fig. 3.155B**). Together, these axons form the anterior root of the spinal nerve in the adult. When the combined root fibres reach the vicinity of the paravertebral ganglia, the autonomic fibres leave the root and enter the ganglia (**Fig. 3.155B**). Since these fibres are entering the paravertebral ganglia from outside or 'before the ganglia', they are known as **preganglionic fibres**. Coming from the central nervous system, they are covered with a myelin sheath which gives them a 'whitish' appearance and hence their other name of **'white rami communicantes'**. On entering the paravertebral ganglia, these axons will synapse with the dendrites of the ganglion cells. The axons of the ganglion cells will then emerge from the paravertebral ganglion and join the nearby mixed spinal nerve to be distributed with the branches of the peripheral nerves (**Fig. 3.155C**). The axons from the paravertebral ganglion cells are called the **postganglionic fibres**. These fibres are devoid of a myelin sheath, giving them a 'greyish' appearance and hence their other name of **'grey rami communicantes'**.

Development of Sympathetic Trunk

On leaving the neural tube and entering a ganglion, preganglionic fibres may synapse with the cells of *that* ganglion, or may pass cranially or caudally to synapse with the cells of other ganglia. It is these 'cranially-' and 'caudally-' running fibres which join the individual paravertebral ganglia to form the **sympathetic trunk** or **'chain'** (**Fig. 3.156**). At whatever level the synapse takes place, the postganglionic fibre will leave *that* segment and join the spinal nerve of *that* segment.

Development of Prevertebral Ganglia

By contrast, there are some preganglionic fibres which enter the paravertebral ganglionated chain and which *do not* synapse with the ganglia at their point of entry but pass through the chain and synapse with **prevertebral ganglia**, such as the **coeliac ganglion** and the **aorticorenal ganglia** (**Fig. 3.156**). In these cases the postganglionic fibres emerge from the *pre*vertebral ganglia. In the dissected adult, these nerve fibres are the

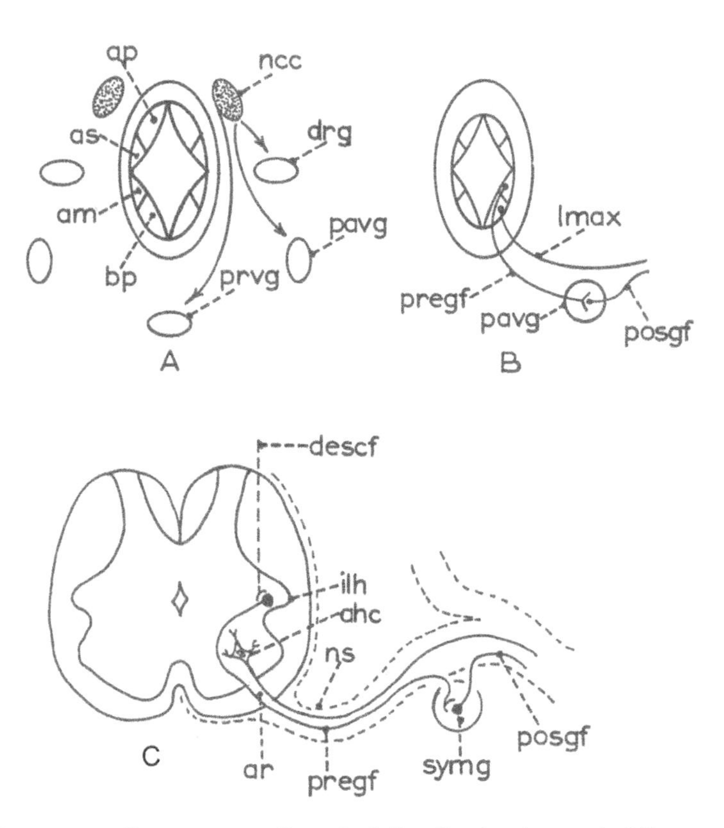

Figure 3.155: Transverse sections depicting the development of the autonomic nervous system including the migration of neural crest cells (ncc). A, B: Embryo. **C:** Adult. ahc = anterior horn cell; ap = alar plate; am = autonomic motor; ar = anterior root; as =autonomic sensory; bp = basal plate; descf = descending fibre; drg = dorsal root ganglion; ilh = intermediolateral horn; lmax = lower motor axon; ns = nerve sheath; pavg = paravertebral ganglion; posgf = post-ganglionic fibre; pregf = preganglionic fibre; prvg = prevertebral ganglion; symg = sympathetic ganglion.

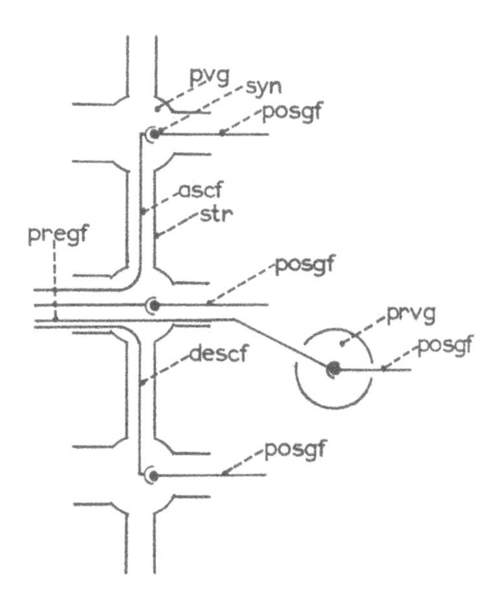

Figure 3.156: Distribution of the sympathetic fibres. ascf = ascending fibres; desf = descending fibres; posgf = postganglionic fibres; pregf = preganglionic fibres; prvg = prevertebral ganglion; pvg = paravertebral ganglion; str = sympathetic trunk; syn = synapse.

greater, lesser and least **splanchnic nerves**, emerging from thoracic segments T6 to T12. These nerves are formed in addition to the ordinary preganglionic fibres.

Development of Primary and Secondary Sympathetic Outflows

Thus far, we have considered the **thoraco-lumbar sympathetic outflow** which may be regarded as the 'primary sympathetic outflow'. Clearly, the spinal cord above T1 and below L2 does not possess individual sympathetic outflow channels although the sympathetic trunks extend into the cervical, lower lumbar and sacral regions. These cervical, lower lumbar and sacral outflows may be regarded as 'secondary outflows' from the primary outflow. The cervical sympathetic ganglia are formed by neural crest cells, but the preganglionic fibres are provided by the T1 to T4 segments of the primary outflow, and their passage from ganglion to ganglion forms the cervical sympathetic trunk. From the cervical ganglia, the postganglionic fibres form the **cardiac** and **pulmonary plexuses** as well as supplying the head and neck structures via the peripheral nerves of the somatic cervical plexus. The structures of the face and interior of the skull are supplied by peri-arterial sympathetic networks, derived from the cervical sympathetic trunk, around the internal and external carotid arteries and the vertebral arteries (**Fig. 3.157**).

The upper limb is supplied with postganglionic fibres (grey rami communicantes) from T1-T5 which enter the C8 and T1 roots of the brachial plexus (these grey rami are referred to as the nerve of Kuntz in the clinic) (**Fig. 3.157**). These postganglionic fibres originate in either the inferior cervical ganglion or the stellate ganglion (which is a combination of the inferior cervical ganglion and the 1st thoracic ganglion).

The preganglionic fibres to the lower lumbar and sacral regions are from the segments T10 to L2 of the primary outflow. These enter the paravertebral ganglia below L2 and form a trunk in the same way as described above. The trunks run inferiorly along the sides of the lower lumbar vertebrae and the anterior surface of the sacrum to end in the **ganglion impar** (unpaired) which lies on the anterior surface of the coccyx. The postganglionic fibres from this part of the trunk form the **superior** and **inferior hypogastric plexuses** which supply the lower abdominal and pelvic organs as well as the lower limbs (**Fig. 3.158**).

The pelvic viscera are also supplied by parasympathetic fibres emanating from the sacral segments S2, 3 and 4 of the spinal cord. They reach parasympathetic ganglia in the walls of the pelvic viscera.

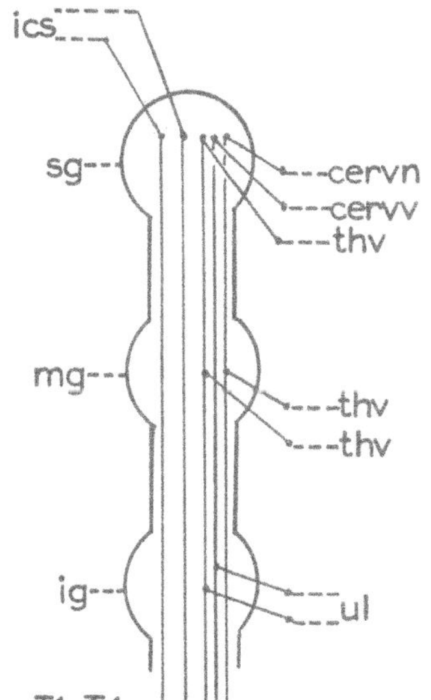

Figure 3.157: Sympathetic outflow from T1-T4. cervn = cervical nerves; cervv = cervical viscera; ics = intracranial structures; ig = inferior ganglion; mg = middle ganglion; sg = superior ganglion; thv = thoracic viscera; ul = upper limb.

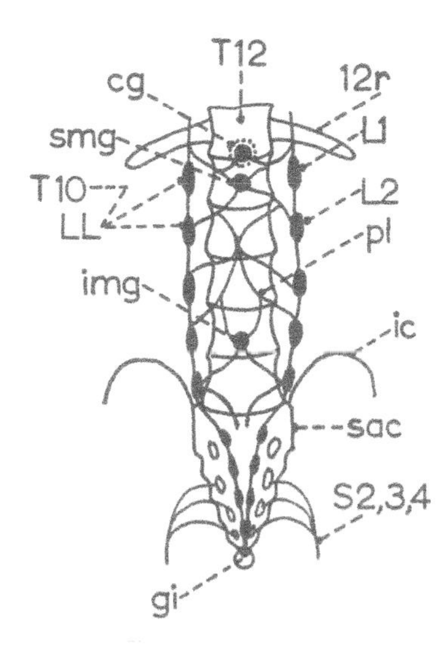

Figure 3.158: The sympathetic system in the lower abdomen and pelvis. 12r = twelfth rib; cg = coeliac ganglion; gi = ganglion impar; ic = iliac crest; img = inferior mesenteric ganglion; L1, L2 = lumbar ganglia; pl = preaortic plexus; S2, 3, 4 = pelvic splanchnic nerves; sac = sacrum; smg = superior mesenteric ganglion.

THE NEURAL CREST

Throughout the history of mankind, it has been recorded that children have been born with physical and/or mental defects. With more accurate recording and study of the defects, it has become apparent that some are due to genetic defects while others may be due to maternal problems such as smoking, alcoholism or drug abuse. Others are due to the mother being infected with a virus (e.g. German measles) while in others, no offending factor may be found.

Careful study of the defects has also shown a number of syndromes (concurrence of symptoms and signs) where the defective parts of the body are found at a distance from one another. If the child is born with such a syndrome, it may be inferred that the causative agent/s were active during the embryonic or fetal period of development. Occasionally, the syndrome may become manifest only in later life. The presence of a particular syndrome does not, in itself, explain the causation for the defects occurring in disparate regions of the body simultaneously. Although there may be no apparent anatomical or physiological relationship between the affected organs, it does raise the possibility that some **common factor** may, at one time, have existed between them.

When one realises that the embryonic disc is about 1mm in diameter, it becomes clear that the only satisfactory way of studying changes at this early stage of development is by the use of stained sections which are viewed under the microscope. In fact, it is this technique which has given answers to some of the problems posed above. The painstaking examination of sections of embryos has shown, amongst numerous other discoveries, that the development of the nervous system is associated with the formation of a 'special' group of cells, the **neural crest cells**. These cells are special in the sense that they are capable of **migrating** to various regions of the embryonic body and there, taking part in the formation of particular structures or organs. Could these cells be the common factor responsible for some of the abnormalities referred to above?

It is common knowledge that salmon, eels and birds migrate seasonally to various parts of the world. These migrations are relatively easy to study as the media through which the animals pass are clear and the direction and destination of the animals is easily determined. This ease of identification and movement is, unfortunately, denied to the microscopist, who is required to recognise and identify a cell (or group of cells) moving along an ill-defined pathway, surrounded by cells having almost the same characteristics as the migrating cell. In addition, discovery of the destination of the migrating cell/s is a problem of major magnitude.

From this it is evident that the proper and secure identification of a migrating cell requires the use of special techniques, since ordinary staining methods cannot distinguish a migrating cell from the similar surrounding cells. One of the successful techniques which have been applied is to use antibodies to existing antigens in or on the migrating cells. The antibody is linked to a specific molecule which can be visualised by staining or by fluorescence. This enables the microscopist to track the course of the migrating cells by examining sections of embryos at different ages. Another commonly used technique is to substitute cells with a known characteristic in place of cells which are thought to migrate. Such cells are available from the embryo of the quail (a type of bird, allied to the partridge). The cells of this bird possess a **large stainable nucleolus**. By substituting embryonic quail neural crest cells for embryonic chick neural crest cells, it is possible to follow the migration of the cells by observing the presence of the large nucleolus. This method is called the chick/quail **chimaera**, which is simply another word for 'substitution'.

In this way embryologists have mapped out the migration of neural crest cells in the chick embryo and in certain amphibian and mammalian embryos. Much of the work has been performed in the chick as these embryos are easy to obtain. The use of mammalian embryos presents considerable difficulties.

Of course, we shall always find the sceptics who say that because the cells migrate in the chick, this is no proof that the same occurs in the human embryo. However, the observations in the chick and other lower vertebrates have enabled embryologists to lay down certain general principles and these apply very well to the human. Recent work in the human has confirmed these results.

The concept of neural crest cells and their numerous derivatives has become so important in embryology and especially in the elucidation of abnormalities, that it has been suggested that these cells be regarded as the fourth germ layer of the embryo. Because of the increasing importance of the neural crest, the authors have felt that it should be considered in some detail.

Origin of Neural Crest Cells

The identification and separation of these cells occurs during the process of development of the neural tube when the lateral walls of the neural plate, which are derived from the dorsal surface ectoderm, elevate to give rise to the **neural folds** (Fig. 3.120A,B,C). A population of ectodermal cells in the **crests** of the folds separate from other ectodermal cells in the crests, taking on a specialised existence of their own. These cells then begin to emerge into the surrounding mesoderm and are called **neural crest cells.** In the cranial region of the embryo, migration of the neural crest cells from the folds occurs before fusion of the folds has taken place. However, in the trunk region of the embryo, this migration from the folds occurs after fusion of the folds. While still in the crests of the neural folds, these cells change their shape and properties so that they become 'different' from the other neuroepithelial cells. They stop expressing cell adhesion molecules which characterise the other neuroepithelial cells, and thus lose cell-to-cell adhesiveness. (When these cells eventually complete their migrations and reach their destinations, the cell adhesion molecules reappear.) On emerging from the neuroepithelium, the neural crest cells change from epithelial to mesenchymal cells (this involves the '*slug*' gene) and are referred to as **ectomesenchymal cells** to indicate their origin from ectoderm and their change to mesenchymal cells. They are then free to migrate extensively through the mesenchyme of the body to reach specific destinations where they give rise to a variety of important derivatives.

Several complex molecules are believed to be important in neural crest cell formation such as fibroblast growth factors (FGF's) and members of the transforming growth factor-b (TGF-b) superfamily (e.g. dorsalin-1, BMP-4).

Migration of Neural Crest Cells

The neural crest cells migrate through an environment which is rich in extracellular matrix molecules. They also encounter certain barriers, for example the surface ectoderm and the posterior halves of the somites. Tracing experiments have shown that neural crest cells follow well-defined pathways. Certain extracellular matrix molecules such as laminin, fibronectin, and type IV collagen are supportive (permissive) of migration. Chondroitin sulphate proteoglycans (which, for example, occurs in the posterior half of the somites) appears to be inhibitory to the migration of neural crest cells, as fully developed basement membranes. Hyaluronic acid 'opens' spaces for the migrating crest cells. These migrations are not only determined by the extracellular environment which the cells encounter, but also by properties inherent in the crest cells themselves.

In mammals, it is thought that the **cranial neural crest** cells arise from the neural tube between the levels of the mid-diencephalon and the fifth somite (in the cervical region). **Trunk neural crest** cells arising from the neural tube, extend from the level of the fifth somite to the caudal region of the embryo. Derivatives of the cranial and trunk neural crest form two distinct groups (see Table 3.4).

Cranial Neural Crest

The cranial neural crest cells, arising between the mid-diencephalon and the level of the fifth somite, leave the neural folds well before fusion of the tube has occurred. The cells migrate in diffuse streams through the cranial mesoderm (which is at this stage composed of lateral plate and paraxial mesoderm) to reach their final destinations. While most of these crest cells shift to the ventral aspects of the embryonic head and neck during a brief period, some remain dorsal to the hindbrain and aggregate as cranial sensory ganglia. Following the initial migration of the crest cells, an **interface** (imaginary line) that separates neural crest-derived mesenchymal tissues from those of general head mesoderm is established (Fig. 3.159A,B,C). This interface is said to occur at the mesencephalic–prosencephalic junction and passes dorsal to the hypophysis cerebri. It then extends caudally close to the dorsal margin of each pharyngeal pouch. Neural crest mesenchyme extends caudal to, but does not include, the laryngotracheal diverticulum. Any cartilage or bone that forms dorsal or caudal to this interface is thus said to develop from general mesoderm because this would be the only mesenchymal population available. Conversely, all skeletal and connective tissue structures which develop ventral to the interface are derived from the neural crest. Thus, even when epithelial structures such as the adenohypophysis, thymus and salivary glands secondarily shift their positions across this interface, they retain the neural crest-derived ectomesenchymal characteristic of the gland's site of origin. It is now evident that the neural crest in the cranial region gives rise to most structures of the ventral head and neck, with the

Table 3.4: Derivatives of the neural crest

System	*Cranial Neural Crest*	*Trunk Neural Crest*
1. **Nervous system:** Sensory nervous system	Ganglia of the trigeminal (V), facial (VII), glossopharyngeal (IX), and vagus (jugular ganglion) (X) nerves Bipolar neurons of the vestibular and cochlear ganglia of the VIII nerve	Spinal ganglia (pseudo-unipolar neurons of the dorsal root)
Autonomic nervous system	Parasympathetic ganglia: sphenopalatine, submandibular and ciliary	Parasympathetic ganglia: visceral and pelvic plexus
Sympathetic nervous system	Peri-arterial plexuses (possibly)	Sympathetic trunk (ganglion cells)
Non-neuronal cells	Satellite cells of sensory ganglia, Schwann cells of peripheral nerves, Oligodenrocytes, leptomeninges	Satellite cells of sensory ganglia, Schwann cells of peripheral nerves, enteric glial cells
2. **Endocrine and para-endocrine cells**	Carotid body (type I cells and possibly type II), Calcitonin cells (C cells) of thyroid	Chromaffin (phaeochromocytes) cells of the suprarenal medulla, endocrine cells of heart and lungs (possibly), extra-suprarenal chromaffin cells of the paraganglia (of Zuckerkandl)(para-aortic)
3. **Pigment cells**	Melanocytes, also possibly of iris but not of retina	Melanocytes
4. **Skeletal elements**	Cranial vault: squamosal and part of frontal, part of sphenoid, nasal, orbital, part of otic capsule and trabeculae, Meckel's cartilage, sphenomandibular ligament, anterior ligament of the malleus, maxillary and palate, ear ossicles, part of external ear cartilages, visceral cartilages and bones (hyoid, thyroid and laryngeal cartilages)	None
Connective tissues	Dermis and fat of skin; fibroblasts of stroma of cornea and corneal endothelium; sclera and sheaths of extrinsic ocular muscles; all of the tooth (odontoblasts, cementoblasts, associated connective tissue) except enamel and endothelium of blood vessels of pulp; connective tissue of parathyroid, thyroid, thymus, salivary and lacrimal glands and hypophysis cerebri; connective tissue of tongue; outflow tract of heart (ascending aorta and pulmonary trunk), semilunar valves, walls of pharyngeal arch arteries	None
Muscle	Dermal smooth muscle, vascular smooth muscle, ciliary muscles	None

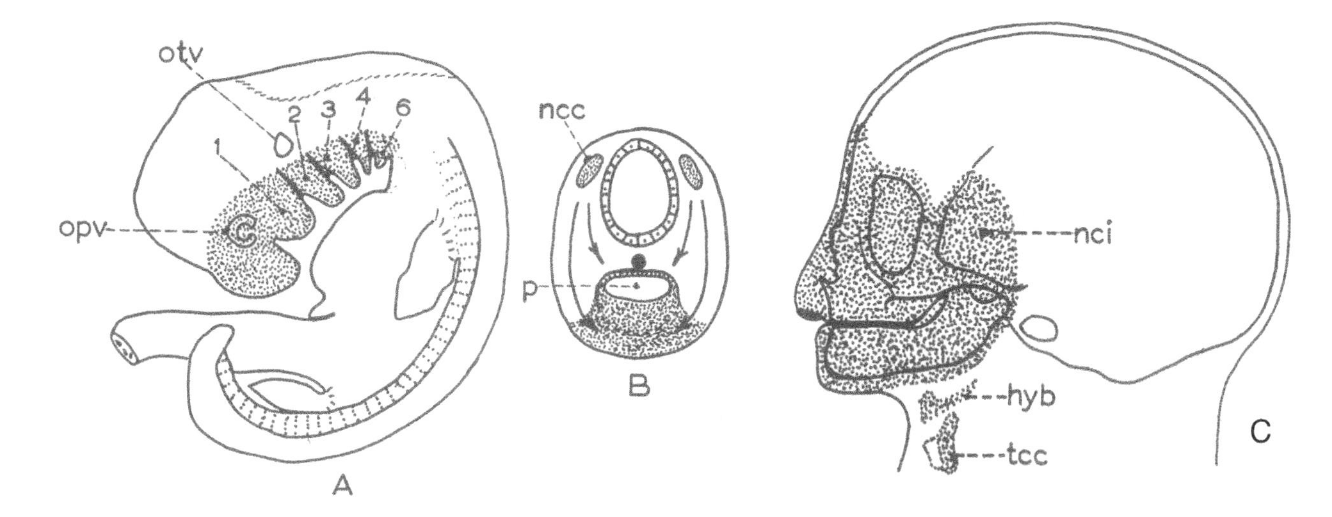

Figure 3.159: Migration of neural crest cells to form an interface between the general mesenchyme and the ectomesenchyme in the head and neck. A-C: Lateral views. **C:** Neonate. 1, 2, 3, 4, 6 = pharyngeal arches; arrows = ventral migration of neural crest cells; hyb = hyoid bone; ncc = neural crest cells; nci = neural crest cell infiltration; opv = optic vesicle; otv = otic vesicle; p = pharynx; tcc = thyroid-cricoid complex.

exception of the retina and lens, the enamel of teeth, epithelial tissues, vascular endothelia and skeletal (voluntary) muscle.

The most rostral region of the cranial neural crest contributes to the tissues of the face. e.g. bone, cartilage, dermis, the major parts of the teeth (except enamel), connective tissues of glands and certain tissues associated with the eye.

In relation to the pharyngeal arches, it is now known that neural crest cells arising from specific rhombomeres (segments of the hindbrain) migrate into specific arches. Neural crest cells from rhombomere 2 migrate into the first pharyngeal arch. Pharyngeal arch 2 derives its neural crest cells mainly from rhombomere 4. Pharyngeal arch 3 derives its neural crest cells mostly from rhombomere 6.

Cardiac (vagal) neural crest is derived from neural tube of the region bounded roughly by the otic placode and the third somite. Migrating crest cells from this region aggregate in the circumpharyngeal tract. Streams of cells from the vagal region migrate into the outflow tract of the heart.

Trunk Neural Crest

The trunk neural crest is derived from neural tube from the level of the sixth somite caudally. Three main migratory pathways are seen in the trunk region (**Fig. 3.160**):

(a) A **dorso-lateral pathway** between the ectoderm and the somites;

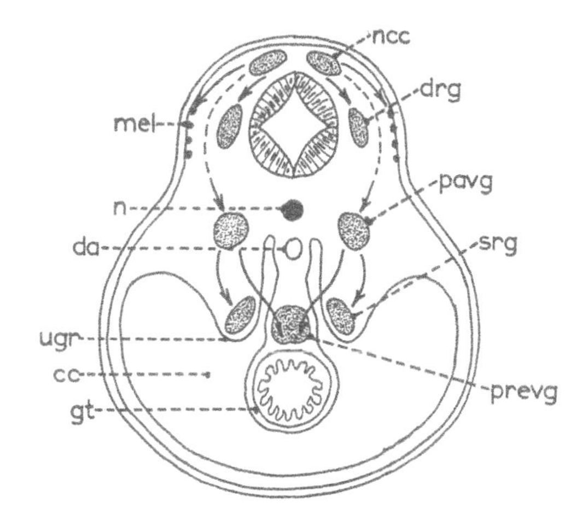

Figure 3.160: Transverse section of neural crest cell migration in the trunk region of the embryo. cc = coelomic cavity; da = dorsal aorta; drg = dorsal root ganglion; gt = gut tube; mel = melanocytes; n = notochord; ncc = neural crest cells; pavg = paravertebral ganglion; prevg = prevertebral ganglion; srg = suprarenal gland; ugr = urogenital ridge.

(b) A **ventral pathway** between the somites and the neural tube, which continues to the dorsal aortae;

(c) A **ventrolateral pathway** in which the crest cells migrate into the rostral halves of the somites.

Cells which elect the dorso-lateral pathway come to lie beneath the surface ectoderm. Ultimately they will enter the ectoderm and form the melanocytes of the skin.

Cells which select the ventral pathway belong to the sympatho-adrenal lineage. They will ultimately contribute to the cells of the suprarenal medulla (chromaffin cells) as well as parts of the sympathetic nervous system (small cells of the sympathetic ganglia and carotid body), adrenergic sympathetic neurons and a small population of cholinergic sympathetic neurons.

The gut is populated by neural crest-derived parasympathetic neurons and enteric glial cells. These are derived from the vagal and sacral levels of the neural tube.

The cells which follow the third migratory route (ventro-lateral into the rostral halves of the somites) form the segmentally arranged sensory dorsal root ganglia of the spinal nerves. The cells of these ganglia are pseudo-unipolar neurons.

The Ailing Neural Crest

There exists a large constellation of developmental anomalies, clinical conditions and syndromes which involve one or more of the derivatives of the neural crest. In 1951 Professor P.V. Tobias suggested the term the 'Sick' Neural Crest, while in 1974 Bolande classified a number of neoplastic conditions as 'neurocristopathies'. Some of the more important conditions are listed below.

- **Von Recklinghausen's multiple neurofibromatosis**
  The affected cells are the Schwann cells. The patient develops schwannomata and café-au-lait spots from disturbance of the melanocytes.
- **Phaeochromocytoma**
  This is a tumour of the catecholamine – producing phaeochromocytes of the suprarenal medulla which produce large amounts of norepinephrine.
- **Hirschprung's Disease**
  Aganglionic or hypoganglionic colon, which, as a result of lack of peristalsis, causes megacolon.
- **Peutz-Jegher syndrome**
  The syndrome consists of patchy pigmentation of the skin and mucous membranes associated with gastro-intestinal polyposis.
- **Medullary carcinoma of the thyroid**
  This is the result of cancer of the parafollicular cells or C-cells of the thyroid gland.
- **Di George's Syndrome**
  The outflow tract of the heart is affected. The thymus and parathyroid glands are affected.
- **Klein-Waardenburg Syndrome**
  This is characterised by deafness and a depigmented forelock.
- **Albinism**
  Inability to convert tyrosine to melanin in cutaneous melanocytes, but not in the substantia nigra.

In addition to the above conditions other congenital abnormalities may result. Examples of these are: **cyclopia, hypertelorism, midline clefts of the face, clefts of the lip and palate** and certain **dental anomalies.**

Development of the Eye

The eye is derived from three main embryonic sources:

(a) **A protrusion from the prosencephalic vesicle of the brain**; this provides the nervous (seeing) elements of the eye and will become the **optic vesicle**;

(b) **A thickening of the surface ectoderm overlying the optic vesicle**; this forms the lens and is known as the **lens placode**;

(c) **Neural crest cells** (ectomesenchyme) are a prominent feature of development in this region of the embryo.

General Development

At an early stage in the development of the central nervous system (about 28 days pf), bilateral tubular evaginations extend from the lateral prosencephalic wall – these are the **optic stalks.** When the prosencephalon divides into telencephalon and diencephalon, the stalks remain as protrusions from the diencephalon (Fig. 3.161A). The optic stalks will later become the optic nerves. Each stalk develops a dilatation at its distal end – these are the **optic vesicles,** which will form part of the eyeballs (Fig. 3.161B). At the same time, the ectoderm over the optic vesicles thickens to form **lens placodes** (Fig. 3.161C).

Note: It seems likely that the proximity of the nervous elements to the ectoderm 'induces' the placode formation.

The optic vesicles grow into the vascular mesenchyme surrounding the central nervous system. Later, this mesenchyme will give rise to the dura mater (pachymeninx), the arachnoid

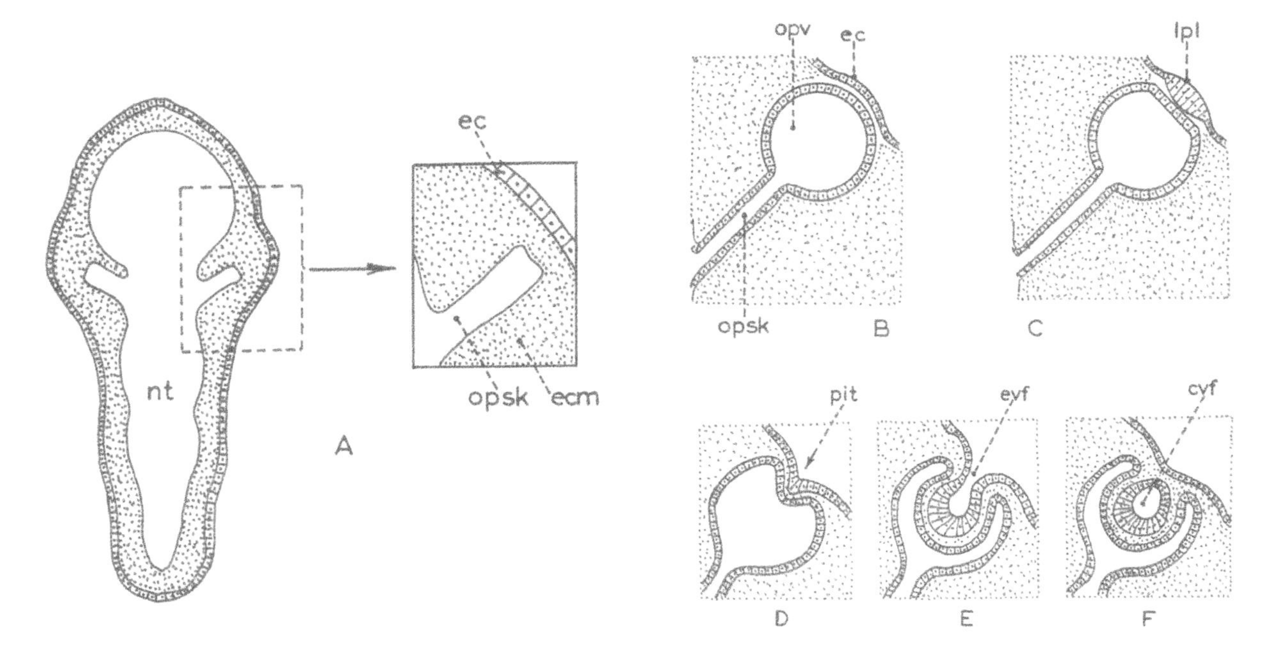

Figure 3.161: A series of views showing the early develoment of the eye. cvf = complete vesicle formation; ec = ectoderm; ecm = ectomesenchyme; evf = early vesicle formation; lpl = lens placode; nt = neural tube; opsk = optic stalk; opv = optic vesicle; pit = initiation of lens vesicle.

mater and pia mater (leptomeninges) as well as the outer layers of the optic vesicle.

The external surface of the lens placode develops a deep pit (Fig 3.161D,E) in its centre and this initiates an invagination of the placode to form a cavity which ultimately becomes the **lens vesicle**. The resulting vesicle will eventually separate from the surface ectoderm to form the **lens** (Fig. 3.161F). Each optic vesicle is then induced to undergo an invagination to form an **optic cup** (Fig. 3.161E,F). The lens then comes to lie completely surrounded by the optic cup and separated from it by a layer of mesenchyme (Fig. 3.161F). Later in development, the rim of the optic cup will grow ventrally to participate in the development of the ciliary body and iris. At the same time, an infolding develops on the inferior surface of the optic cup and stalk resulting in the formation of a slit, the **choroidal fissure** (Fig.3.162A). This fissure will later allow access of blood vessels and nerves to the interior of the eye. The edges of the slit fuse on about the 36th day pf forming a canal in the optic stalk (Fig. 3.162B,C).

Condensations of ectomesenchyme surrounding the optic cup will later give rise to two layers, an inner vascular layer, the **choroid**, and an outer fibrous layer, the **sclera**. The choroid will extend ventrally to participate in the development of the ciliary body, the iris and the cornea. The sclera ends by attaching to the conjunctiva externally and internally to the rim of the cornea.

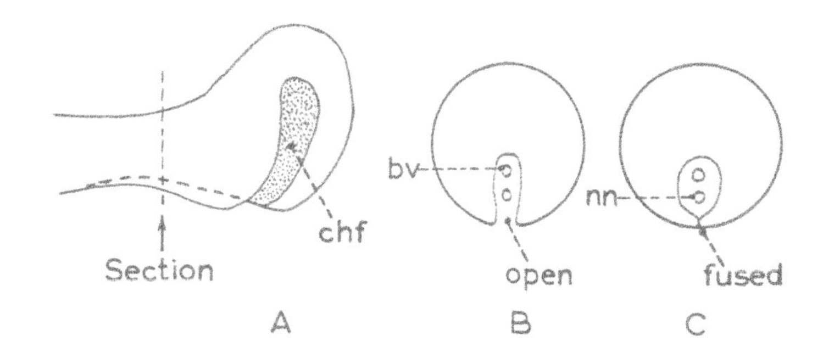

Figure 3.162: Formation of the optic stalk and choroid fissure (chf). The vertical interrupted line indicates the position of the transverse section in B and C. bv = blood vessels; nn = nerves.

Development of the Optic Cup

The optic stalk and cup form the optic nerve and retina respectively of the eyeball. When the optic vesicle undergoes invagination, two layers of cells are formed – the outer and inner layers of the optic cup. These are the primitive layers of the future retina. Most of the retina forms from the inner layer of the cup (Fig. 3.163A,B), while the outer layer of the cup is thinner and becomes pigmented. These pigmented cells are not of neural crest origin. Due to the invagination of the optic cup by the lens vesicle there is a space between the outer and inner layers of the cup – the **intraretinal space**. As the lens enlarges, the intraretinal space is reduced in size so that the layers make contact with each other and the very presence of the developing lens induces the cells of the inner layer of the cup to proliferate and form the nervous elements of the retina (**Fig. 3.163B**).

The space between the layers of the optic cup begins to disappear so that the pigmented layer comes to lie directly on the inner neural layer of the cup. However, the two adjacent layers of the cup are separated from each other by a thin **outer limiting membrane** (**Fig. 3.164A**). Occasionally, the space between the pigmented layer and the inner neural layer persists and results in **congenital detachment of the retina.**

There is also a space between the optic cup and the lens vesicle – the **lentoretinal space** (**Fig. 3.164A**). The lentoretinal space is not a true space as it will become occupied by a fine cellular network derived from the adjacent surfaces of the retina and lens (**Fig. 3.164A**). This network will give rise to the **vitreous body**, which will come to occupy the lentoretinal space. As the neural retina develops, its layers too are separated from the underlying vitreous body by an **inner limiting membrane,** while the vitreous body itself is surrounded by the **hyaloid membrane**.

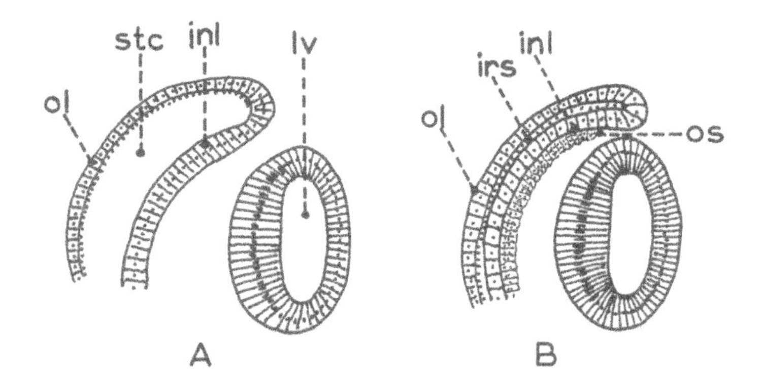

Figure 3.163: Sagittal section of the developing eye showing the optic cup and lens. inl = inner layer; irs = intra-retinal space; lv = lens vesicle; ol = outer layer; os = ora serrata; stc = stalk cavity.

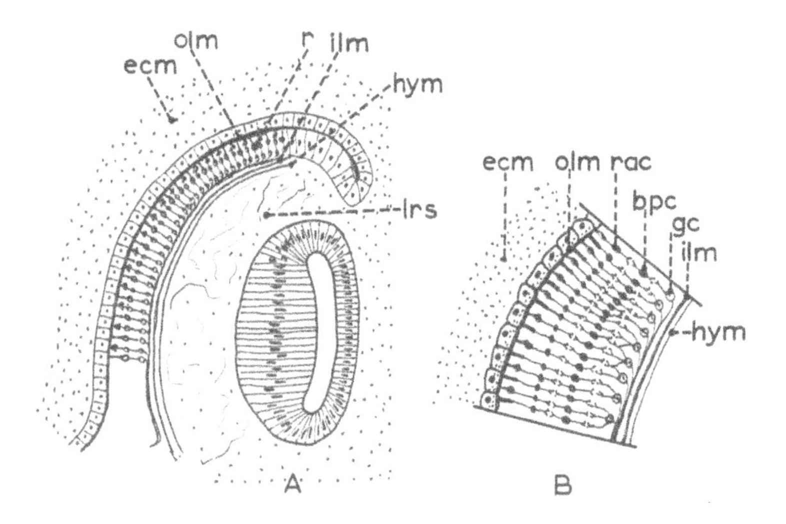

Figure 3.164: Sagittal sections to illustrate the development of the retina. bpc = bipolar cells; ecm = ectomesenchyme; gc = ganglion cells; hym = hyaloid membrane; ilm = inner limiting membrane; lrs = lento-retinal space; olm = outer limiting membrane; r = retina; rac = rods and cones.

In structure, the neural retina possesses three fundamental groups of cells. The outer layer consists of rods and cones (bipolar neurons, receptors), the middle layer consists of connector neurons, and the inner layer consists of ganglionic neurons (effector cells), the axons of which carry impulses into the central nervous system. Note the similarity to the anatomical structure of the peripheral nervous reflex arc which has receptor, connector and effector components. Because the optic vesicle becomes invaginated, the layers of the retina appear to be 'inside out' (**Fig. 3.164B**).

Although from its structure the retina would appear to be a fairly thick membrane it is, in fact, quite thin and offers little or no obstruction to the passage of light.

Since the optic stalk and cup are 'offshoots' of the neural tube it might be expected that the cells of the retina would be approximately similar to those of the neural tube. The receptor cells of the retina (rods and cones) form the lining layer of the optic cup, resembling ependyma. The next layer (connector neurons and ganglion cells) resemble the mantle layer. The final layer, the axonal fibres of the ganglion cells, resembles the marginal layer of the neural tube.

Development of the Ciliary Body

Up until now the retina has extended to the level of the angle of the lens where it ends in a circular fringe, the **ora serrata** (**Fig. 3.163B**). Further growth of the neural and pigmented layers of the retina result in it extending ventral to the angle of the lens (**Fig. 3.164A**). The neural part of the retina loses its 'neural' properties whereas the pigmented layer remains as such. These two layers together come to overlie the inner surface of the choroid where they are known as the ciliary part of the retina (**Fig. 3.165A,B**). The ciliary part of the retina and the underlying core of choroid are thrown into a series of irregular elevations called the **ciliary processes** (**Fig. 3.166A**). The lens is attached to the ciliary processes by a circular set of fine suspensory (zonular) fibres. The ciliary body contains muscle fibres (**Fig. 3.166B**) which are derived from local ectomesenchyme and are disposed as meridional, radial and circular groups. The pupillary and ciliary muscles are involuntary (smooth muscle) and are, therefore, innervated by the autonomic nervous system.

Development of the Iris

The ciliary part of the retina grows more ventrally as the iridial part of the retina to cover the inner surface of the choroid (**Fig. 3.166A,B**). The extension of the two combined layers forms the **iris** of the eye (**Fig. 3.166A**). The pigmented layer of the iridial part of the retina accounts for the pigmentation of the iris. Since the choroid is formed from neural crest cells, the muscle fibres developed within it are of ectomesenchymal origin. These muscles are the circular (constrictor) and radial (dilator) muscles of the iris and are responsible for alteration of the size of the pupil. They too are smooth muscle.

Mesenchyme will persist for a short while between the edges of the iris to form the **pupillary membrane** (**Fig. 3.167A**). When this membrane breaks down the pupil is formed. The remainder of the mesenchyme both on the outer and inner surface of the iris will degenerate, thus creating a space which will form the anterior and posterior chambers of the eye respectively (**Fig. 3.167B**) in the adult. The outer layer (ventral part) of the mesenchyme will participate in the development of the inner corneal epithelium.

It should be noted that the optic cup is a spherical structure and the extension of the combined retinal layers ventral to the lens will form a spherical cover so that the iris is a circular structure and the pupil is likewise a circular aperture.

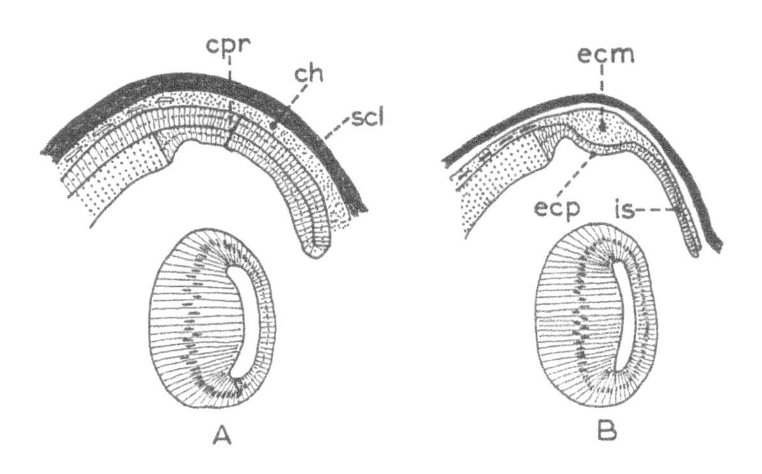

Figure 3.165: Sagittal sections depicting the development of the ciliary part of retina (cpr). ch = choroid; ecm = ectomesenchyme; ecp = early ciliary process; is = iris; scl = sclera.

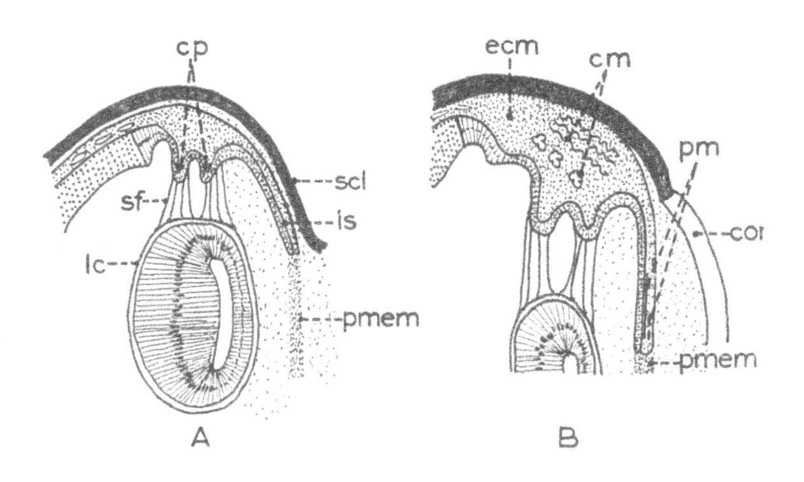

Figure 3.166: Sagittal sections to show the development of the ciliary processes and suspensory fibres. cm = ciliary muscles; cor = cornea; cp = ciliary processes; ecm = ectomesenchyme; is = iris; lc = lens capsule; pm = pupillary muscles; pmem = pupillary membrane; scl = sclera; sf = suspensory fibres.

Development of the Lens

Having been derived from surface ectoderm, the lens consists of a vesicle lined by cuboidal epithelium (**Fig. 3.168A,B**). The epithelium on the dorsal wall of the vesicle proliferates, the cell bodies extending ventrally, steadily reducing the size of the lens cavity (**Fig. 3.168C,D**). As the proliferation occurs, the cells lose their nuclei so that they become transparent. These cells are called the **primary lens fibres**. The ventral wall of the vesicle, also covered with cuboidal epithelium, undergoes no change and later becomes the subcapsular epithelium of the lens (**Fig. 3.168E**). Note that the ventral cuboidal cells extend dorsally as far as the equator of the lens. Viewed ventrally, the cuboidal cells surround the equator and produce **secondary lens fibres** which reinforce the perimeter of the primary lens fibres (**Fig. 3.168F**). Since the depression in the optic cup is filled with mesenchyme it follows that the lens is surrounded by mesenchyme. This mesenchyme is highly vascular and the lens is supplied by the **hyaloid artery** in the form of a surrounding vascular network – the **tunica vasculosa lentis**. Later, the blood vessels and the tunica degenerate, leaving a lens capsule which may be regarded as the basement membrane of the original ectodermal cells which gave rise to the lens vesicle. The lens will gradually change its shape from a sphere to an elliptical structure. The lens is suspended from the ciliary processes by the **suspensory ligament of the lens** (Fig. 3.166A,B).

Development of the External Layers of the Eye

The mesenchyme surrounding the optic vesicle condenses and divides into an inner pigmented and vascular layer, the **choroid,** and an outer dense fibrous layer, the **sclera** (Fig. 3.165A,B). The sclera corresponds to (is homologous with) the layer of mesenchyme which will form the dura mater (pachymeninx) of the central nervous system. The choroid corresponds to the layers which will form the arachnoid mater and the pia mater (leptomeninges).

The choroid will give rise to a continuous layer which forms the vascular layer on the outer surface of the retina. It also gives rise to the ciliary body and its processes, as well as the stroma of the iris. The sclera similarly is a continuous layer over the posterior aspect of the retina and passes ventrally as a thick fibrous layer which provides an attachment for the extra-ocular muscles and for the cornea.

That part of the choroid which lies ventral to the lens undergoes attenuation in such a way that the iris forms a partition between the anterior and posterior compartments of the adult eye (Fig. 3.167B).

The space between the iris and the lens is the **posterior chamber** while that between the cornea and the iris is the **anterior chamber**. The two chambers communicate with one another through a deficiency in the iris known as the **pupil**.

The surface epithelium of the cornea is formed from surface ectoderm. This ectoderm will secrete the underlying stromal components. Waves of neural crest cells will migrate into the developing cornea to form fibroblasts as well as the inner corneal epithelium.

The surface epithelium of the cornea folds deeply to form the **conjunctival sac** over the surface of the cornea (Fig. 3.167B). These folds become infiltrated with mesenchyme to form the eyelids, the edges of which fuse with one another and remain so, often until the 7th month of fetal life, when they separate to form the **palpebral fissure** (Fig. 3.167B).

After the formation of the conjunctival sac, the **lacrimal gland** forms from a series of ectodermal outgrowths in the upper outer part of the sac (Fig. 3.169). The **nasolacrimal duct** forms during the development of the face by fusion of the maxillary process with the lateral nasal process. At its superior end, it becomes dilated as the lacrimal sac and from the sac, ducts connect with each eyelid.

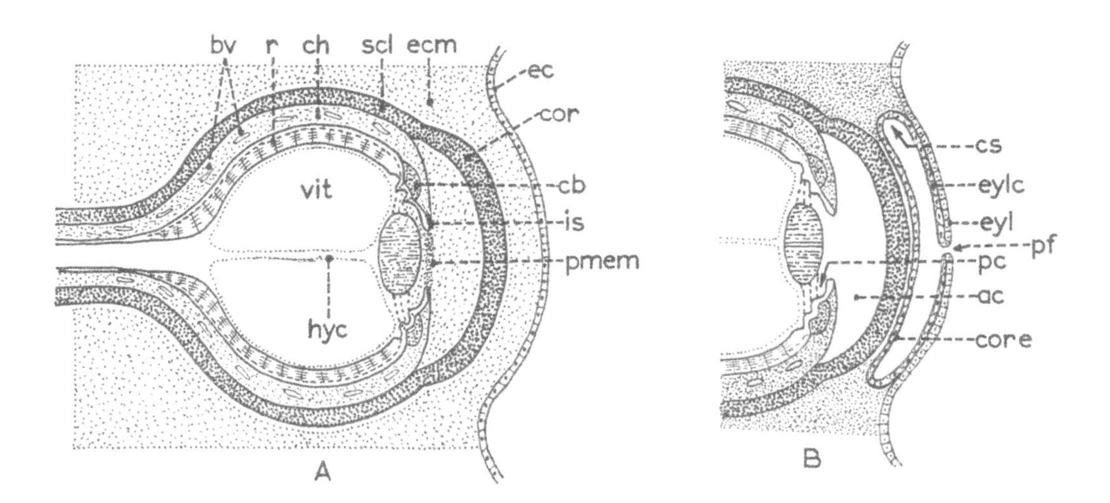

Figure 3.167: Sagittal sections showing the development of the chambers of the eye. ac = anterior chamber; bv = blood vessels; cb = ciliary body; ch = choroid; cor = cornea; core = corneal epithelium; cs = conjunctival sac; ec = ectoderm; ecm = ectomesenchyme; eyl = eyelid; eylc = eyelid conjunctiva; is = iris; pc = posterior chamber; pf = palpebral fissure; pmem = pupillary membrane; r = retina; scl = sclera.

POSSIBLE COMMON ABNORMALITIES

- Degeneration of the optic vesicle/s may result in unilateral or bilateral **anopthalmos.**
- Infection of the mother with rubella may result in **congenital cataract** in the infant.
- **Cyclopia** (median eye) results from the total or partial fusion of the eyes. This is associated with holoprosencephaly (non-separation of the forebrain). This abnormality is thought to result from an associated neural crest deficiency. The proboscis (tubular nose) usually lies above the eye.

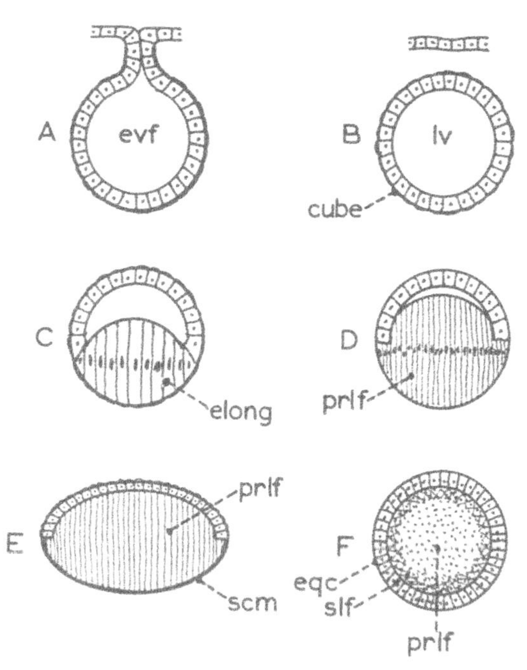

Figure 3.168: Formation of the lens.
cube = cuboidal epithelium; elong = elongation of posterior cells; eqc = equatorial cells; evf = early vesicle formation; lv = lens vesicle; prlf = primary lens fibres; scm = subcapsular membrane; slf = secondary lens fibres.

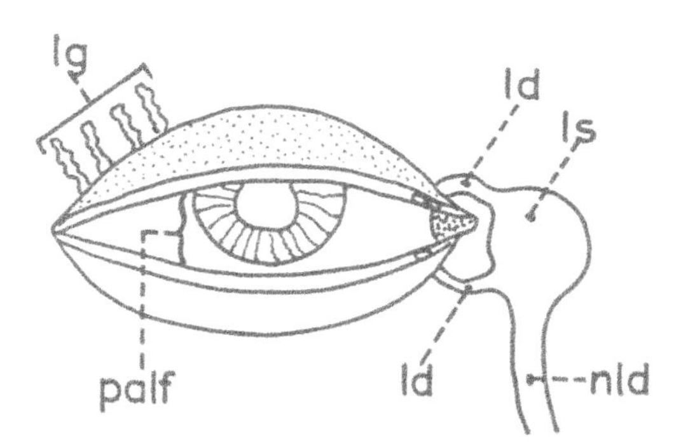

Figure 3.169: Lacrimal apparatus. ld = lacrimal duct; lg = lacrimal gland; ls = lacrimal sac; nld = nasolacrimal duct; palf = palpebral fissure.

Development of the Ear

The ear consists of three parts:

(a) the **external ear** – in the embryo situated around and between the external surfaces of the 1st and 2nd pharyngeal arches (1st pharyngeal groove);

(b) the **middle ear** – a cavity originating from the 1st pharyngeal pouch and containing the three ear ossicles;

(c) the **inner ear** – a complex series of passages called the labyrinth, developing from the ectodermal otic placode.

To understand the positioning of the three components, consider Figures 3.170 and 3.171.

External Ear

This consists of the auricle (pinna) and the external auditory canal leading to the **tympanic membrane** (ear drum). The auricle arises from four to six surface elevations produced by underlying ectomesenchymal condensations in the first and second pharyngeal arches (Fig. 3.170 and 3.172). These elevations, covered with ectoderm, fuse with one another to form the pinna of the ear. The elevations first appear in a more ventral and caudal position and are 'pushed' to a cranial and lateral position by the developing mandible. As they 'ascend' cranially the elevations fuse to give rise to a smooth contour which forms the pinna around the opening of the 1st pharyngeal groove.

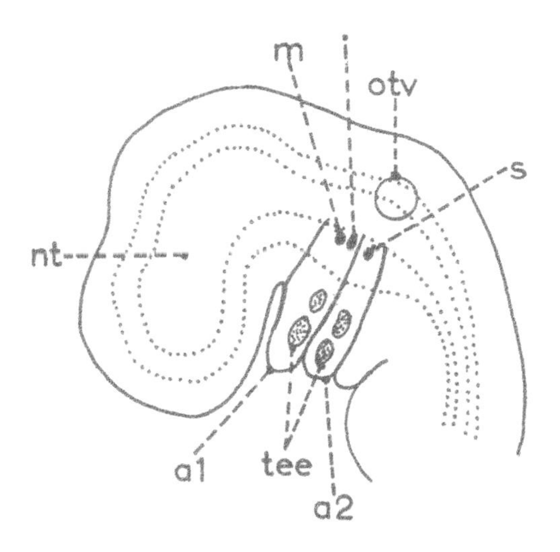

Figure 3.170: Lateral view to indicate the positions of ear structures. a1, a2 = 1st and 2nd pharyngeal arches; i = incus; m = malleus; nt = neural tube; otv = otic vesicle; s = stapes; tee = tubercles of external ear.

The 1st groove between the two arches deepens to form the **external auditory canal** which terminates at the tympanic membrane (**Fig. 3.172**). This membrane is covered on the outer surface by ectoderm (later, skin) and on the inner surface of the 1st pharyngeal pouch by endoderm (later, pharyngeal epithelium). Between the two lies a layer of ectomesenchyme (later, connective tissue).

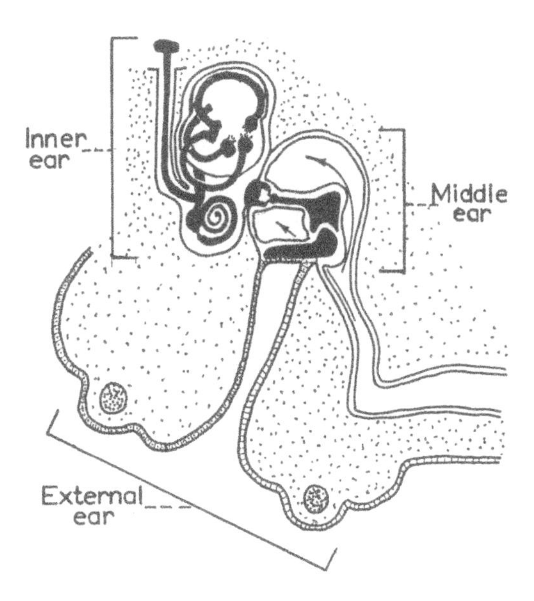

Figure 3.171: General disposition of the external, middle and inner ear.

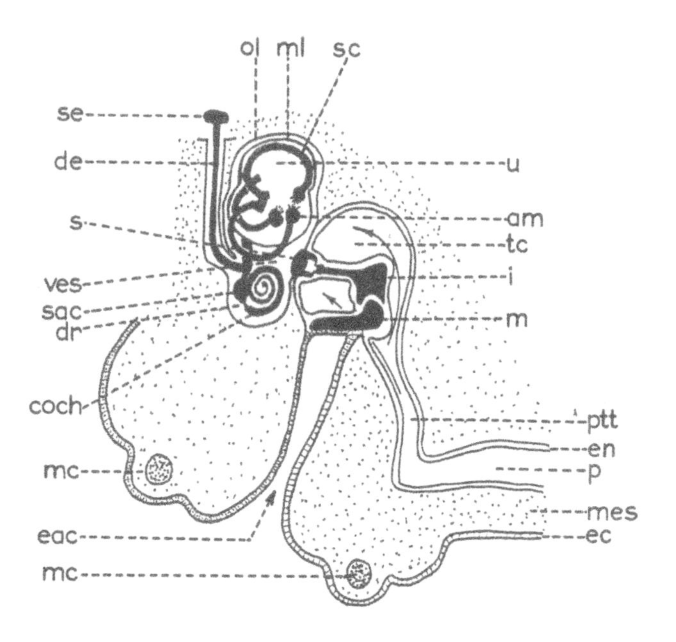

Figure 3.172: Detailed structure of the external, middle and inner ear. am = ampulla of semicircular canal; coch = cochlea; de = endolymphatic duct; dr = ductus reuniens; eac = external auditory canal; ec = ectoderm; en = endoderm; i = incus; mc = mesodermal condensation; m = malleus; mes = mesoderm; ml = membranous labyrinth; ol = osseous labyrinth; p = pharynx; ptt = pharyngotympanic tube; s = stapes; sac = saccule; sc = semicircular canal; se = endolymphatic sac; tc = tympanic cavity; u = utricle; ves = vestibule.

POSSIBLE COMMON ABNORMALITIES

- The external ear may fail to ascend to its normal position. The pinna may thus be found anywhere along its pathway of normal ascent.
- The abnormal shape of the pinna may be caused by aberrant fusion of the ear hillocks.
- **Anotia** or the absence of the pinna may occur.
- Cervically placed, ventrally located pinnae are referred to as **synotia**.

Middle Ear

While the processes stated above are taking place, a tubular extension of the 1st pharyngeal pouch, the **tubotympanic recess** (**Fig. 3.173A**) grows into the mesenchyme surrounding the hindbrain. This tubular extension has a dilated lateral component which will become the middle ear. The tubular part of the tubotympanic recess forms the **pharyngotympanic tube** (of Eustachius). Three spots of mesenchyme condense to form two of the ossicles; the **malleus** and **incus** arise in the mesenchyme of the first pharyngeal arch while the **stapes**, a further condensation of mesenchyme, arises in that of the second pharyngeal arch (**Fig. 3.173A**). The ossicles become jointed together by synovial joints forming a chain, the outer end (handle of the malleus) being attached to the tympanic membrane, the incus being the intermediate link and the footplate of the stapes being indirectly related to the cochlear duct. The greater part of the development of the middle ear takes place in the fetal period of life. Due to the retained connection between the middle ear and the pharynx, infections of the pharynx may spread to the middle ear (otitis media).

POSSIBLE ABNORMALITY

Abnormalities of the malleus and incus may be associated with the **1st arch syndrome.**

Note. The 1st arch syndrome manifests as micrognathus, a cleft of the palate, hypoplasia of the mandible or maxilla and may be seen as a complex of abnormalities in Pierre-Robin Syndrome and Mandibulofacial Dysostosis.

Inner Ear

At about the 27th or 28th day pf, the surface ectoderm opposite the hindbrain becomes thickened to form the **otic placode.** Its external position is approximately opposite the first pharyngeal groove (**Fig. 3.173A**). The placode becomes deeply dimpled in the centre resulting in an invagination to form the **otocyst** (**Fig. 3.173B**). The otocyst lies deeply within the mesenchyme of the region (**Fig. 3.173C**) where it undergoes a series of complex changes to form the **membranous labyrinth** (**Fig. 3.174a-g**). The vesicle then becomes elongated dorsoventrally and develops a constriction in its middle from which emanates a long narrow tube, the **endolymphatic duct** (**Fig. 3.174c**). The **endolymphatic sac** forms at its termination. The dorsal dilated part of the vesicle is called the **utricle** and the ventrally placed elongated part will become the **cochlea** (**Fig. 3.174f**).

At this stage, three 'flange-like' extensions emanate from the utricle to form the three **semicircular canals** (ducts) and these will later be set in the three planes of space. Below the original constriction, a further dilatation occurs and this

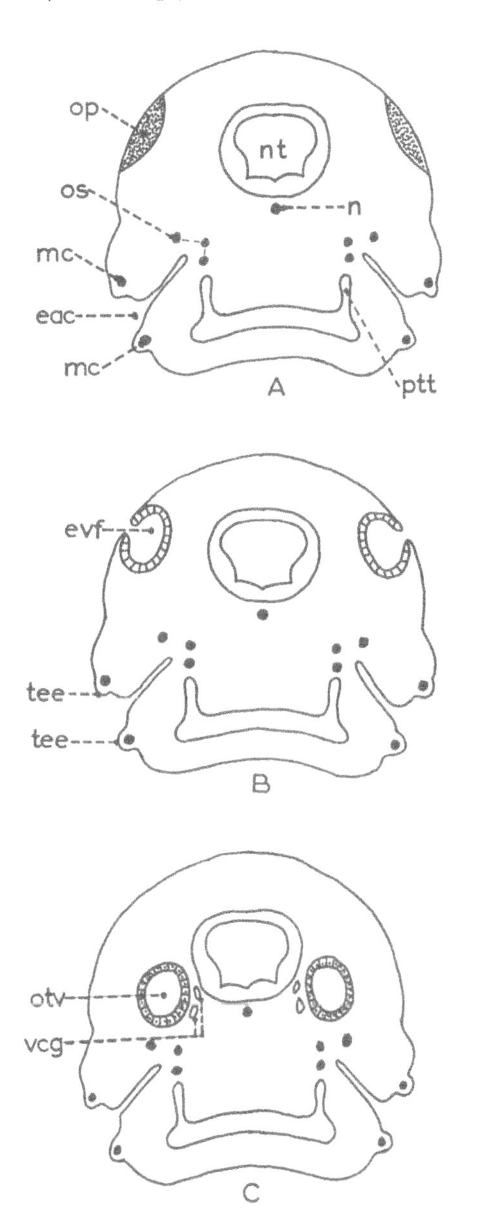

Figure 3.173: Transverse sections of the head of the embryo showing the development of the various parts of the ear. eac = external auditory canal; evf = early vesicle formation; mc = mesodermal condensation; n = notochord; nt = neural tube; op = otic placode; os = ossicles; ptt = pharyngotympanic tube; otv = otic vesicle; tee = tubercles of external ear; vcg = vestibulocochlear ganglia.

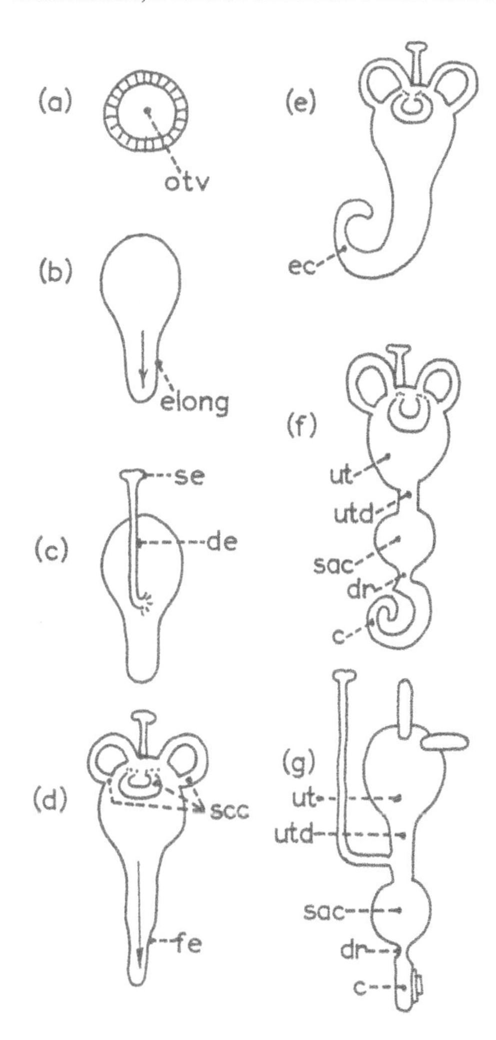

Figure 3.174: Formation of the semicircular canals and membranous labyrinth. Note: Figure (**g**) is at right angles to (**f**). c = cochlea; de = endolymphatic duct; dr = ductus reunions; ec = early cochlea; elong = elongation of vesicle; fe = further elongation; otv = otic vesicle; sac = saccule; scc = semicircular canals; se = endolymphatic sac; ut = utricle; utd = utricular duct.

is known as the **saccule** and below this, the cochlear tube is attached by a narrow connection, the **ductus reuniens** (Fig. 3.174f). To conserve space, the cochlear tube undergoes about 2¾ turns of 'spiralisation' to form the definitive cochlea.

The semicircular canals, the utricle, saccule and cochlea are formed of membrane and are collectively known as the **membranous labyrinth.** These contain a fluid called **endolymph** which is said to be secreted by the **stria vascularis** (Fig. 3.175) within the cochlear duct. As the mesenchyme condenses to form the cartilage of the **otic capsule** and then the bone of the **petrous part of the temporal bone,** a bony cavity is formed around the existing membranous labyrinth. Bone forms the **osseous labyrinth** and it has a shape similar to that of the membranous labyrinth. The space between the bony and membranous labyrinths is filled with **cerebrospinal fluid,** obtained from the subarachnoid space to which the membranous labyrinth is attached by the **perilymphatic duct.** At one place, an oval opening (**fenestra vestibuli**) occurs in the surrounding bone. It is to the edge of this opening that the footplate of the stapes is fitted. Thus, the movement of the tympanic membrane due to sound is transferred to the stapes which, in turn, agitates the fluid surrounding the membranous labyrinth and this, in turn, agitates the fluid in the cochlear duct, containing the hearing organ (of Corti) (Fig. 3.175).

The otocyst (inner ear) is closely related to the neural crest of the hindbrain ganglia. The neuroblasts of the accoustico-facial ganglion will separate into two groups and the accoustic (auditory) group will grow into the inner ear structures to supply the receptor cells of the ampullae, the maculae and the hair cells of the organ of Corti.

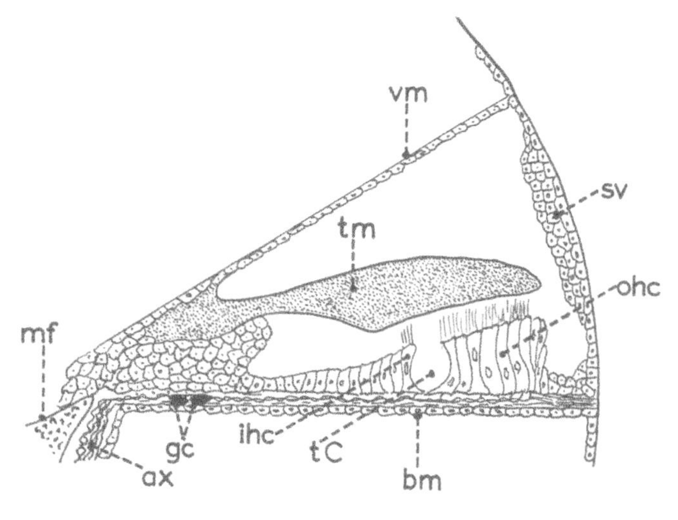

Figure 3.175: The organ of Corti. ax = axons of ganglion cells; bm = basilar membrane; gc = ganglion cells; ihc = internal hair cells; mf = modiolar flange; ohc = outer hair cells; sv = stria vasculosa; tC = tunnel of Corti; tm = tectorial membrane; vm = vestibular membrane.

Structures Developing in the Membranous Labyrinth

(a) Each semicircular canal develops an ampulla at one end which contains a group of ampullary crest cells. These are responsible for responding to movement of the head.
(b) The utricle and saccule develop maculae in their walls upon which are the otolith organs which are gelatinous and contain crystals of calcium carbonate. These crystals are responsible for responding to the effects of gravity.
(c) The cochlear duct contains the organ of hearing (the organ of Corti) which arises from the proliferation of cells on the basilar membrane of the duct. These form two sets of 'hair cells' (inner and outer) extending between the basilar membrane and the tectorial membrane (Fig. 3.175).

POSSIBLE COMMON ABNORMALITY

Congenital hearing loss may result from damage to the spiral organ (of Corti) after infection of the mother with the rubella virus.

Skeletal and Muscular Systems

Development of a Typical Vertebra

It is common zoological knowledge that all vertebrates are segmental in nature. This segmentation arises early in embryonic life. One of the best examples of segmentation is to be found in the vertebral column.

The development of each sclerotome will enclose the notochord. The sclerotomes fuse around the notochord to form a series of segments, each of which is known as a **centrum** (later the body of the vertebra) (Fig. 3.176a). Because the myotome arises from the same segment as the centrum, it clearly lies in the same plane. This defeats the object of a muscle which must act across a joint to be effective.

To achieve this physiological state, an alteration takes place very early in the arrangement of the scleratomic masses and this is known as **resegmentation**. In this process, a split occurs in the middle of the sclerotome (Fig. 3.176b). This is called the sclerotomic fissure and it divides the sclerotome into cranial and caudal parts. This fissure indicates the position of the future intervertebral disc or joint. The caudal part of each sclerotome fuses with the adjacent cranial part of each successive sclerotome to form a resegmented centrum of the vertebra. With the development of the transverse processes (Fig. 3.176c,d), it becomes evident that the muscle now exists across the intervertebral joint.

At the sclerotomic fissure, the adjacent mesenchyme condenses to form a **perichordal disc** (surrounding the notochord) which will ultimately form the **anulus fibrosus** of the intervertebral disc. The remnant of the notochord forms the **nucleus pulposus** of the disc.

After each resegmented sclerotome has enclosed the notochord, the sclerotome of each side will give rise to one side of the neural arch (Fig. 3.177). The fused neural arch will ultimately surround the developing neural tube (spinal cord) and its membranes. Each half of the neural arch is attached to the single centrum at the neurocentral junction. The neural arches normally fuse dorsally to form the **spinous process** of the vertebra (Fig. 3.178B).

POSSIBLE COMMON ABNORMALITIES

- If one or both parts of the neural arch are deficient, the condition of **rachischisis** or **spina bifida** occurs. Nervous elements with their membranes may prolapse through the resulting opening (Fig. 3.179A,B,C).
- When only membranes prolapse, the condition is called **spina bifida meningocoele** (Fig. 3.179A).
- When membranes and nervous elements prolapse, the condition is called **spina bifida meningomyelocoele** (Fig. 3.179B);
- Occasionally, there is lack of fusion of the neural arches but the opening is so small that no prolapse occurs. This is called **spina bifida occulta** (hidden split vertebral spinous process) (Fig. 3.179C).
- Clearly, if one sclerotome is absent, deformed or deficient, the vertebral column will be angulated at that site, e.g. **hemivertebra** (Fig. 3.177B).
- Other deformities may result from fusion of mis-shapen vertebral segments, e.g. **Klippel-Feil Syndrome.**

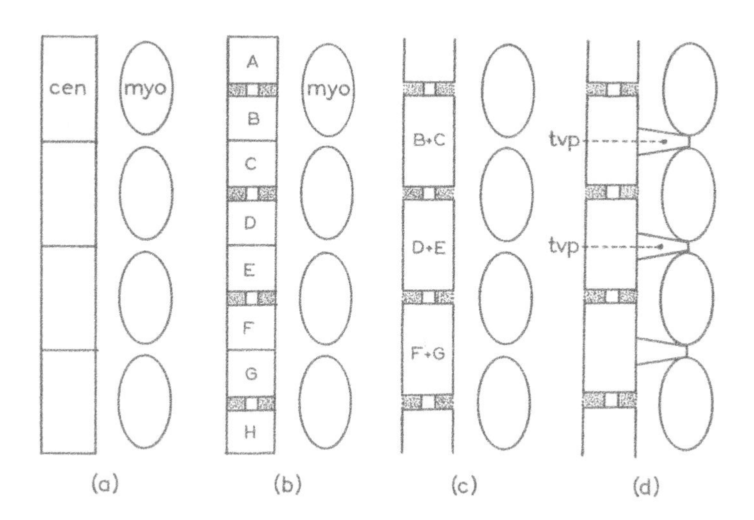

Figure 3.176: a –d: Segmentation and resegmentation of the vertebral column. cen = centrum; myo = myotome; tvp = transverse process.

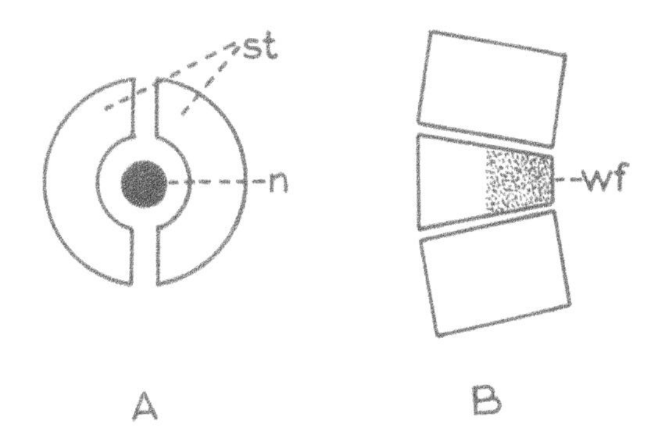

Figure 3.177: An early stage in the development of a vertebra. A: Transverse section. **B:** Lateral view. st = sclerotome; wf = wedge formation.

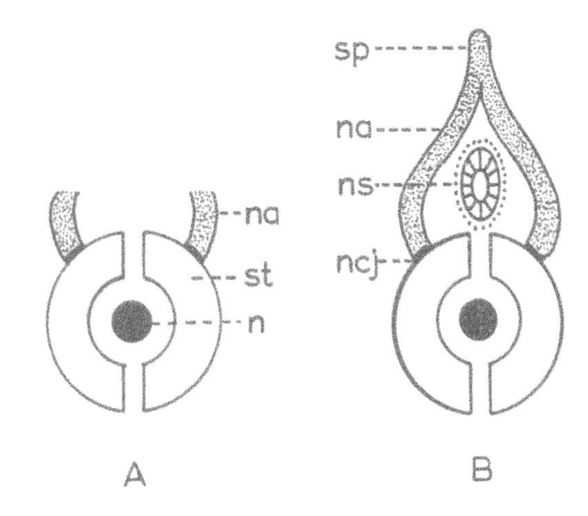

Figure 3.178: Formation of the neural arch (na). **A** and **B** are both transverse sections. n = notochord; ncj = neurocentral junction; ns = nervous system; sp = spinous process; st = scelrotome.

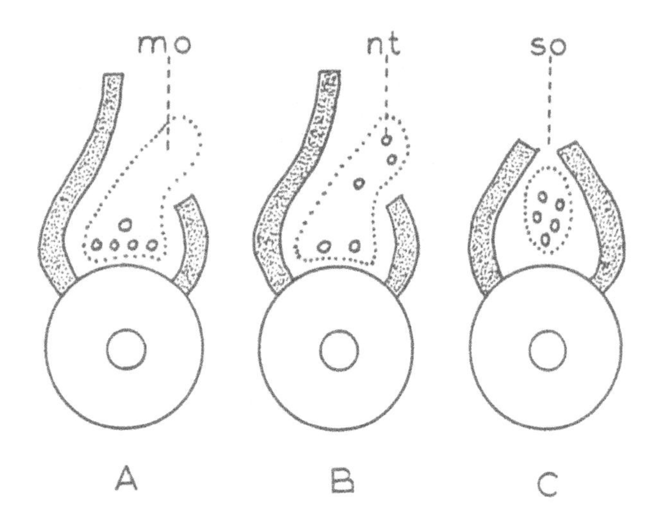

Figure 3.179: Transverse sections illustrating spina bifida. mo = membranes only; nt = neural tube elements; so = spina bifida occulta.

Development of the Myotome and Associated Nervous Elements.

By the time the vertebra has developed, the myotome has appeared lateral to the vertebra (Fig. 3.180A) and the nervous system has provided a nerve which will supply the muscle mass as well as the overlying skin.

Note. The nerve supply is both motor (to the muscle mass) and sensory (from the skin) (Fig. 3.180A).

A transverse process grows laterally from the neural arch and splits the myotome into dorsal (**epaxial**) and ventral (**hypaxial**) parts. This means that the nerve must now supply two masses of muscle by giving two branches –a **dorsal primary ramus** (branch) and a **ventral primary ramus** (Fig. 3.180B). This is part of the formation of the anatomical spinal nerve. The dendrites coming from the skin are the processes of pseudo-unipolar neurons located in the dorsal root ganglion and are derived from neural crest cells. At this very early stage of vertebral development, the vertebra already has all its adult features namely, a body (centrum), pedicles, transverse processes, laminae and spine (Fig. 3.181).

Clearly, the end result of the muscle formation is a mass dorsal to the transverse processes of the vertebrae and a mass ventral to the transverse processes so that the transverse processes separate the masses. The line marking the tips of the transverse processes from atlas to coccyx is known as the **coronal morphological plane** and is an undulating plane (not truly coronal, in that it follows the curvatures of the vertebral column) (Fig. 3.182A,B).

Development of the Limbs

In the normal adult individual, common observation reveals that the cranially placed limb consists of the **arm, forearm** and **hand** and is attached to the trunk by the **shoulder girdle**. The caudally placed limb consists of the **thigh, leg** and the **foot** and is attached to the trunk by the **pelvic girdle**.

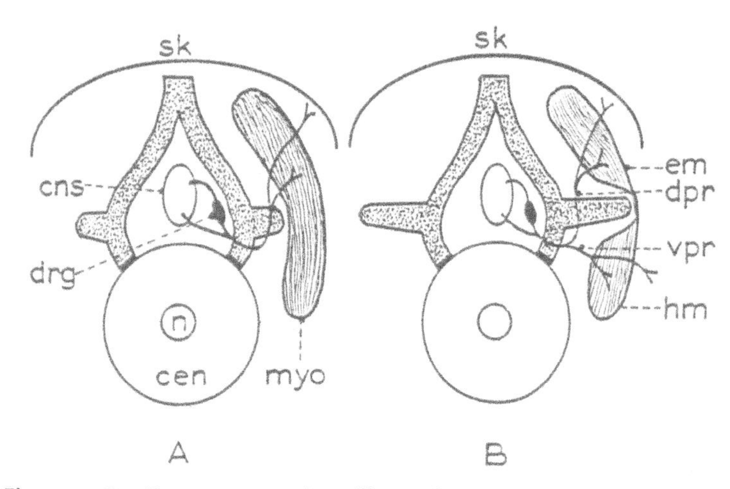

Figure 3.180: Transverse sections illustrating the innervation of muscle masses. cen = centrum; cns = central nervous system; dpr = dorsal primary ramus; drg = dorsal root ganglion; em = epaxial muscle; hm = hypaxial muscle; myo = myotome; n = notochord; sk = skin; vpr = ventral primary ramus.

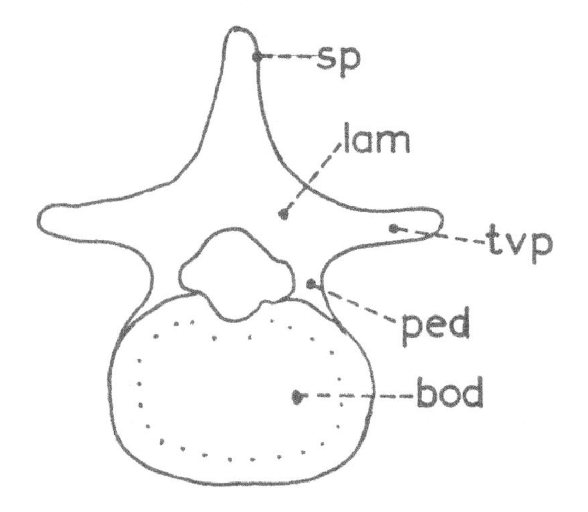

Figure 3.181: An adult vertebra. bod = body; lam = lamina; ped = pedicle; sp = spinous process; tvp = transverse process.

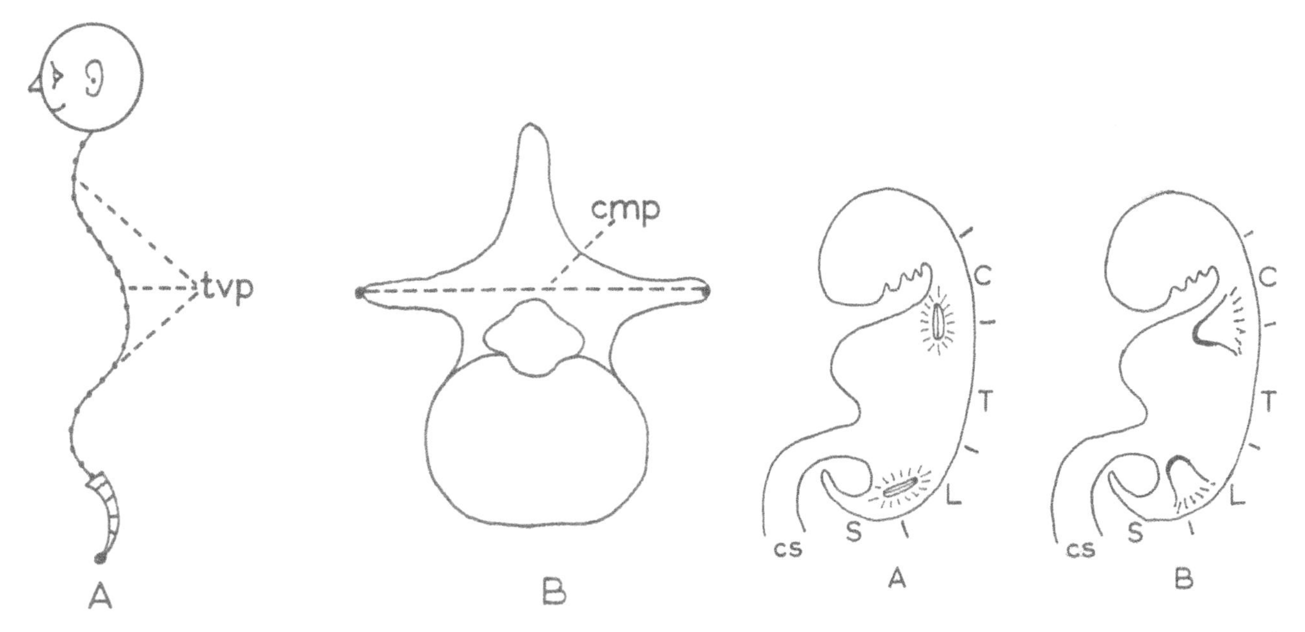

Figure 3.182: Figure A depicts the vertebral curvatures, and the interrupted line in figure B indicates the coronal morphological plane (cmp). tvp = transverse process.

Figure 3.183: Lateral views to show the positions of the developing limb buds. C, T, L, S = cervical, thoracic, lumbar and sacral; cs = connecting stalk.

Note. Some textbooks refer to the upper limb as the 'arm' and to the lower limb as the 'leg'. These terms as used are anatomically incorrect since 'arm' and 'leg' refer to parts of the limbs and not to the whole limb.

From their function, the humans limbs are complex mechanical structures. It follows therefore, that abnormalities in the limbs will result in considerable disability.

The study of the development of the limbs may be considered from two aspects:

- The development of their external features;
- The development of their internal structure.

Development of the External Features of the Limbs

From a general standpoint, the upper and lower limbs emerge from the trunk as anterolateral structures. The forelimbs lie at the level of the pericardial cavity opposite the segments C4, 5, 6, 7, 8, T1 and T2. The lower limbs lie just inferior to the attachment of the connecting stalk to the trunk and opposite segments L2 to S3 (**Fig. 3.183A,B**). The limbs emerge as 'buds' and consist of a core of mesoderm covered with surface ectoderm. The mesoderm is derived from the somatic component of the lateral plate of intra-embryonic mesoderm (**Fig. 3.184A,B**). The buds

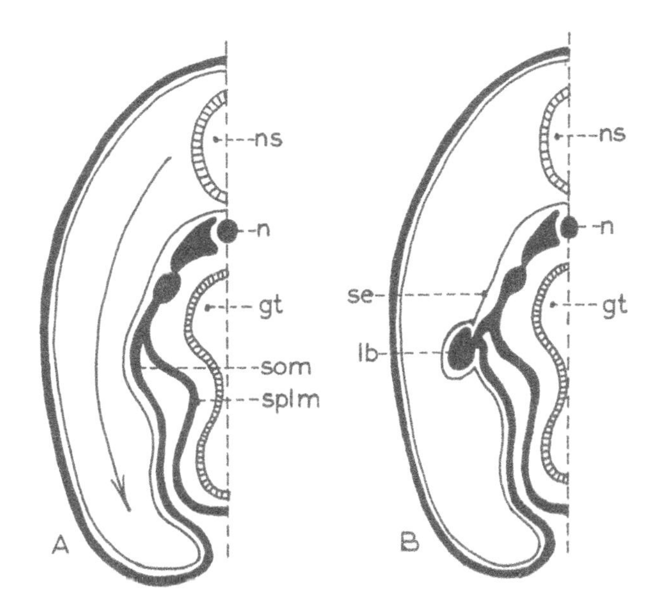

Figure 3.184: Transverse sections to show the derivation of the limb bud from lateral plate mesoderm. Arrow = direction of growth of the amniotic sac. gt = gut tube; lb = limb bud; n = notochord; ns = nervous system; se = surface ectoderm; som = somatic mesoderm; splm = splanchnic mesoderm.

are flattened mediolaterally and the vertical plane of the buds lies in the long axis of the embryo (**Fig. 3.185A**). The transverse plane of the bud passes at right angles to the vertical plane. As the limb extends ventrally, the transverse plane of the limb keeps pace with the extending limb (**Fig. 3.185B**). The transverse plane which is regarded as the axis of the limb bud, divides the limb into a cranial (**pre-axial**) part and a caudal (**postaxial**) part. The cranially placed edge of the flattened bud is called the 'pre-axial' border and the lower (caudal) edge is the 'post-axial' border. The apex of the pre-axial and the postaxial border is marked by a

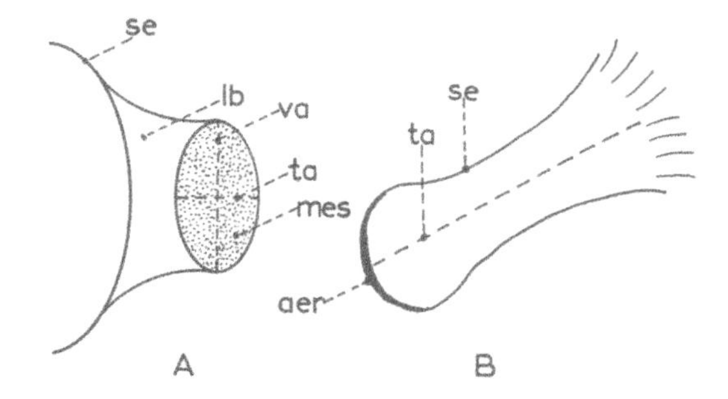

Figure 3.185: Axes of limb buds in transverse and longitudinal view. aer = apical ectodermal ridge; lb = limb bud; mes = mesoderm; se = surface ectoderm; ta = transverse axis; va = vertical axis.

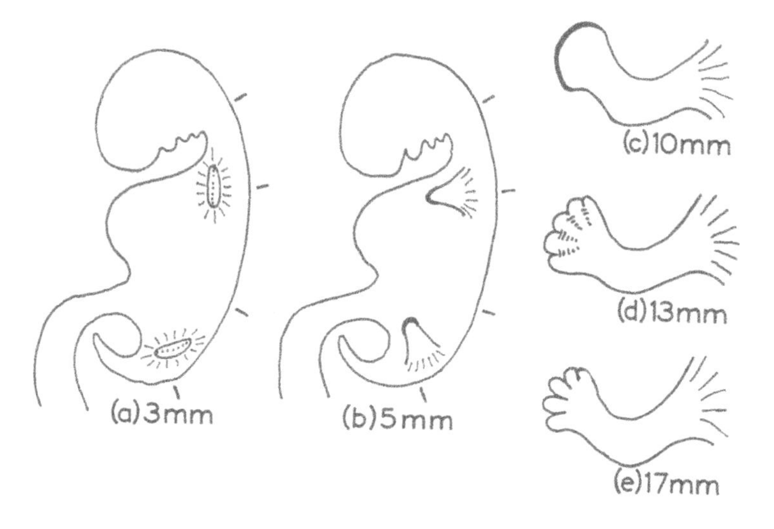

Figure 3.186: Stages in the development of the external features of the limb. The length of the bud is shown in embryos of various stages of development (mm's indicate crown–rump length of embryo).

ridge of thickened epidermis called the **apical ectodermal ridge** (**Fig. 3.185B**). Important inductive interactions take place between the apical ectodermal ridge and the underlying mesenchyme during the development of the limbs.

Stages in the development of the external features:

Stage 1: At about 3mm crown-rump length (CRL), the upper and lower limb positions appear as bulges at the sides of the pericardial cavity and just inferior to the attachment of the connecting stalk to the trunk (**Fig. 3.186a**).

Stage 2: At about 5mm CRL, the upper and lower limbs protrude as flattened 'buds' from the trunk (**Fig. 3.186b**).

Stage 3: At about 10mm CRL, the forelimb shows separation into arm, forearm and hand while the hindlimb, although enlarged, still retains its 'bud' shape (**Fig. 3.186c**).

Stage 4: At about 13mm CRL, the hand has 'flared' out and digital clefts are appearing; the hindlimb now shows division into thigh, leg and foot (**Fig. 3.186d**).

Stage 5: At about 17mm CRL, the digits of the hand are undergoing separation but those of the foot are still fused (**Fig. 3.186e**). Separation of the digits is initiated by disintegration of the apical ectodermal ridge. The ridge disappears where the digital clefts will appear and remains where the digits will develop. This is followed by programmed cell death (apoptosis) of the mesenchymal cells between the digits producing digital clefts, while the epidermal cells grow into the clefts to separate the digits (**Fig. 3.187A,B,C**).

Stage 6: At about 30mm CRL, the fore and hind limbs have developed all their normal external features with the elbows and knees slightly flexed and pointing laterally.

Stage 7: The upper limbs rotate laterally and the lower limbs rotate medially to bring the

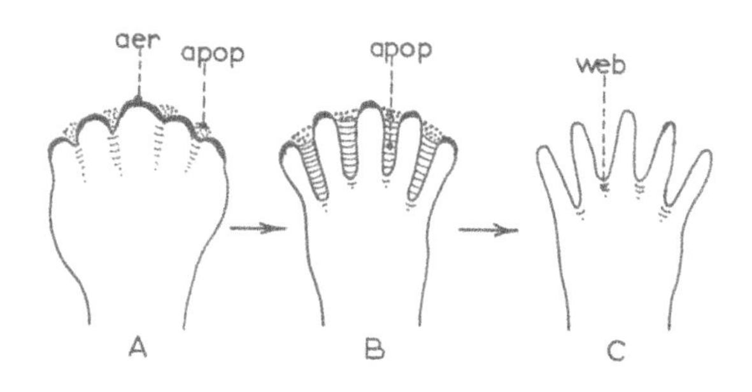

Figure 3.187: Formation of the digits. aer = apical ectodermal ridge; apop= apoptosis.

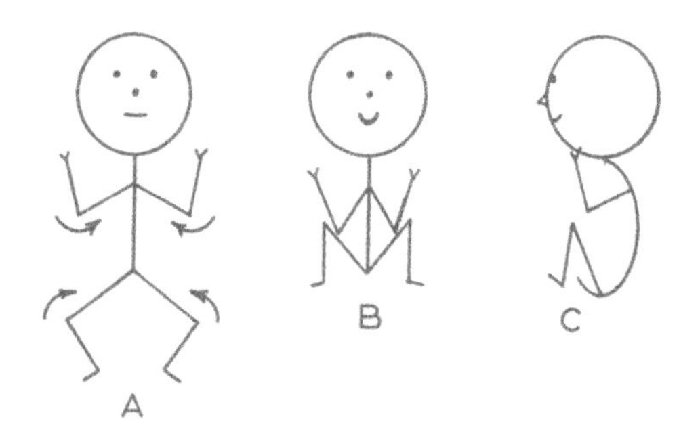

Figure 3.188: Rotation of the limbs to their anatomical positions.

elbows and knees into alignment with the trunk. This is called the **fetal position** in which the elbow points caudally and the knee, cranially (Fig. 3.188A,B,C).

Development of the Internal Structures of the Limbs

The internal structures may be considered as bones, joints, muscles, nerves and blood vessels.

Bones: The limb mesenchyme is at first a homogeneous mass but soon condensations occur in it and these chondrify to form cartilaginous models of the various bones. Each **cartilage model** is surrounded by perichondrium which is a condensation of mesenchyme (Fig. 3.189). An ossific centre (**primary ossification centre**) is formed upon each cartilage model by the ingrowth of osteoblasts (bone-forming cells) from the surrounding mesenchyme (Fig. 3.189). The surrounding mesenchyme is now termed the periosteum (surrounding the bone). Osteoblasts now produce the **bone** which will give rise to the skeletal elements of the limbs. Around the diaphysis of each cartilage model, a 'collar' of ossification forms. This is called the **periosteal collar** which forms by intramembranous ossification (Fig. 3.189). The **'primary ossification centre'** for each long bone forms within the diaphysis (Fig. 3.189). This leaves the bone with two cartilaginous ends (epiphyses). Secondary centres of ossification occur later in the epiphyses (Fig. 3.204). This leaves a plate of cartilage, the **epiphyseal plate**, sandwiched between each epiphysis and the diaphysis. Growth of bone takes place longitudinally, by extension from one or both of the epiphyseal plates. Union of the epiphyses and diaphysis takes place later with disappearance of the epiphyseal plates.

Growth in width of the bone takes place by the circumferential deposition of bone from the osteoblasts in the surrounding periosteum.

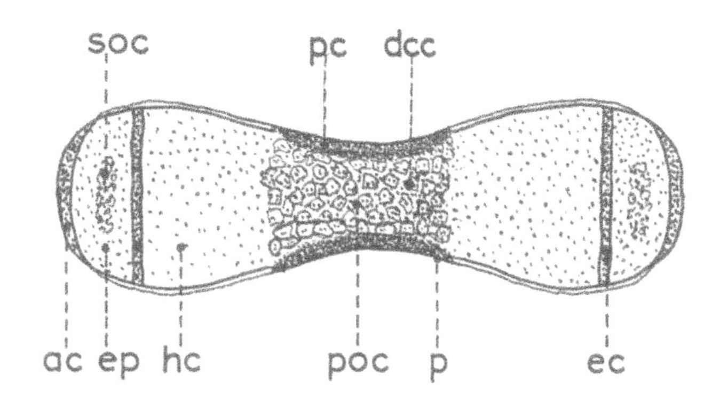

Figure 3.189: The regions of a long bone and its ossification.
ac = articular cartilage; dcc = degenerating cartilage cells; ec = epiphyseal cartilage; ep = epiphysis; hc = hyaline cartilage; p = perichondrium; pc = periosteal collar; poc = primary ossification center; soc = secondary ossification center.

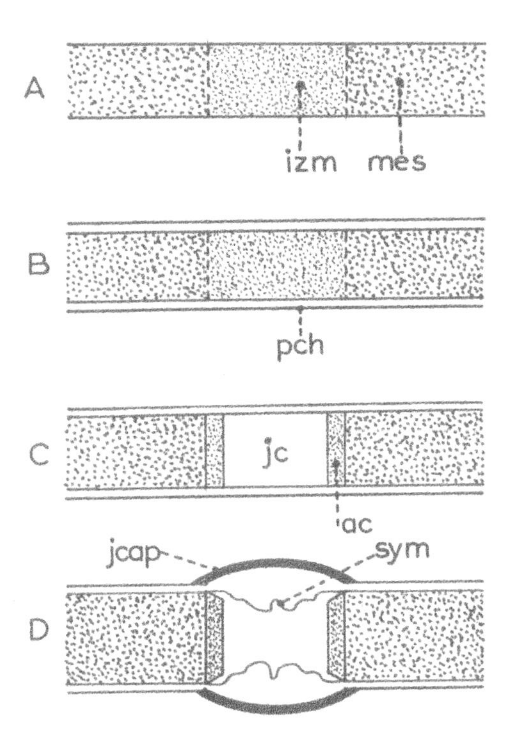

Figure 3.190: The formation of joints.
ac = articular cartilage; izm = interzonal mesoderm; jc = joint cavity; jcap = joint capsule; mes = mesoderm; pch = perichondrium; sym = synovial membrane.

Development of Limb Joints

At places where joints are to occur, cartilage and bone development is interrupted by the formation of a plate of **interzonal mesenchyme**. Depending upon the type of joint to be formed, the interzonal mesenchyme is transformed into **hyaline** or **fibrocartilage** which links the bones.

In the case of the upper limb, all the joints, including those of the shoulder girdle, are **synovial joints** whereas in the lower limb, all the joints, including those of the pelvic girdle, are of the synovial variety, except one, the **symphysis pubis**, which is cartilaginous in nature. In the formation of a synovial joint, the interzonal mesenchyme forms intra-articular structures such as ligaments and menisci (e.g. cruciate ligaments of the knee joint) (**Fig. 3.190A-D**). The central layer of the interzonal mesenchyme temporarily retains its attachment to the surrounding vascular mesenchyme. Proliferation and condensation of mesenchymal cells then seals off the gap between the cartilaginous models to form the **capsule** of the joint. This separates the joint from the surrounding mesenchyme so that the internal vascular structure now becomes the **synovial membrane** of the joint. At this time, the remaining central tissue develops a number of clefts which coalesce to form the joint cavity.

Development of the Limb Musculature

For many years it was believed that the limb muscles differentiated *in situ* from limb mesenchyme. It is now known that **myogenic cells** invade the limb buds from the somites at the 'roots' of the buds. However, it seems that tendons and other connective tissue elements of the muscle (e.g. endomysium, perimysium and epimysium) are formed *in situ* from the limb bud mesenchyme.

Soon after the cartilaginous models of the bones have been established, the myogenic cells, which have now become myoblasts, aggregate to form muscle masses on the ventral and dorsal aspects of the limbs. These muscle masses, in the relevant compartments, form the **flexors** and **extensors** of the joints (**Fig. 3.191**). Rotator muscles are also formed so that flexors and pronators are related and extensors and supinators are related.

Note. Pronation is a form of flexion in bringing a part nearer to the body, while supination is a form of extension in moving a part away from the body.

Ultimately, the muscles and tendons become attached to the bony structures so that they produce their actions across the joints.

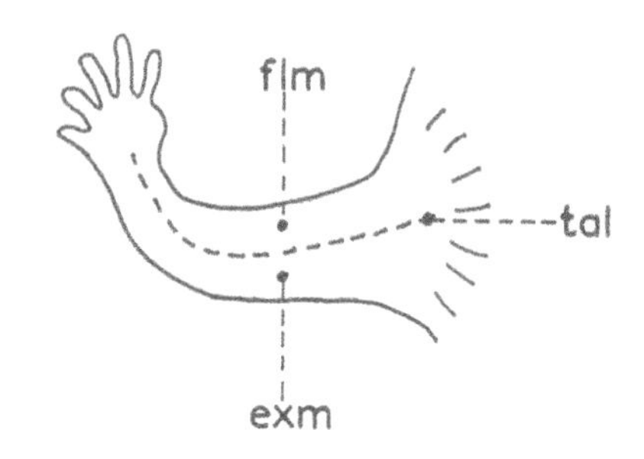

Figure 3.191: Flexor (flm) and extensor (exm) muscles of the limb develop in relation to the transverse axial line (tal).

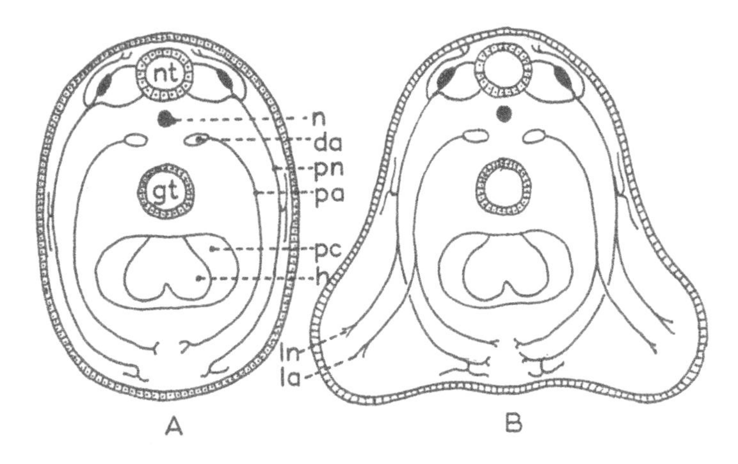

Figure 3.192: Transverse section of an embryo (A) and fetus (B) to show the nerve and blood supply of a limb. da = dorsal aorta; h = heart; la = limb artery; ln = limb nerve; n = notochord; nt = neural tube; pa = peripheral artery; pc = pericardial coelom; pn = peripheral nerve.

Innervation of the Limbs

This may be considered in the light of the construction of a spinal nerve (**Fig. 3.192A,B**). As explained previously, the limbs are supplied by nerves which arise from the segments opposite to which the limb bud originates. The upper limb is supplied by segments C5 to T1, but it should not be forgotten that the tip of the shoulder is supplied by C4 and the upper medial part of the arm is supplied by the lateral cutaneous branch of T2 (intercostobrachial nerve).

Note. Knowledge of the area supplied by C4 is important in sub-diaphragmatic problems and that of T2 in anaesthesia produced by blocking the brachial plexus.

The remainder of the limb is supplied by the branches of the brachial plexus formed by the **anterior primary rami** of the spinal nerves C5 to T1. These rami entering the upper limb must negotiate a narrow opening at the apex of the axilla and to do so, the nerves come together to form a consolidated mass known as the **brachial plexus.** When they have entered the limb, the nerves spread out again into their segmental distribution and since they are mixed nerves (motor and sensory), both muscle and skin are innervated.

In the lower limb, the anterior primary rami emanate from the segments L2 to S3 (lumbosacral plexus). We should not forget however, that the upper part of the thigh is also supplied by L1 which is of importance if it is damaged in the operation for indirect inguinal hernia. If L1 is damaged the area may become anaesthetic or pruritic. In the case of the lower limb, the nerves must find exits from the pelvis. Those nerves from the upper segments (femoral and obturator) in their growth emerge from under the inguinal ligament and through the obturator foramen respectively, while the sciatic nerve from the lower segments escapes via the greater sciatic foramen. These three nerves are responsible for the supply of the muscles of the lower limb and for most of the cutaneous supply.

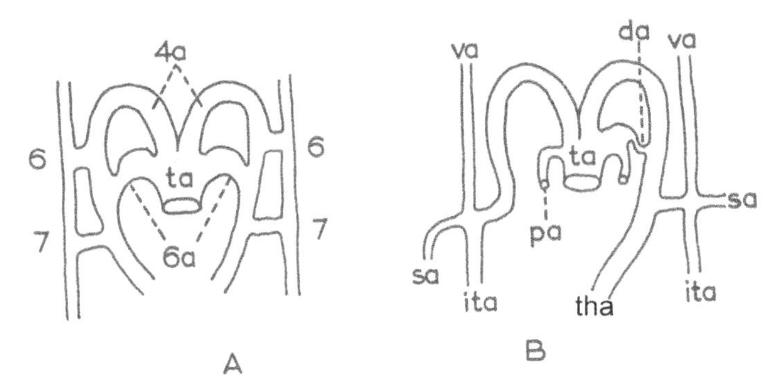

Figure 3.193: Origin of the blood supply to the upper limb. 4a = fourth arch arteries; 6, 7 = intersegmental arteries; 6a = sixth arch arteries; da = ductus arteriosus; ita = internal thoracic artery; pa = pulmonary artery; sa = subclavian artery; ta = truncus arteriosus; tha = thoracic aorta; va = vertebral artery.

Development of the Arteries of the Limbs

As mentioned previously, angioblasts in the blood islands spread through the entire mesoderm of the embryo forming an extensive capillary network. Preferential channels develop between the somites to form the intersegmental arteries. In the cervical region, these are involved in the formation of the vertebral artery; in the thoracic region, branches of these form the posterior intercostal arteries and in the lumbar region, the upper four form the lumbar arteries, while the last forms the common iliac artery.

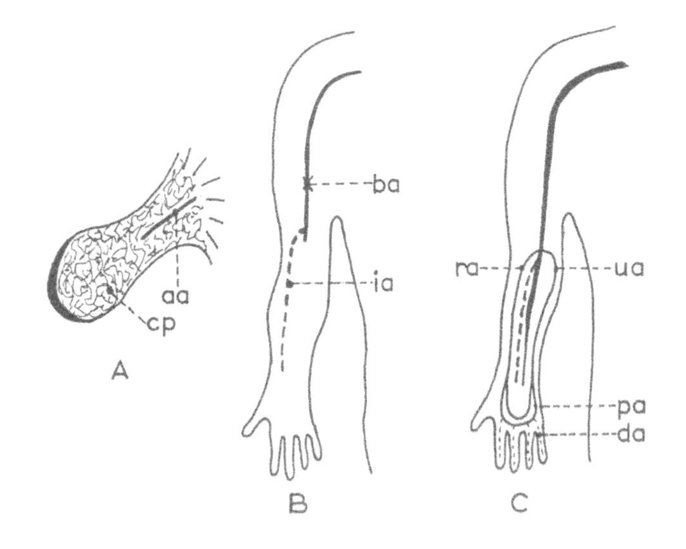

Figure 3.194: Development of the blood supply to the upper limb. aa = axial artery; ba = brachial artery; cp = capillary plexus; da = digital arteries; ia = interosseous artery; pa = palmar arches; ra = radial artery; ua = ulnar artery.

Upper limb: There is a difference in the origin of the major vessel to the upper limb on each side of the body. On both sides of the body the **7th intersegmental artery** is the 'root' artery of the upper limb. On the left side, the 7th intersegmental artery **opens into** the left 4th pharyngeal arch artery (which will become the definitive aortic arch). Thus, the left 7th intersegmental artery forms the **entire left** subclavian artery. On the right side, the right **7th intersegmental artery** taps into the right 4th pharyngeal arch artery which will become the proximal part of the right subclavian artery **(Fig. 3.193A,B)**. The right 7th intersegmental artery thus forms the **distal** part of the right subclavian artery. These subclavian arteries give rise to the **axial artery** of the upper limb bud which is surrounded by a capillary network extending to the distal end of the limb bud (hand). The axial artery consists of a proximal part, the **brachial artery** and a distal part, the **interosseous artery** **(Fig. 3.194A,B,C)**. The **radial** and **ulnar arteries** branch from the brachial and join to form the **arches of the palm,** while preferential channels in the capillary network form the **digital vessels.** Occasionally, a central branch of the brachial artery persists as the **median artery**, running with the median nerve.

Lower limb: The axial artery is an outgrowth of the **5th lumbar intersegmental artery**. It accompanies the sciatic nerve and is thus called the **sciatic artery** (arteria ischiadica). It forms a capillary plexus in the distal part of the limb-bud (foot) **(Fig. 3.195A,B,C)**. As the hip joint develops and undergoes flexion, the sciatic artery, lying dorsal to the joint, is at risk of compression when the fetal position (flexion of hip joint) is adopted. Thus, a ventral artery, the **femoral**

artery, a continuation of the common iliac artery, develops. This artery joins the sciatic artery via the deep femoral network (profunda femoris) which then supplies the dorsal (posterior) aspect of the thigh. The middle part of the sciatic artery then disappears, while the femoral makes contact with the remains of the sciatic artery forming the popliteal artery. This ultimately divides into three branches – the **anterior tibial, posterior tibial** and **fibular** arteries. The digital arteries are formed in preferential channels from the capillary network in the foot.

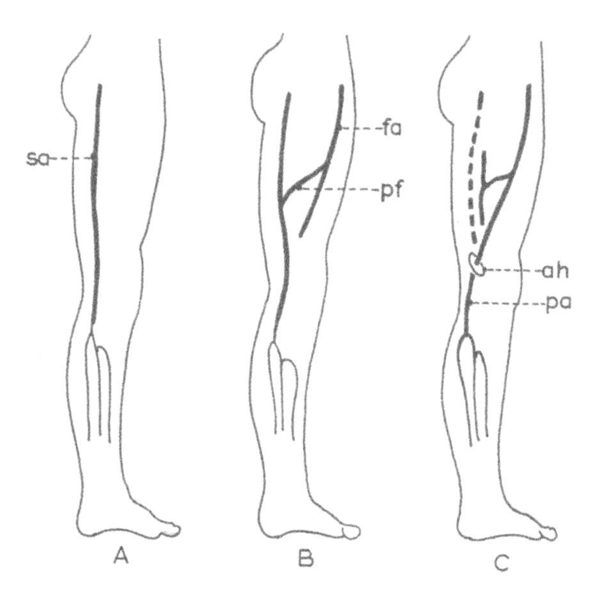

Figure 3.195: Development of the blood supply to the lower limb. ah = adductor hiatus; fa = femoral artery; pa = popliteal artery; pf = profuda femoris arterty; sa = sciatic artery.

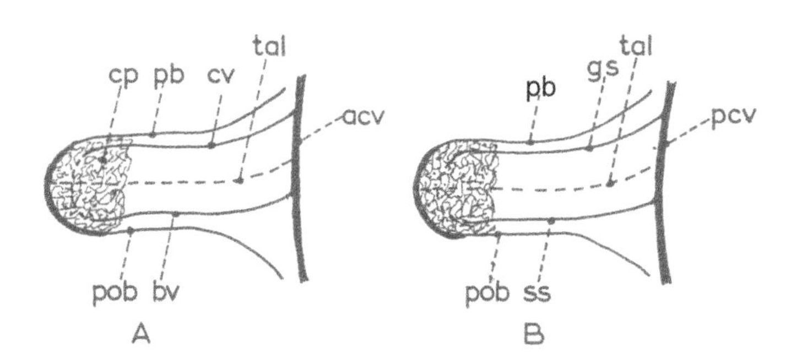

Figure 3.196: Development of the venous system of a limb. Figure A: Upper limb. **Figure B:** Lower limb. acv = anterior cardinal vein; bv = basilic vein; cp = capillary plexus; cv = cephalic vein; gs = great saphenous vein; pb = pre-axial border; pob = postaxial border; ss = small saphenous vein; tal = transverse axial line.

Venous Drainage of the Limbs

The superficial veins of the upper and lower limbs are largely comparable in that the pre-axial border and postaxial border of each limb are drained by a large vein. In the case of the upper limb, the pre-axial border is drained by a vein which will develop into the **cephalic vein** and the postaxial border by the a vein which will develop into the **basilic vein.** Both these veins drain into the **anterior cardinal vein.** In the case of the lower limb, the pre-axial border is drained by a vein which will become the **great saphenous vein** whereas the postaxial border is drained by the **small saphenous vein.** Both these drain into the **posterior cardinal vein** (Fig. 3.196A,B).

Rotation of the Limbs

From the fetal position, the upper limb rotates **laterally** so that the pre-axial border comes into a lateral position and the medial surface of the limb becomes anterior (Fig. 3.197A,B). The radius and the thumb thus assume a lateral position.

The hind limb rotates **medially** so that the pre-axial border comes into a medial position and the medial surface of the limb becomes posterior (Fig. 3.198A,B). The tibia and great toe thus assume a medial position.

The position of the segmental nerves is related to these rotations. In the upper limb, when in the fetal position, the segments are placed from pre-axial to postaxial and after rotation, they are disposed from lateral to medial. The nerve arrangements are not as clear in the lower limb as in the upper limb because of the greater degree of 'twisting' in the lower as compared with the upper limb, to achieve the position suitable for locomotion. A good example is shown by the fact that the nerves from segments L2, 3, and 4 have anterior and posterior divisions which are placed in the pre-axial and post axial compartments respectively, of the limb. After rotation, the posterior divisions come to lie anteriorly as the femoral nerve, while the anterior divisions reach a medial position as the obturator nerve.

Osteogenesis

Osteogenesis refers to the formation of osseous (bony) tissue. Bones are generally derived from somitic mesoderm and in some cases from neural crest cells. In the embryo, bone develops either by **intramembranous** ossification or by **endochondral** ossification:

- **Intramembranous** (within membrane) ossification occurs within a layer (membrane) of mesenchyme.
- **Endochondral** (upon cartilage) ossification takes place on a pre-existing (hyaline) cartilage model, which has formed from mesenchyme.

Osteogenesis (ossification) is a two-phase process:

Phase 1 is when an organic matrix consisting of collagen fibres and ground substance is laid down.

Phase 2 occurs when mineralisation of the organic matrix takes place.

In both intramembranous and endochondral ossification the first bone tissue that appears is **primary** or immature. It is a temporary tissue which is soon replaced by definitive lamellar bone.

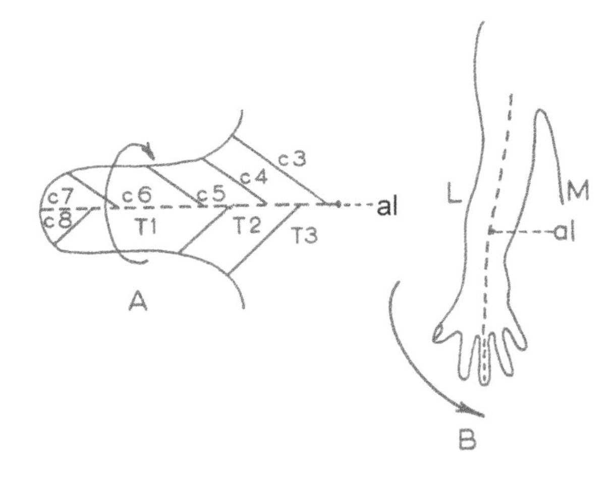

Figure 3.197 A: Rotation and final position of the upper limb indicating the distribution of segmental nerves (C3 – T3) to the limb. **B:** Lateral rotation of the upper limb showing the lateral and medial aspects of the limb with the axial line (al) running in an anterior plane. al = axial line; L and M = lateral and medial borders respectively.

Intramembranous Ossification

Intramembranous ossification is the source of most flat bones. Bones such as the clavicle, frontal, parietal, as well as parts of the occipital and temporal bones, mandible and maxilla are formed in this way. Intramembranous ossification is also responsible for the growth of short bones and thickening of long bones. In this form of ossification a pre-existing cartilage model does ***not*** exist.

Osteogenesis will begin at a **primary ossification centre**. This is usually near the centre of the area to be occupied by the adult bone (**Fig. 3.199A**). Mesenchyme in this area condenses into a richly vascularised layer in which the mesenchymal cells (which will form osteoprogenitor cells) are in contact with one another by long tapering processes (**Fig. 3.199B**). The intercellular spaces are occupied by randomly orientated delicate bundles of collagen fibrils embedded in a thin gel-like ground substance. In sections of developing embryos stained with haematoxylin and eosin, the first appearance of bone development is the appearance of thin strands or bars of denser eosinophilic matrix (**Fig. 3.199C**). This is known as **osteoid** (collagen and ground substance) and refers to the unmineralised bone that has been laid down by **osteoblasts** (which have developed from the osteoprogenitor cells). Thus, prior to calcification the unmineralised bone is known as osteoid. The bars of osteoid tend to be deposited equidistant between neighbouring blood vessels and since

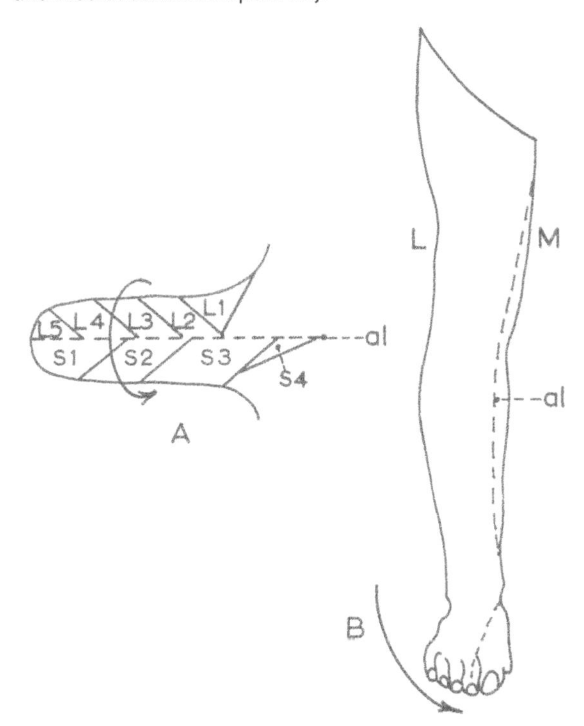

Figure 3.198 A: Rotation and final position of the lower limb indicating distribution of the segmental nerves (L1- S3) to the limb. **B:** Medial rotation of the hind limb. al = axial line.

the blood vessels form a network, the earliest bars of bone matrix also develop in a branching and anastomosing pattern.

The osteoblasts (transformed mesenchymal cells) produce the extracellular matrix (collagen and ground substance) which is able to bind calcium and phosphate salts. These salts are brought to the region through the capillaries of the vascular network. Mineralisation of the osteoid occurs by the deposition of the mineral salts **(Fig 3.199D)**.

Thus, numerous isolated areas of differentiating bone exist in the form of **spicules** (irregularly shaped pieces of bone) **(Fig. 3.199E)**. These will coalesce as differentiation spreads randomly in all directions. The result is a network of spicules of bone which extend radially from the primary ossification centre. The fusion of spicules to form a network of trabeculae (a beam-like structure) gives the bone a spongy nature **(Fig. 3.199F)**. Osteoblasts form a unicellular layer on the surface of this bone. Some of the osteoblasts are entrapped in the osteoid to form osteocytes. The connective tissue which differentiates from mesenchyme remains between the bony trabeculae and is penetrated by blood vessels. Undifferentiated mesenchymal cells give rise to bone marrow cells **(Fig. 3.199F)**.

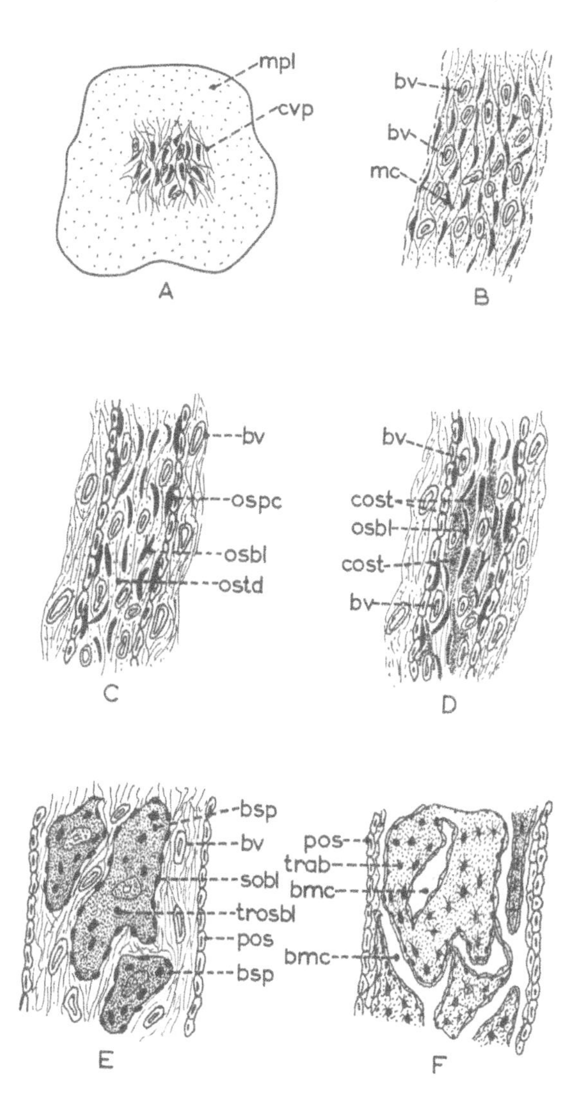

Figure 3.199: Process of intramembranous ossification. bmc = bone marrow cavity; bsp = bone spicule; bv = blood vessel; cost = calcified osteoid; cvp = central vascular proliferation; mc = mesenchymal cell; mpl = mesenchymal plate; osbl = osteoblast; ospc = osteoprogenitor cell; ostd = osteoid; pos = periosteum; sobl = surface osteoblast; trab = trabecular bone; trosbl = trapped osteoblast.

As rapidly as the layer of osteoblasts on the surface of the bones is depleted by their incorporation into the bone to form osteocytes, new osteoblasts are formed from osteoprogenitor cells in the mesenchyme. Mitotic division is frequent in the precursor cells (osteoprogenitor cells) but is rarely observed in the osteoblasts themselves.

That portion of the connective tissue which does not undergo ossification gives rise to the **endosteum** lining the marrow cavities and **periosteum** surrounding the bone. Both endosteum and periosteum have the potential to produce bone as they contain undifferentiated mesenchymal cells (osteoprogenitor cells).

Endochondral Ossification

Endochondral ossification takes place on a piece of hyaline cartilage whose shape resembles a small version of the bone to be formed. Bones at the base of the skull, in the vertebral column, the pelvis and the extremities (for example the long bones) are loosely called 'cartilage bones' because they are formed on cartilage which is then replaced by bone by the process of endochondral ossification.

For ease of description, this form of ossification will be described in a long bone. The shaft of the model is referred to as the diaphysis and the two ends as the future epiphyses (sing. epiphysis). The shaft of the model is surrounded by a condensation of vascular mesenchyme called the **perichondrium**. The perichondrium terminates a short distance from the epiphyses leaving the hyaline cartilaginous ends exposed **(Fig. 3.200)** as joint formation will occur in these regions. The first indication of the formation of a primary centre of ossification is an enlargement of the chondrocytes in the middle of the shaft of the cartilaginous model **(Fig. 3.201A)**. The cytoplasm of the chondrocytes in the diaphysis becomes vacuolated indicating degeneration **(Fig. 3.201B)**. Enlargement of the lacunae of the chondrocytes

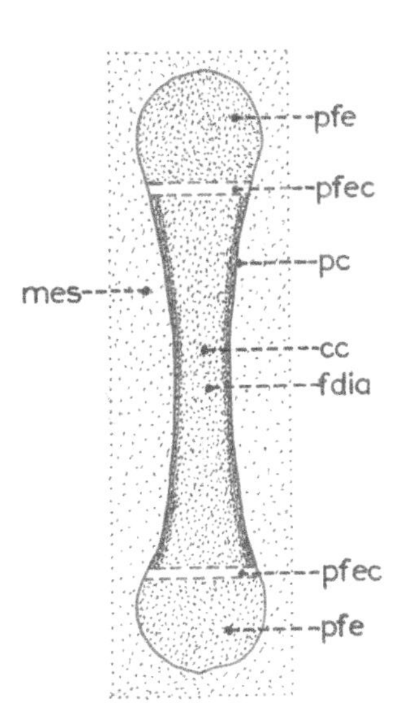

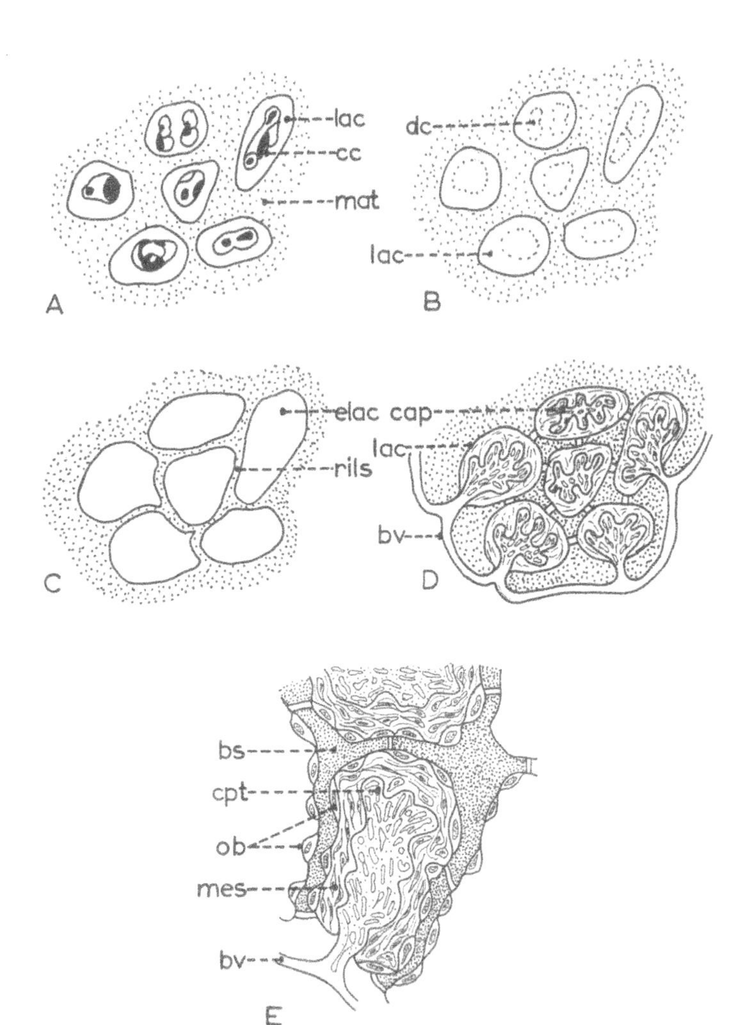

Figure 3.200: A cartilage model indicating the position of the future regions of developing bone. cc = cartilaginous condensation; fdia = future diaphysis; mes = mesenchyme; pfe = position of future epiphysis; pfec = position of future epiphyseal cartilage; pc = perichondrium.

Figure 3.201: Process of endochondral ossification. bs = bone spicule; bv = blood vessel; cap = capillary; cc = cartilage cell; cpt = capillary tuft; dc = degenerating cell; elac = enlarged lacuna; lac = lacuna; mat = matrix; mes = mesenchyme; ob = osteoblasts; rils = reduced interlaminar matrix.

takes place and results in narrowing of the intervening cartilage matrix (**Fig. 3.201C**), which is gradually reduced to thin septae and spicules. This reduced hyaline matrix becomes calcified and the chondrocytes degenerate completely and die. One might pose the question: 'Why do the chondrocytes undergo degeneration and death?' Could it be that the diffusion of nutrients from the perichondrial vascular network is severely reduced? Is it possible that the formation of the periosteal collar (see next paragraph) interrupts the diffusion gradient? This matter has not yet been elucidated fully by embryologists.

While degeneration of the cartilage in the diaphysis is going on, mesenchymal cells in the perichondrium are stimulated to form osteoblasts which secrete osteoid (**Fig. 3.202**). This newly formed osteoid soon becomes mineralised to form a thin layer of **intramembranous** bone around the midportion of the cartilaginous diaphysis. This is the **periosteal collar**. It must be emphasised that the periosteal collar is formed by intramembranous ossification and *not by* endochondral ossification. The enlarged lacunae of the dead chondrocytes in the diaphysis become confluent and these are invaded by blood vessels from the perichondrium which is now called **periosteum**.

After entering the cartilaginous model, the periosteal vessels continue to form branches and

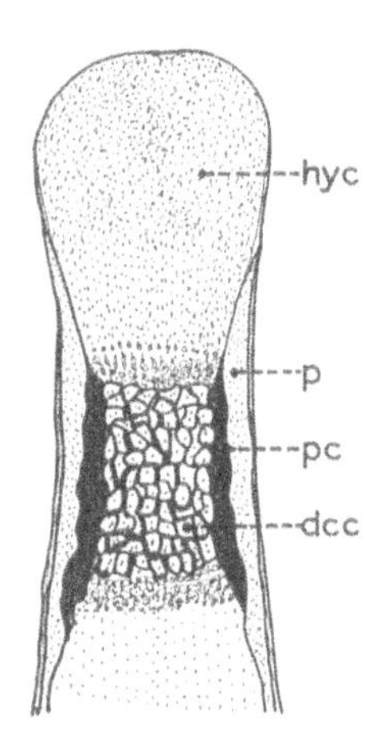

Figure 3.202: Formation of the periosteal collar in relation to endochondral ossification taking place in the diaphysis of a long bone. dcc = degenerating cartilage cells; hyc = hyaline cartilage; p = periosteum; pc = periosteal collar.

penetrate the spaces previously occupied by the chondrocytes (**Fig. 3.201D**). Mesenchymal cells accompanying the vessels transform into osteoblasts and begin to form bone (**Fig. 3.201E**).

The penetrating periosteal vessels grow towards either end of the cartilage model (future epiphyses), forming capillary loops that extend into the blind ends of the lacunae (**Fig. 3.201E**). Mesenchymal cells with differentiating potencies are thus carried into the interior of the cartilage by the perivascular connective tissue. Some of these cells will differentiate into haemopoeietic elements of the bone marrow. Other undifferentiated mesenchymal cells (osteoprogenitor cells), coming into contact with the calcified cartilage, differentiate into osteoblasts. These osteoblasts will align themselves on the outside of the spicules of calcified cartilage matrix and start to secrete osteoid (**Fig. 3.203B**). The earliest trabeculae formed in the diaphysis thus have a core of calcified cartilage covered by a layer of osteoid.

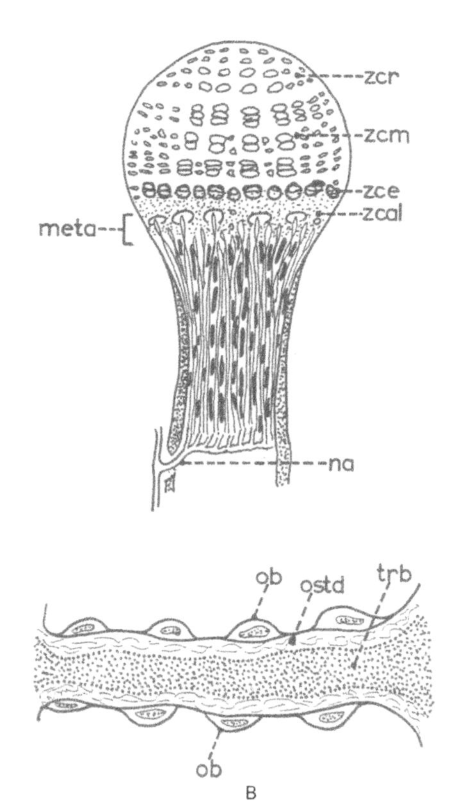

Figure 3.203: Endochondral ossification in the region of the epiphysis and showing the metaphyseal region. na = nutrient artery; ob = osteoblast; ostd = osteoid; meta = metaphysis; trb = trabecula of bone; zcal = zone of calcification; zce = zone of cell enlargement; zcm = zone of cell multiplication; zcr = zone of cell rest.

Following the appearance of the primary ossification centre in the cartilaginous shaft, ossification of the model continues towards the ends.

After the appearance of the primary ossification centre in the diaphysis, the cartilaginous model continues to grow in length and the chondrocytes in the future epiphyses become arranged in longitudinal columns. Microscopic examination of the developing epiphysis at successive stages reveals that fundamentally it goes through the same series of events that have occurred in the shaft, but in a more ordered fashion.

At first, the ends consist of a mass of normal resting chondrocytes, especially near the articular surface. This may be regarded as the **zone of cell rest** (**Fig. 3.203A**).

On the deep surface of this outer zone the chondrocytes multiply to form longitudinal columns. This may be regarded as the **zone of multiplication.**

As these columns elongate, the deeper (older) cells stop proliferating and undergo enlargement. This may be regarded as the **zone of cell enlargement.**

Vacuolation and degeneration of the subsequent chondrocytes leads to enlargement of the lacunae. This reduces the thickness of the cartilaginous matrix between the lacunae which undergoes calcification. This may be regarded as the **zone of calcification.**

Now the lacunae are invaded by perivascular mesenchyme containing osteoprogenitor and haemopoeietic cells. The perivascular mesenchyme originates from around the invading periosteal vessels. Some of the osteoprogenitor cells give rise to osteoblasts, which lay down osteoid, which mineralises to form bone. This may be regarded as the **zone of ossification.** The first bone that is laid down is known as primary or woven bone. This will later be replaced by lamellar or compact bone. The haemopoeietic cells occupy the lacunar spaces and generate cells of the blood series.

The transitional zone between diaphysis and the future epiphysis, where the cartilage is being replaced by advancing bone, is called the **metaphysis** (**Fig. 3.203A**).

Primary centres of ossification appear in the diaphyses of each of the principal bones of the body by the 3rd month pf. Much later, usually following birth, the appearance of **secondary ossification centres** occur in the epiphyses and become vascularised by the ingrowth of blood vessels from the periosteum (**Fig. 3.204**).

Ossification spreads peripherally in all directions until there is replacement of all cartilage by bone except in two places:

(a) Cartilage will remain over the free ends of the epiphyses as articular cartilage.

(b) Cartilage will remain as a plate between each epiphysis and the diaphysis. These are known as the **epiphyseal plates** (Fig. 3.204). Subsequent growth of the long bone is thus possible at the surface of the epiphyseal plate facing the diaphysis following the appearance of the secondary ossification centre. When the epiphyseal plates are replaced by spongy bone, longitudinal growth of the bone ceases. This is known as **closure or union of the epiphyses.**

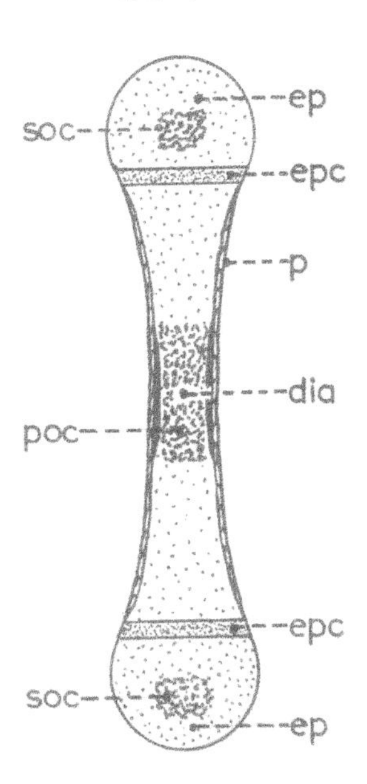

Figure 3.204: Model of a bone to demonstrate the position of secondary ossification centres and the epiphyseal plate. dia = diaphysis; ep = epiphysis; epc = epiphyseal cartilage (plate); p = periosteum; poc = primary ossification centre; soc = secondary ossification centre.

Remodelling of Bone

Remodelling may be defined as the selective deposition and removal of bone in specific places, for example in developing bone and in fractures. This process is performed by osteoblasts and **osteoclasts** respectively.

Note. Osteoclasts are multinucleated giant cells derived from the mononuclear-phagocytic series. Osteoclasts occupy depressions on the side of the bone known as osteoclastic crypts (Howship's lacunae). This remodelling is exemplified in Wolff's Law which states that the shape and architecture of a bone will depend upon the external forces applied to it.

POSSIBLE ABNORMALITIES

- **Achondroplasia** results from a genetic mutation and is characterised by deficiencies in bone formed by endochondral ossification.
- **Osteogenesis imperfecta** is also genetically linked and is characterised by the formation of very thin and delicate bones.

The Development of the Skull

The adult skull may be divided into a number of parts which are joined to one another. The anatomical parts may be classified as follows:

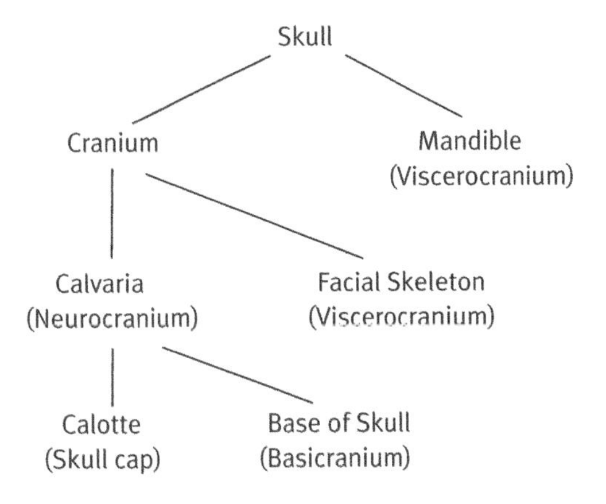

The major components of the cranium are the **neurocranium** and the **viscerocranium.** These two major components are developed from different primordia. The neurocranium consists of 'membrane' bones (**desmocranium**) and 'cartilage' bones (**chondrocranium or basicranium**). The desmocranium is developed by intramembranous ossification, the basicranium by endochondral ossification and the viscerocranium by intramembranous ossification (that is, in mesenchyme derived largely from neural crest cells) and partly endochondral ossification.

Note. A viscus is an 'organ' and it follows, therefore, that the viscerocranium contains a number of structures which may be classed as organs. These are: the lacrimal glands, the eyeballs, the nasal conchae, nasolacrimal ducts, paranasal sinuses, teeth, gums, palate, tongue and salivary glands. On the other hand, the neurocranium contains the brain and the organs of equilibrium and hearing, the vestibulo-cochlear organs.

Recall that the notochord forms at an early stage from the primitive pit mesoderm and extends cranially to the prechordal plate. With the longitudinal flexion (folding) of the embryo, the prechordal plate is angled ventrally to form the oropharyngeal membrane. Just cranial to this point the hypophysis cerebri (pituitary gland) develops and the cranial end of the notochord lies immediately caudal to it (**Fig. 3.205A**). In the adult, this position is in the dorsum sellae of the sphenoid bone.

At this time the neural tube is forming, lying dorsal to the notochord and surrounded by mesenchyme from the paraxial band of mesoderm. It is thus evident that the bones of the skull will form in the mesenchyme surrounding that region of the neural tube.

Development of the Neurocranium

Since the mesenchyme forms a 'membrane' around the neural tube and also forms the 'dermis' of the skin on the dorsal aspect of the tube, the bones which form in the dorsal mesenchyme are called 'membrane' or 'dermal' bones. Collectively these bones are said to form the 'desmocranium' which really is the vault of the skull or the neurocranium.

On the interior of the developing skull bones, the mesoderm, which is intermingled with neural

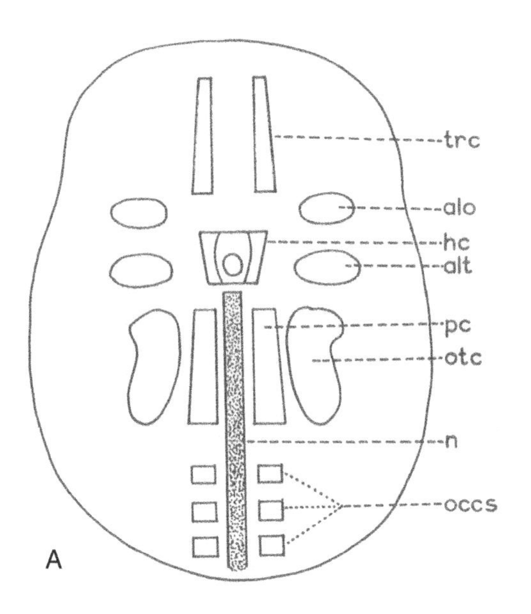

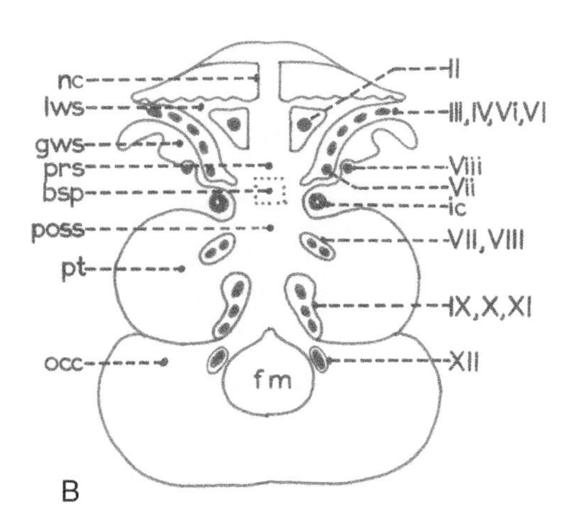

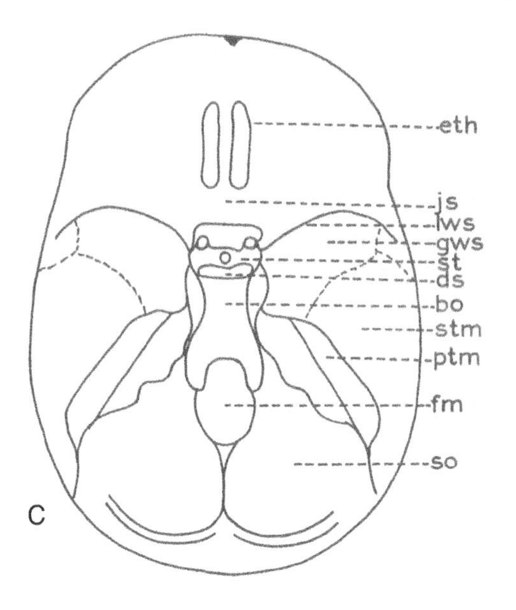

Figure 3.205: Development of the base of the skull. **A:** Cartilaginous centres. **B:** Spread of cartilage and resulting foramina. **C:** Adult, indicating bones of skull. I-XII = cranial nerves passing through specific foramina; alo = ala orbitale (orbitosphenoid); alt = ala temporale (alisphenoid); bo = basi-occiput; bsp = basisphenoid; ds = dorsum sellae; eth = ethmoid; fm = foramen magnum; gws = greater wing of sphenoid; hc = hypophyseal cartilage; ic = internal carotid artery; js = jugum sphenoidale; lws = lesser wing of sphenoid; n = notochord; nc = nasal capsule; occ = occipital cartilage; occs = occipital somites; otc = otic capsule; pc = parachordal cartilages; poss = post-sphenoid; prs = presphenoid; pt = petrous temporal cartilage; ptm = petrous temporal bone; so =squamous occipital bone; st = sella turcica; stm = squamous temporal bone; trc = trabeculae cranii.

crest cells, is called the **primitive meninx.** It later divides into outer and inner layers which are the **ectomeninx** and **endomeninx** respectively. Each of these layers again divides into two, the ectomeninx forming the **dura mater** and the so-called **superficial membrane,** the cells of which have osteogenic potential. At about the 8th week pf, this membrane undergoes ossification to form the bones of the vault of the skull, namely the frontal, parietal, membranous (squamous portion) occipital and the squamous temporal bones. The inner layer of the superficial membrane forms the periosteum (endocranium) of the inner surface of the skull.

Since the dura mater and endocranium arise from the same membrane, they are firmly attached to one another. Despite this, they are split apart at certain places to contain the intracranial venous sinuses.

Development of the Basicranium

The bones of the cranial base are developed on a series of independent cartilages by endochondral ossification. These cartilages surround the cranial part of the notochord and extend rostrally to meet the developing viscerocranium and laterally to make contact with the elements of the desmocranium.

Chondrification of the cranial base begins during the 2nd month pf with the appearance of the **parachordal cartilages** on either side of the notochord (Fig. 3.205A).

The parachordal cartilages fuse around the notochord to form the clivus of the basi-occiput bone. The clivus extends from the anterior border of the foramen magnum to the level of the cranial end of the notochord. In the adult cranium, the remnant of the 'vertebral' notochord extends through the body of the axis vertebra into the odontoid process (dens) and on reaching the end of the dens, forms the apical ligament of the dens. This is the region in which a **craniopharyngioma** (a malignant tumour of the notochord) may form. The apical ligament is attached to the anterior margin of the foramen magnum. From there the ligament passes through the clivus to end in the dorsum sellae of the sphenoid bone.

On either side of the parachordal cartilages are the **otic capsules** (Fig. 3.205A) which will form the petrous parts of the temporal bones and will envelope the middle ear and inner ear (vestibulocochlear apparatus) as well as part of the internal carotid artery.

At this point, study Figure 3.205C and a prepared skull from which the calotte has been removed. Note particularly the 'central' elements of the base – the **body of the sphenoid bone** forming the 'central pole' of the base. Identify the curved **lesser wings of the sphenoid** joined by the jugum and the hollowed-out greater wings lying inferior to the lesser wings. Anterior to the jugum sphenoidale, identify the cribriform plates with the perpendicular plate of the ethmoid bone, surmounted by the crista galli, between them.

At the junction of the body of the sphenoid bone and the clivus of the occipital bone, a plate of cartilage persists after ossification of the bones. This is the **spheno-occipital synchondrosis** and is responsible for the lengthening (growth) of the skull base. This represents a type of epiphyseal plate. The cartilage disappears at about the age of 25 years. The sphenoid and basi-oocipital bones then unite and this terminates the growth of the base of the skull.

The sphenoid bone is developed from rostral and dorsal components. The cartilages (Fig. 3.205A) which will form the anterior and posterior parts of the body of the sphenoid are bilateral (as are all other skull cartilages) and they fuse across the midline. The **trabecular** cartilages fuse to form the **presphenoid** cartilage, while the **polar** cartilages fuse to form the **post-sphenoid** cartilage (Fig. 3.205B). Lateral to the polar cartilages, two other bilateral cartilaginous condensations occur (Fig. 3.205A). The rostral of the two is the **orbitosphenoid cartilage** which forms the lesser wing of the sphenoid, and the dorsal of the two is the **alisphenoid cartilage** which forms the greater wing of the sphenoid bone. The presphenoid cartilage will extend rostrally to fuse with cartilage around the nasal pits. The most rostral part of this cartilage forms the nasal bones.

The orbitosphenoid cartilages send medial extensions to form the anterior clinoid processes and to enclose the optic nerves, thus forming the optic canals (Fig. 3.205B). Between the orbitosphenoid and alisphenoid cartilages (that is, between the lesser and greater wings of the sphenoid) a slit is formed which gives access to the orbit. This is the **superior orbital fissure** through which pass the ophthalmic, oculomotor, trochlear and abducent nerves, as well as the superior and inferior ophthalmic veins. More dorsally the alisphenoid cartilages surround the maxillary nerves, the mandibular nerves and the middle meningeal arteries forming the **foramina rotundum, ovale** (Fig. 3.205B) and **spinosum** respectively. The internal carotid arteries emerge into the interior of the skull through openings which are surrounded by the body of the sphenoid, the apices of the petrous temporal bones and the basi-occiput. This aperture is the internal opening of the carotid canal (Fig. 3.205B). On the medial sides of the otic capsules, the facial and vestibulo-cochlear nerves enter the capsules through the internal auditory meatus of each side (Fig. 3.205B).

Between the parachordal cartilages and the otic capsules large spaces give rise to the **jugular foramen** on each side, through which pass the internal jugular veins together with the glosso-pharyngeal, vagus and accessory nerves (Fig. 3.205B).

The sclerotomes of the occipital somites extend around the neural tube to form the occupital bone surrounding the **foramen magnum**, and the occipital condyles (Fig.3.205B). The occipital myotomes are said to migrate forwards to form the intrinsic muscles of the tongue. In doing so, they take with them their motor supply (hypoglossal nerve) which leaves the skull through the **hypoglossal canal** situated in the condylar mass (Fig. 3.205B).

Ossification of the cartilaginous basicranium begins in the region of the parachordal cartilages and proceeds rostrally. Note that foramina will occur from ossification surrounding pre-existing nerves.

Development of the Viscerocranium

This is intimately associated with the development of the face. The viscerocranium is formed of the bars of cartilage of the first two pharyngeal arches. These mesodermal arches have been extensively infiltrated with neural crest cells and their cartilaginous precursors are thus ectomesenchymal in origin. Cartilaginous condensations occur in the ectomesenchyme to form Meckel's cartilage in the mandibular (1st) arch, and Reichert's cartilage in the hyoid (2nd) arch. Both these cartilages are mostly replaced by membranous bone.

At a very early stage of viscerocranial and facial development the mandibular arch splits horizontally, giving rise to a maxillary prominence. As the maxilla increases in size it forms part of the floor of the orbit and then extends medially to meet the nasal bones, the ethmoid bones and the frontal bones. In other words, it links up with an area of the basicranium which is known as the interorbitonasal cartilage.

Growth of the Skull

Since the skull houses the brain as well as a number of other organs such as those of the special senses, its growth must keep pace with the enlargement of these organs. Up to the age of about 25 years the growth of the base is at the spheno-occipital synchondrosis. The growth of the neurocranium and viscerocranium is by the deposition of bone by 'moulding'.

Note. Moulding is the term applied to the deposition of bone by osteoblasts and the removal of bone by osteoclasts; e.g. enlargement of the vault occurs by deposition of bone on the external surface (osteoblasts) and removal of bone from the internal surface (osteoclasts).

Diploë

At birth the vault consists of one layer of bone. At about the fourth year, a diploë appears within the bones creating outer and inner tables of compact bone. The diploic layer is the bone marrow and trabecular bone sandwiched between the tables of compact bone.

Sutural Growth

On the outside of the developing bones of the skull the condensed mesenchyme forms the outer periosteum (pericranium) which consists of two layers. These are an outer fibrous layer and an inner cellular (osteogenic) layer. On the inner surface of the skull a reversed situation occurs, and the double layered periosteum is referred to as the endocranium. The outer layer of the endocranium is the cellular (osteogenic) layer and the inner layer is a fibrous layer to which the dura mater is attached (Fig. 3.206A).

Where 'gaps' exist between bones, the endocranium and pericranium on the inside and the outside of the bone respectively, will pass into the gap forming a sutural ligament. This ligament comprises a central fibrous band on either side of which is an osteogenic membrane (Fig. 3.206B). The osteogenic layers between the bone ends will form bone which 'pushes' the bone edges away from one another during growth of the bones.

Sutures 'close' from inside to outside at a fairly advanced adult age.

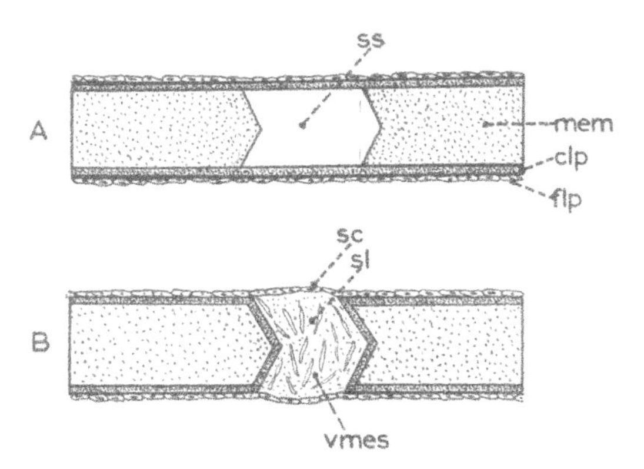

Figure 3.206: Formation of a suture. clp = cellular layer of periosteum; flp = fibrous layer of periosteum; mem = membrane; sc = sutural capsule; sl = sutural ligament; ss = sutural space; vmes = vascular mesenchyme.

Fontanelles

In particular places in the skull during development, the bones are widely separated forming fontanelles. The two most important fontanelles are the anterior and posterior which are 'exaggerated' sutures. These allow for 'moulding' of the vault bones during passage of the fetus through the birth canal. The fontanelles begin to close soon after birth and the procedure is completed in the case of the anterior fontanelle by 24 months and in the case of the posterior fontanelle in three months. Developmentally the frontal bones, like all other bones of the skull, are bilateral. If the two parts of the frontal bone do not unite, a **metopic suture** will result.

POSSIBLE COMMON ABNORMALITIES

- **Acrania** results from the lack of development of the neurocranium. This is often associated with meroencephaly (anencephaly).
- **Craniosynostosis** results from premature closure of the sutures. **Scaphocephaly** is caused by the early closure of the sagittal suture, while **oxycephaly** results from the early closure of the coronal suture.
- **Microcephaly** (small skull) is not caused by early synostosis, but rather by an underdeveloped brain.

The Development of the Temporomandibular Joint

The temporomandibular joint of the adult consists of the head of the condyle of the mandible articulating with the glenoid fossa of the temporal bone. These two articulating surfaces are separated by a fibrocartilaginous disc. The joint is a synovial joint and differs from other synovial joints in the following respects:

(a) It develops from two blastemata (blastema – centre of differentiation): a condylar blastema and a temporal blastema (other synovial joints develop from one blastema).
(b) The joint begins to form a joint cavity at the 9th week pf (other synovial joints have completed their cavity formation by the 7th week pf).
(c) It has fibrocartilage rather than hyaline cartilage on its articular surfaces.

The condylar and temporal blastemata appear as highly condensed mesenchymal masses dorsal to the developing mandible. The condylar blastema appears first, at the 7th week pf, and is superior and lateral to Meckel's Cartilage. The temporal blastema is situated superior to the condylar blastema, and appears at the eighth week. (At this stage the length of the entire mandibular skeleton is ~ 0.6mm.)

The mesenchymal blastemata soon differentiate into the condylar cartilages and the adjacent temporal bony elements, comprising the glenoid fossa and its surrounding bone. The condylar cartilage, one of the accessory cartilages, is wedge shaped and its dorsal end is capped by a condensed layer of fibrous tissue (Fig 3.207B). The bony temporal elements, which are also covered by fibrous tissue, are situated anterior to the malleus and incus.

While the cartilage and bony elements are forming, the mesenchyme between the two condenses into the articular disc of the joint and the mesenchyme surrounding all the parts of the joint condenses into a capsule. This capsular condensation progressively isolates the joint with its developing synovial membrane from the surrounding tissues. An inferior joint cavity forms at the 8-9th week pf, two weeks before the formation of the superior joint cavity, which occurs by the 11-12th week (Fig. 3.207). By the 15th week pf, the joint cavities are fully formed. Joint cavity formation is said to be induced by 'mouth opening reflexes'. These reflexes may occur as early as weeks 8-9 pf.

The tissue forming the articular disc is continuous with the tendon of the lateral pterygoid muscle. In the 12th week pf, endochondral ossification of the condylar cartilage begins and by the 14th week, full differentiation of the temporal bony elements has occurred. By the 16th week complete development of the joint has taken place. At birth the glenoid fossa is flat and there is no articular eminence. Only after eruption of the deciduous dentition does the articular eminence begin to become prominent.

POSSIBLE ABNORMALITY

Ankylosis (bony or fibrous adhesion of joint surfaces) may result from a variety of causes, one of which may be failure of joint cavity formation due to lack of mandibular movement.

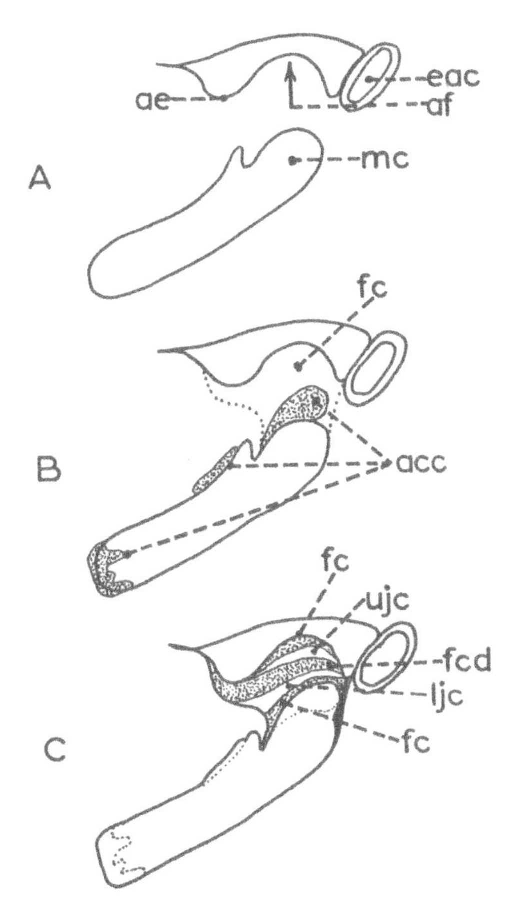

Figure 3.207: Development of the temporomandibular joint. acc = accessory cartilages; ae = articular eminence; af = articular fossa; eac = external auditory canal; fc = fibrocartilage; fcd = fibrocartilaginous disc; ljc = lower joint cavity; mc = mandibular cartilage; ujc = upper joint cavity.

The Integument (The Skin and its Appendages)

In the adult, the skin which covers the entire body is the organ which separates the individual from the environment. It consists fundamentally of two elements, the **epidermis** and the **dermis**.

Epidermis

In the embryo, the 'skin' consists of surface ectoderm coupled with some cells which have been derived from the neural crest. At first the ectoderm consists of cuboidal cells which proliferate to form a more flattened superficial layer called the **periderm** (**Fig. 3.208A,B**). Since the embryo and fetus live in a watery environment, it seems possible that the periderm, an outer layer, may be a protective layer.

By about 12 weeks pf the embryonic epidermis has thickened to three layers, the deepest being the **germinal layer** (stratum basale). External to this is the middle or **intermediate layer** which arises from the germinal layer and the existing peridermal layer which is beginning to undergo a hardening and scaling process (**Fig. 3.208C**). By about 30 weeks pf, the intermediate layer has differentiated into its definitive layers. The stratum basale continues to produce 'daughter' cells, which give rise to the intermediate layers and which ascend through the epidermal layer. As they do so they undergo a process of **cornification** (**Fig. 3.208D**), known as **keratinisation**, in which, as the cells reach the surface, they undergo cytological and chemical changes, and then desquamate. The periderm is a feature of embryonic and fetal life and its desquamating cells will contribute to the formation of the **vernix caseosa** which acts as a protective layer against the amniotic fluid.

In the 6th week pf, neural crest **melanocytes** invade the dermis (the underlying mesodermal layer of the skin) and later invade the stratum basale of the epidermis. These cells produce a dark pigment called **melanin** which gives the skin a dark colour (**Fig. 3.208D**). These cells have long **dendritic processes**, which extend between the cells of the epidermis.

Note. The condition of albinism is not due to an absence of melanocytes, but to a metabolic deficiency in the conversion of tyrosine (an amino acid) to melanin.

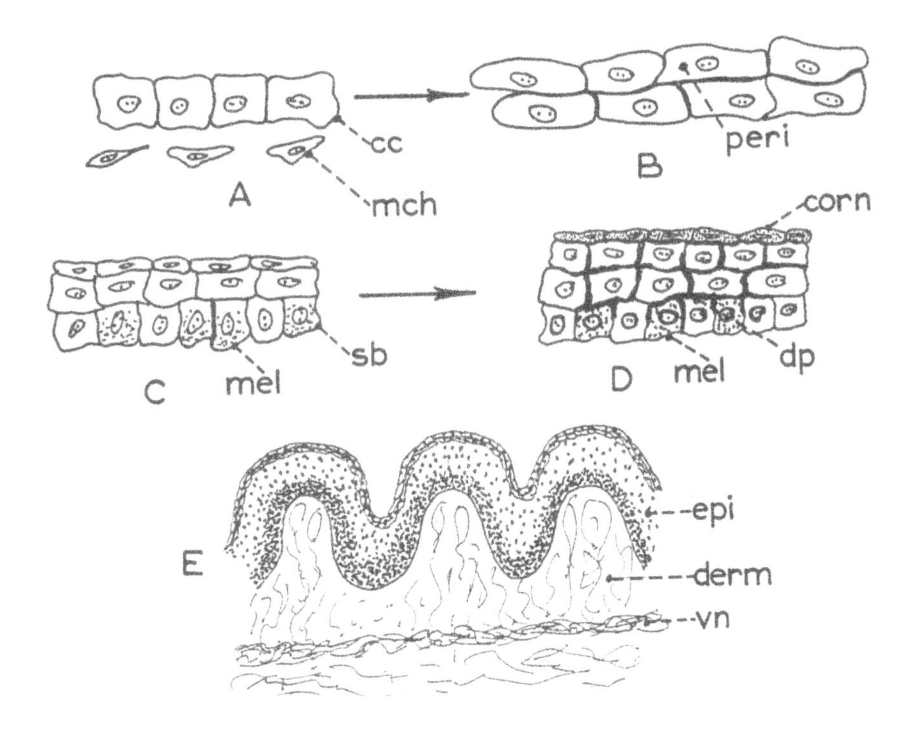

Figure 3.208: Development of the skin. cc = cuboidal cells; corn = keritinisation; derm = dermis; dp = dendritic processes; epi = epidermis; mel = melanocytes; mch = mesodermal cells; peri = periderm; sb = stratum basale; vn = vascular network.

In addition to the acquisition of melanocytes, the basal layer of the epidermis also acquires **tactile epithelial (Merkel) cells** which are located close to free nerve endings. There is some experimental evidence to indicate that the tactile epithelial cells are of neural crest origin. These cells are believed to be **mechanoreceptors** and their attachment to nerve elements supports this view.

The epidermis acquires a third type of cell, the **dendritic (Langerhans) cell.** This cell is said to be associated with the immune activities of the epidermis. The dendritic cells originate in the bone marrow and migrate into the epidermis presumably by first gaining access to blood vessels of the bones and then to dermal blood vessels.

These three cell types (melanocytes, tactile epithelial cells and dendritic cells) are collectively known as *'clear cells'* of the skin.

Dermis

This layer lies immediately deep to the epidermis (**Fig. 3.208E**). It is at first relatively flat but later forms 'bumps and hollows' or interdigitations which are known as papillae and over which the epidermis fits exactly. In the fingers and toes, the epidermal irregularities form finger print ridges on the surface of the epidermis (**Fig. 3.208E**). The dermis is formed partly from dermotomic mesoderm and partly from somatopleuric mesoderm of the lateral plate. Gradually, fat accumulates in the deep part of the dermis resulting in the formation of the **tela subcutanea** (hypodermis, or superficial fascia).

The integument is in itself an organ, and like other organs it contains a group of special smaller organs or appendages. These are, hair, sebaceous glands, sweat glands (eccrine and apocrine), mammary glands and nails. These are dealt with *seriatum*, below.

Hair

During the 3rd month pf, solid rods of epidermal cells penetrate the underlying mesenchyme (dermis) in an oblique direction (**Fig. 3.209A**). These are the **hair buds**. The deepest part of the bud expands, forming the **hair bulb**. At the same time a condensation of the dermal mesenchyme, the **hair papilla,** occurs at the base of the bulb so that the hair bulb grows over it and covers it (**Fig. 3.209B,C**).

The central cells of the hair bud undergo keratinisation to be compressed into the shaft of the hair, while the cells in the bulb proliferate, adding cells to the base of the hair so that it protrudes further and further above the surface of the epidermis (Fig. 3.209D). The melanocytes, which have invaded the epidermis, are also found in the hair follicles and they impart melanin to

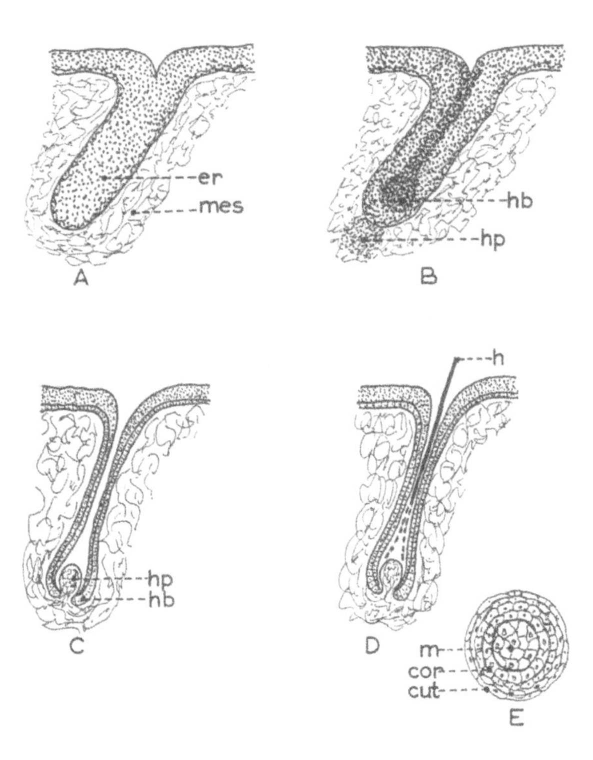

Figure 3.209: Development of hair. A-D: Longitudinal sections. **E:** Transverse section of hair. cor = cortex; cut = cuticle; er = epidermal rod; h = hair; hb = hair bud; hp = hair papilla; m = medulla; mes = mesoderm.

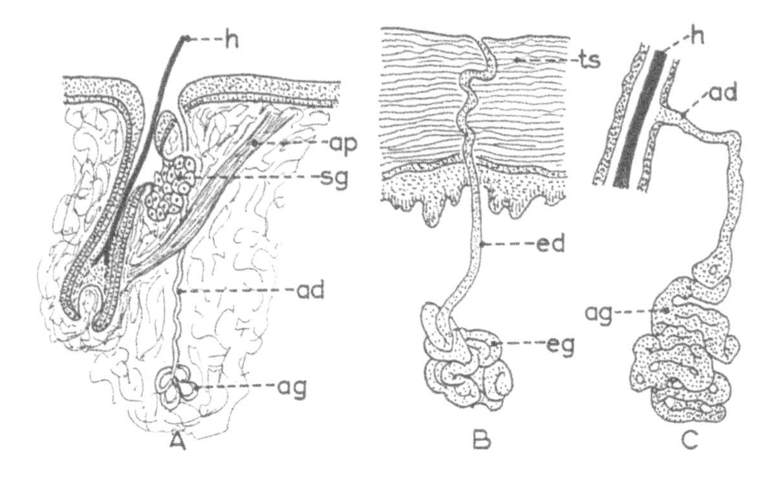

Figure 3.210: Development of the glands of the skin. ad = apocrine duct; ag = apocrine gland; ap = arrector pili muscle; ed = eccrine duct; eg = eccrine gland; h = hair; sg = sebaceous gland; ts = thick skin.

the hair so that its colour may vary from jet black to lighter shades of brown. Gradually, as the hair matures, it has a **medulla** of polyhedral cells separated by air spaces, a **cortex** of elongated cells joined to form fusiform fibres containing melanin and a **cuticle** of a single layer of overlapping flat scales.

The greying of hair is due to the accumulation of air bubbles in the medulla and cortex and to the loss of melanin pigment in the germinal cells of the hair root.

Attached to the root of the hair are a bundle of smooth muscle cells, the **arrectores pilorum**, which differentiate from the mesenchyme of the upper layer of the dermis (Fig. 3.210A). They are supplied by sympathetic nerve fibres and are responsible for the formation of 'goose-pimples'.

At about the 12th week pf, fine hairs known as **lanugo** appear in the region of the forehead and eyebrows. These are shed just before birth.

Sebaceous Glands

Most, but not all, of these glands are associated with hair follicles and lie in the dermis in the angle between the follicle and the arrector pili muscle (Fig. 3.210A). The glands are commonly found in the scalp, face, ear, nose, mouth and anus; they are absent from the thick skin of the palm of the hand and the sole of the foot. Occasionally, the glands open on hairless areas such as the labia minora, glans penis and the angles of the mouth.

The glands develop as a diverticulum from the deep surface of the hair follicle. The duct is fairly wide and leads to a group of acini whose cells consist of foamy cells and whose lumen contains degenerated cells which form part of the secretion. The secreted sebum is said to act as a lubricant for the hair and to keep the skin soft and pliable. These glands are classified as **holocrine glands**.

The secretions of the fetal sebaceous glands coupled with desquamating peridermal cells form a protective layer covering the entire fetus. This layer is known as vernix caseosa.

Sweat (Sudoriferous) Glands

These are normally classified as **eccrine** and **apocrine** glands.

Eccrine glands occur all over the body. They develop first as solid cords of epidermal cells in the skin of the palms and soles at about the 4th month pf. The cords enter the dermis and coil in the deeper layer (Fig. 3.210B,C). The cords then canalise and open on to the epidermal surface at about the 7th month pf. The ducts consist of two layers of cells. A layer of myoepithelial cells surrounds the single layer of acinar cells and aids in the expulsion of secretion by the compression of the acinar cells.

The eccrine glands produce a watery perspiration containing a considerable quantity of sodium chloride and urea, but little protein.

Note. Myoepithelial cells are contractile cells which are related to exocrine secretory tissues (with the exception of the pancreas) and some of their ducts. Myoepithelial cells are thought to be derived from ectoderm.

Apocrine glands: These arise from the same epidermal cores as the hair follicles but are more superficial than the sebaceous glands (Fig.3.210A). The duct of the gland passes deeply into the depths of the dermis where it becomes coiled and may become branched. Myoepithelial cells also surround the acini of these glands. Apocrine glands are found in the axilla, the areola and nipple of the breast, around the anus and in relation to the external genitalia. They produce a somewhat milky proteinaceous secretion. They are fully developed at puberty and are, therefore, responsive to sex hormones.

Mammary Glands

These are thought to be modified sudoriferous (sweat) glands and thus have a type of development similar to that of apocrine sweat glands. At about 5 or 6 weeks pf, epidermal ridges (milk lines) appear on the ventral surface of the embryonic body (Fig. 3.211A). When projected on to the adult body, these lines extend from the axillae to the medial aspect of the thighs (Fig. 3.211B). In lower animals, mammary glands will appear in relation to these lines but in the human embryo only two will arise and these are in the pectoral region. The epidermal ridges disappear in later embryonic life.

In the pectoral regions, solid cords of epidermal cells penetrate the underlying mesenchyme (Fig. 3.212A). These soon develop twelve to twenty radially disposed buds of epidermis (Fig. 3.212B). The cords canalise to form

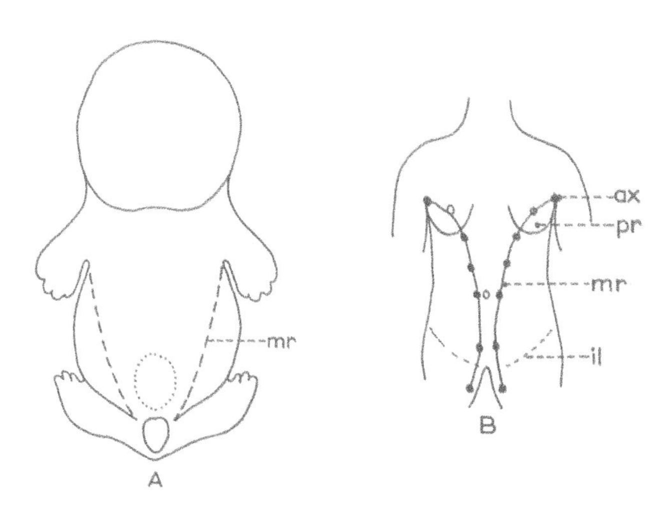

Figure 3.211: Ventral view of an embryo (A) and anterior view of an adult (B) demonstrating the milk line (mr = mammary ridge). In addition, **B** shows the normal position (pr = pectoral) and ectopic positions where breast tissue may occur. ax = axillary; il = inguinal ligament.

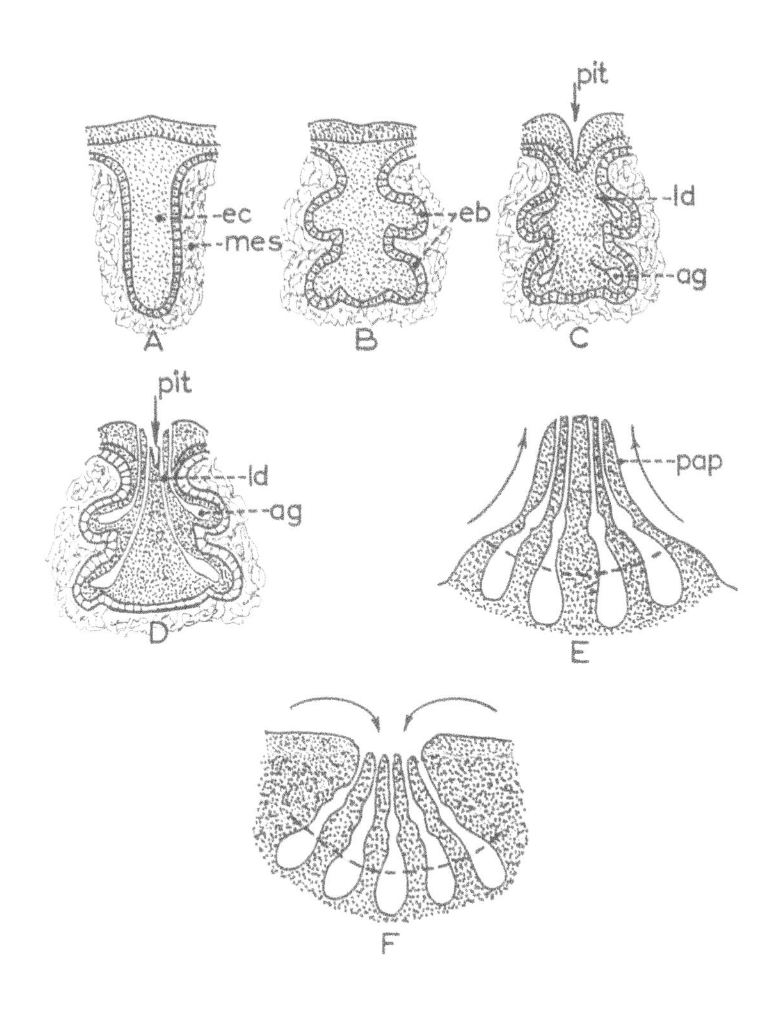

Figure 3.212: Histological development of the mammary gland (breast). Arrows = non-extrusion of nipple resulting in inverted nipple. ag = alveolar gland; eb = ectodernal buds; ec = ectodermal cord; ld = lactiferous duct; mes = mesoderm; pap = papilla (nipple).

lactiferous ducts (Fig. 3.212C) and the buds canalise to form the **alveolar glands** (Fig. 3.212D).

At about 30 weeks pf, after the ducts and glands have completely canalised, a pit forms at the site of the original penetration of the epidermal cord (Fig. 3.212C,D). As the pit deepens, the lactiferous ducts open into it (Fig. 3.212D). After birth, the nipple is formed by eversion of the pit by proliferation of the underlying mesenchyme so that the lactiferous ducts open on the summit of the nipple (Fig. 3.212E). If eversion of the nipple fails to take place, the lactiferous ducts open into separate pits (Fig. 3.212F) and the condition is known as **inverted nipple**.

Nails

At about the 12th week pf, the nails begin to develop as patches of thickened epidermis on the tips of the fingers and toes (Fig. 3.213A,B). These plaques are known as the **primary nail fields**. The proliferation of the epidermis continues, spreading on to the dorsum of the fingers and toes (Fig. 3.213C). Because the epidermal cells of the 'nail fields' grow relatively slowly, the more rapidly proliferating digital epidermis forms a ridge or fold surrounding the 'field'. This fold is called the **nail fold** and is situated at the proximal end and sides of the 'field' (Fig. 3.213D). The proximal part of the fold proliferates to form the **formative zone** (Fig. 3.213C). In the depths of the fold, cellular proliferation takes place and the nail emerges from under the fold. The nail is said to consist of modified **stratum lucidum cells** which first appear from under the nail fold as the **lunule** of the nail, a crescentic pale type of tissue. As the nail grows out, it becomes dessicated and hardened.

At birth, the length of the finger nails has barely reached the end of the fingers and those of the toes, even less.

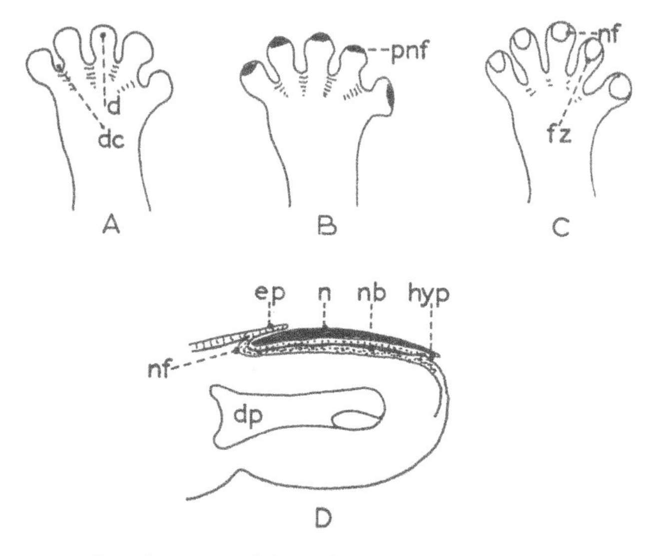

Figure 3.213: Development of the nail. d = digit; dc = digital cleft; dp = distal phalanx; ep = eponychium; fz = formative zone; hyp = hyponychium; n = nail; nf = nail fold; nb = nail bed; pnf = primary nail field.

INDEX

B

C

F

G

H

M

N

O

P

R

S

T

U

Abnormalities of Development

THE CARDIOVASCULAR SYSTEM

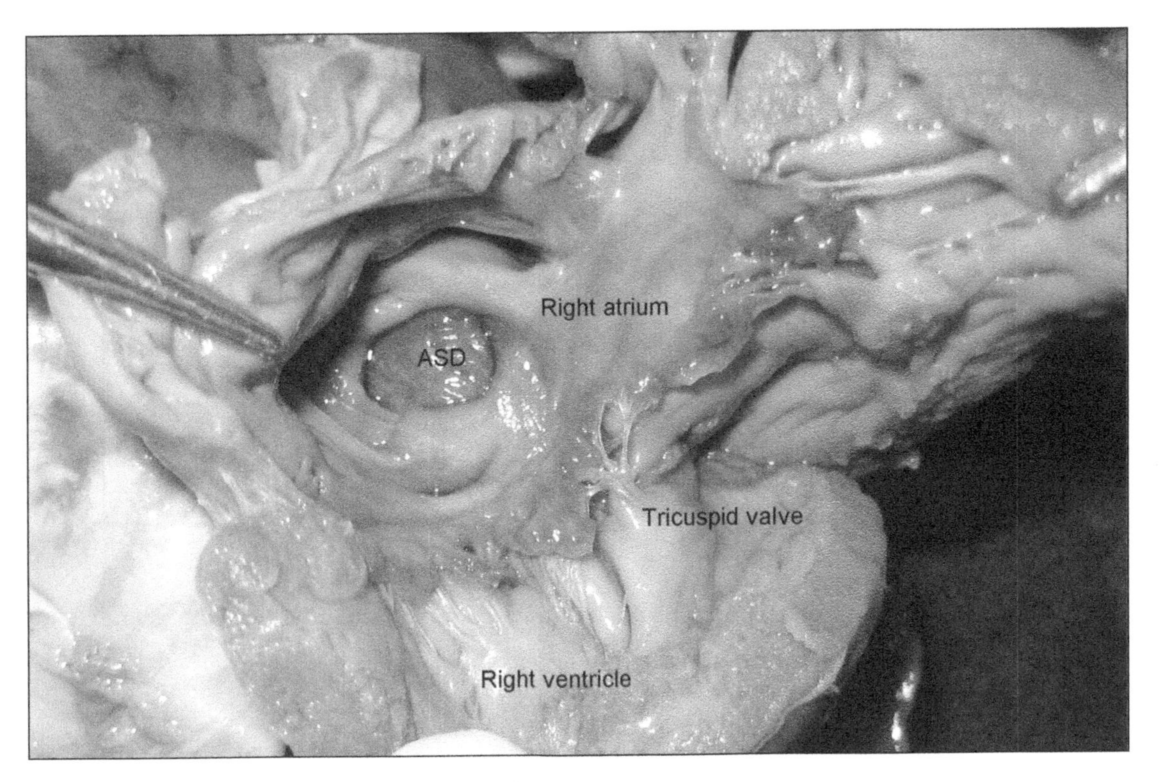

Plate 1 Atrial septal defect. ASD = atrial septal defect.

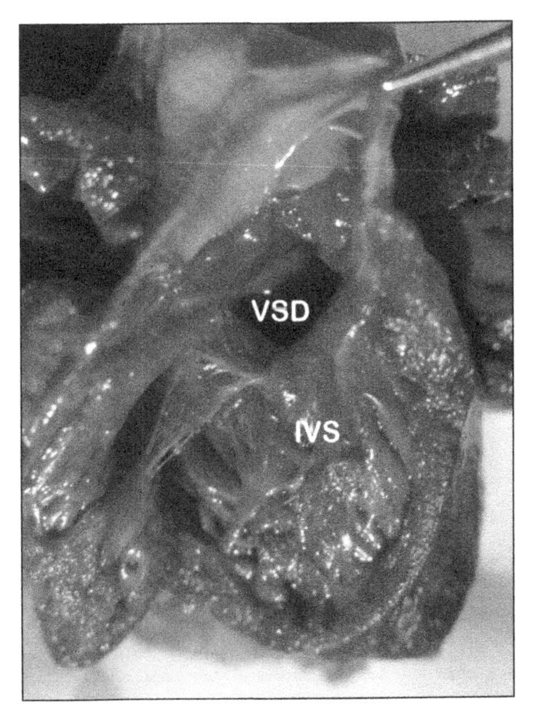

Plate 2 Ventricular septal defect. IVS = interventricular septum; VSD ventricular septal defect.

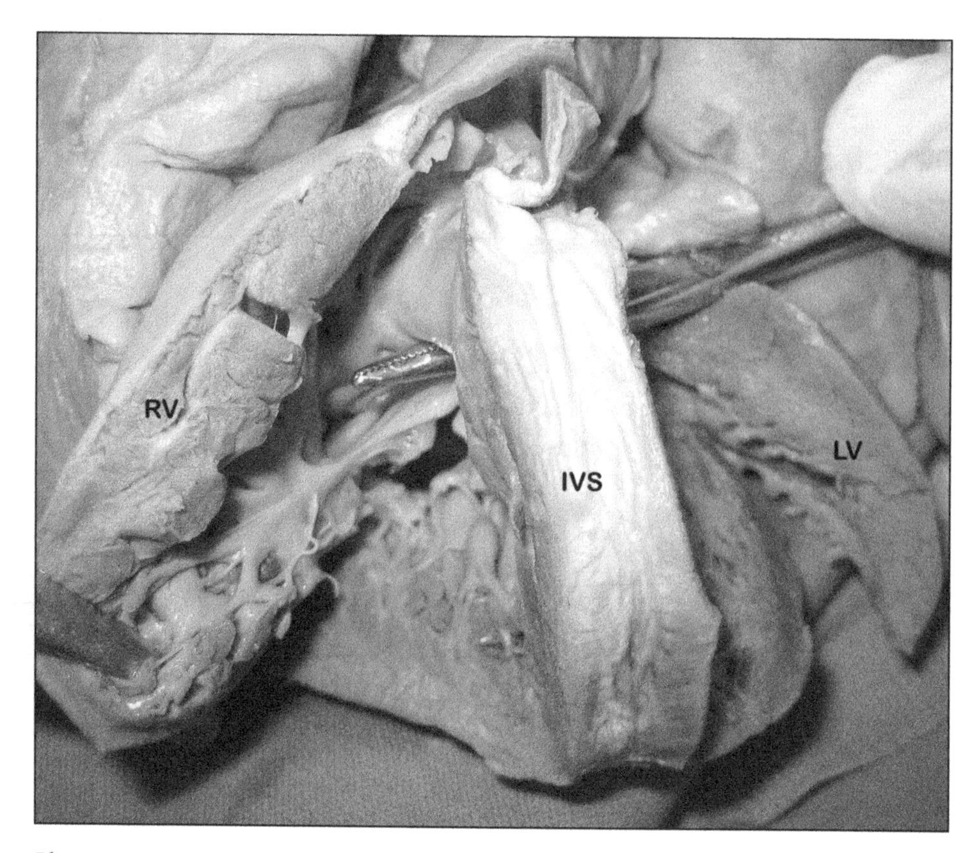

Plate 3 Ventricular septal defect indicated by forceps. RV = right ventricle; IVS = interventricular septum; LV = left ventricle.

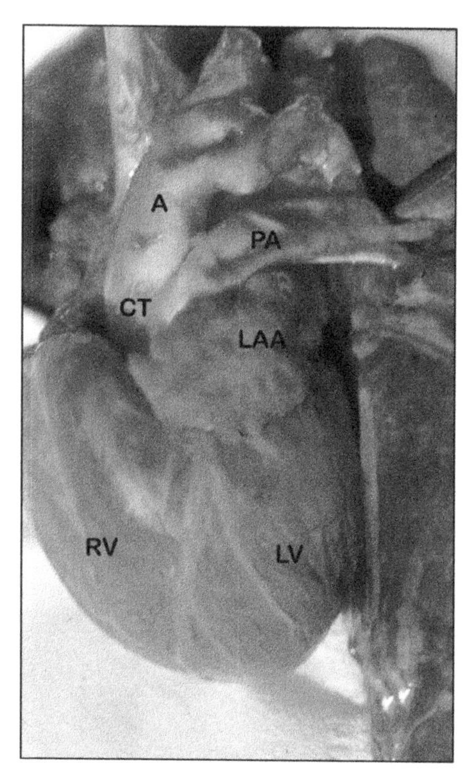

Plate 4 Truncus communis. A = aorta; CT = truncus communis; LAA = left auricular appendage; LV = left ventricle; PA = pulmonary artery; RV = right ventricle.

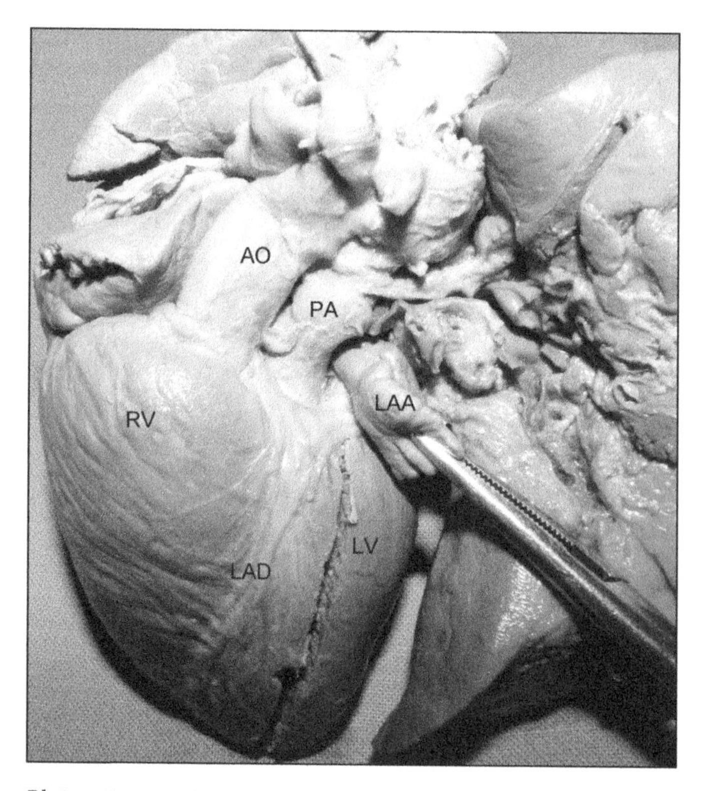

Plate 5 Transposition of the great vessels. AO = ascending aorta; LAA = left auricular appendage; LAD = left interventricular artery; LV = left ventricle; PA = pulmonary trunk; RV = right ventricle.

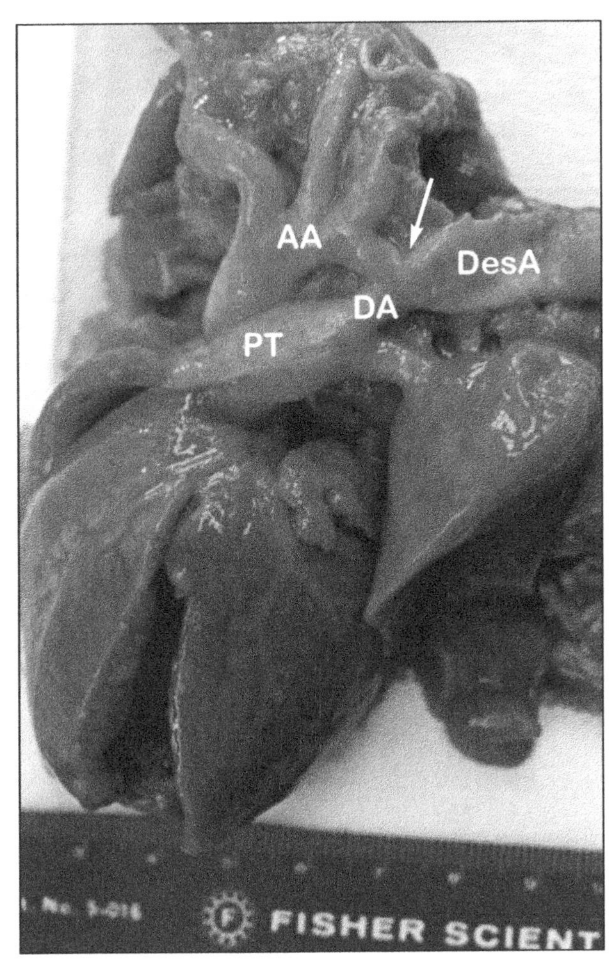

Plate 6 Coarctation of the aorta. AA = aortic arch, DA = ductus arteriosus; PT = pulmonary trunk; DesA = descending aorta pulled to the left side; Arrow = coarctation.

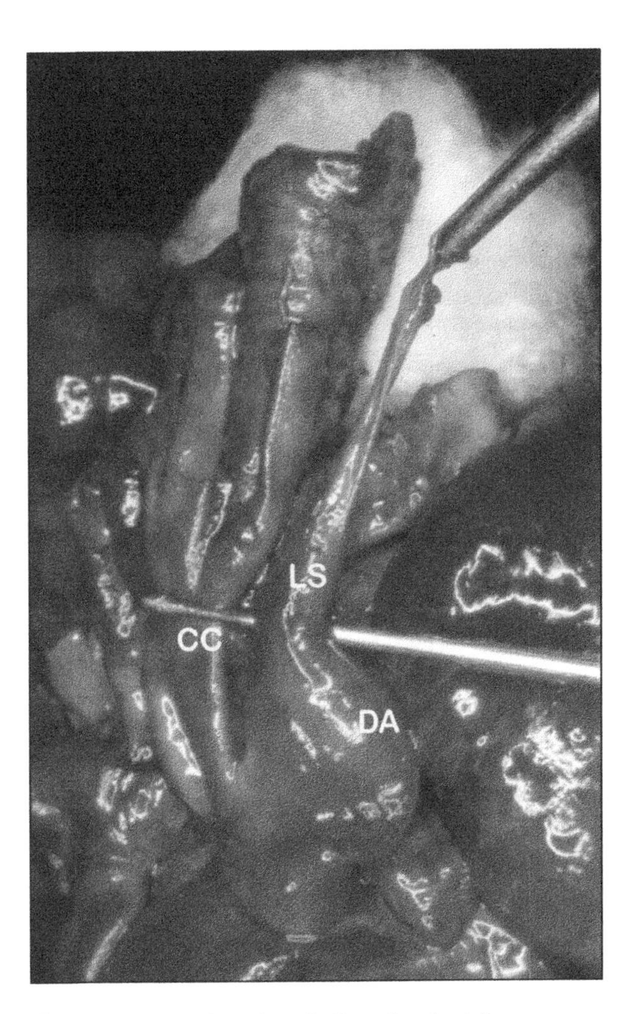

Plate 7 Interrupted aortic arch. Note that the left subclavian artery (LS) takes origin from the descending aorta. CC = common carotid; DA =descending aorta.

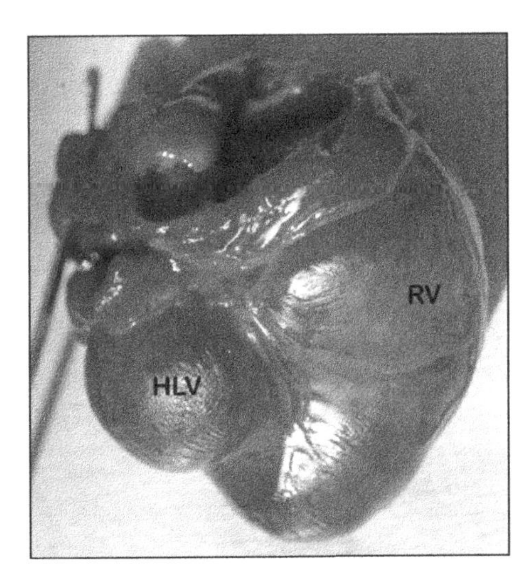

Plate 8 Dorsal view of the heart to show a hypoplastic left ventricle. HLV = hypoplastic left ventricle; RV = normal right ventricle.

The Face

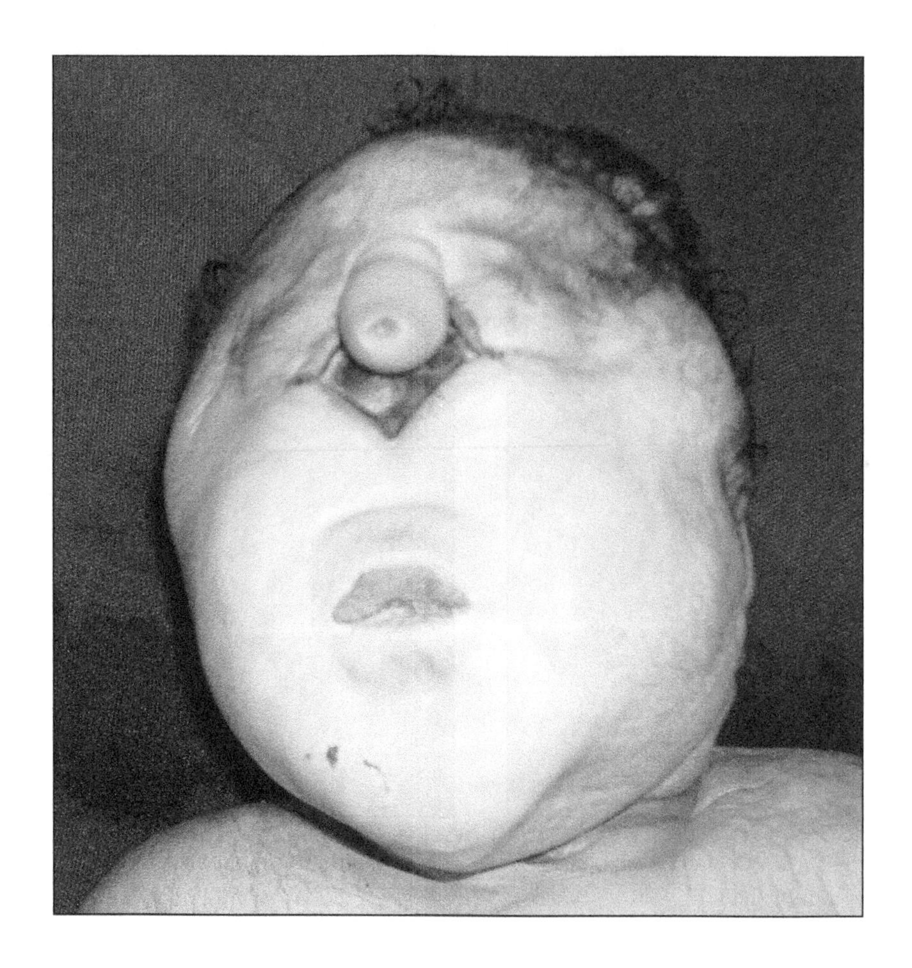

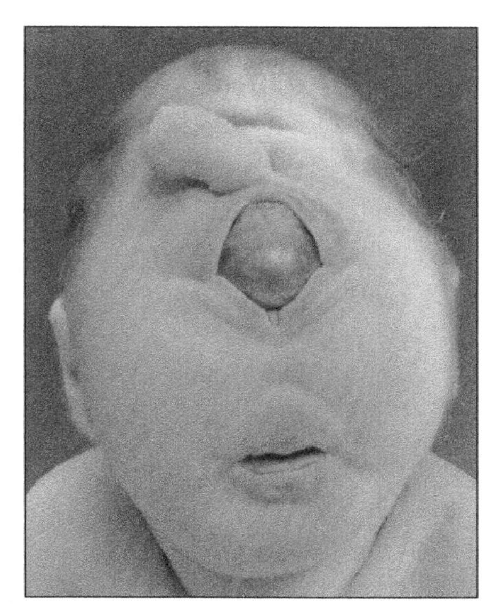

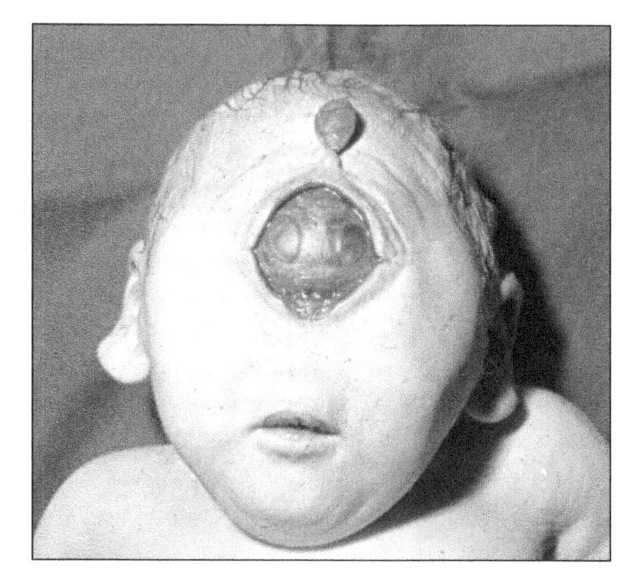

Plate 9a, 9b and 9c View of face indicating the abnormality known as cyclopia. The eyes are fused in the centre of the face, caudal to the nose which exists as a proboscis.

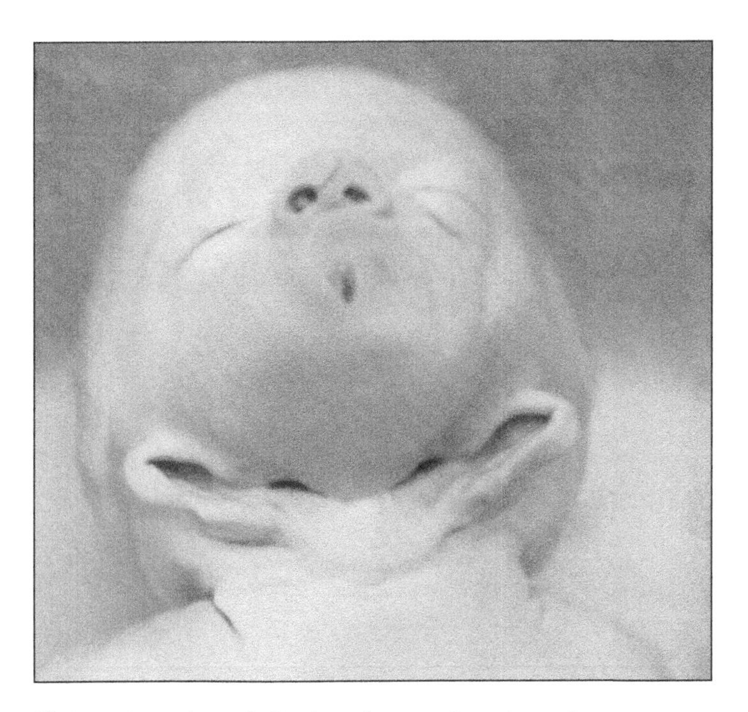

Plate 10 A specimen indicating microstomia and synotia.

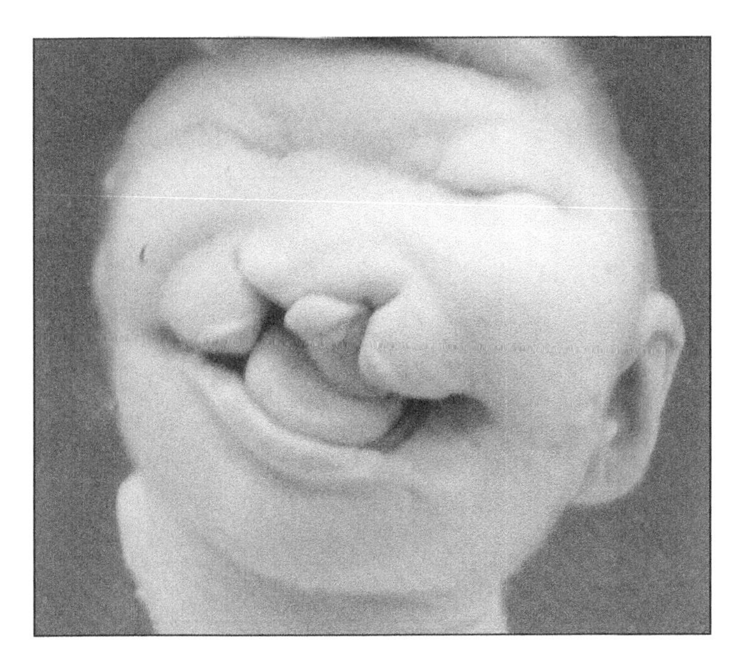

Plate 11 A specimen showing bilateral clefting of the upper lip. Note the low-set ear.

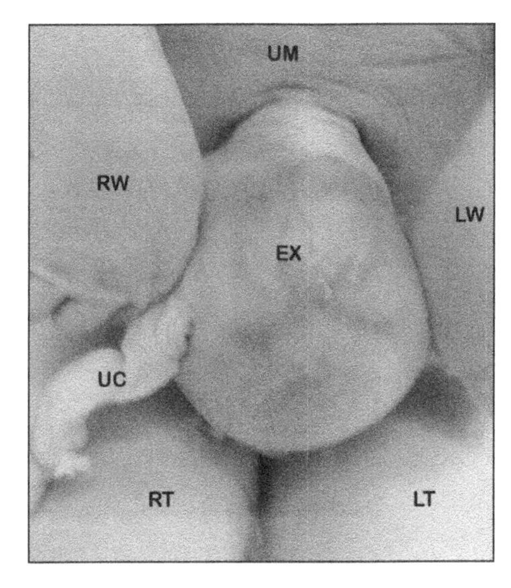

Plate 12 Exomphalos containing loops of bowel and covered by amniotic membrane. EX = exomphalos; LT = left thigh; LW = left wrist; RT = right thigh; RW = right wrist; UC = umbilical cord; Um = umbilicus.

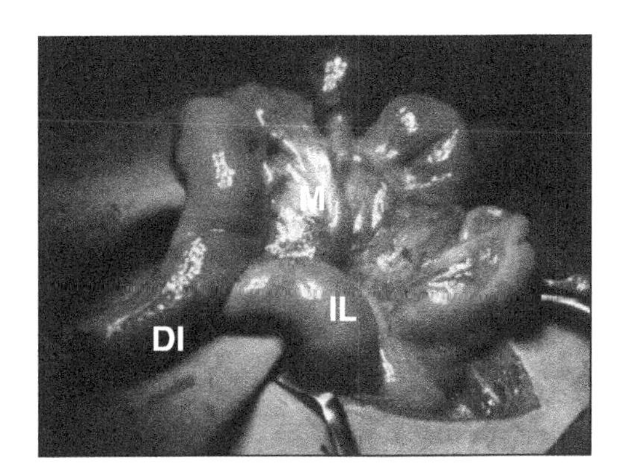

Plate 13 Diverticulum of ileum (DI). IL = ileum; Int = intestine; M = mesentery.

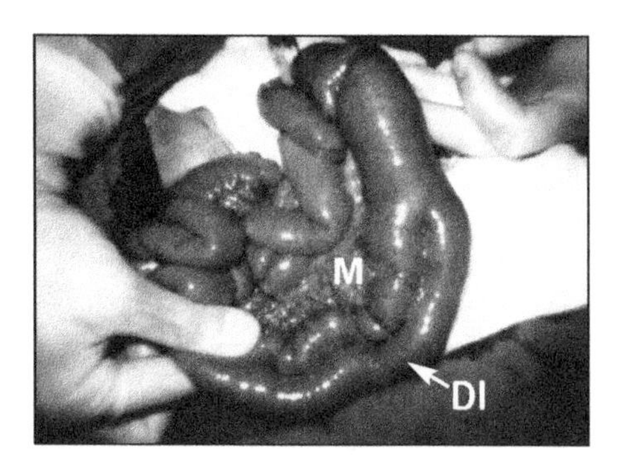

Plate 14 Duplication of small intestine.
DI = duplicated intestine; M = mesentery.

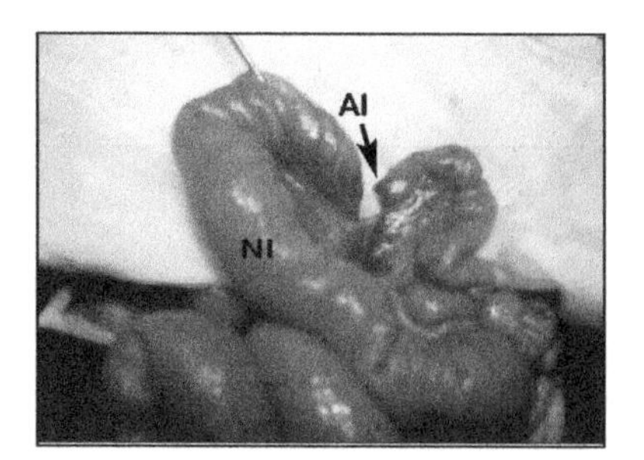

Plate 15 Atresia of intestine (AI).
NI = normal intestine.

THE UROGENITAL SYSTEM

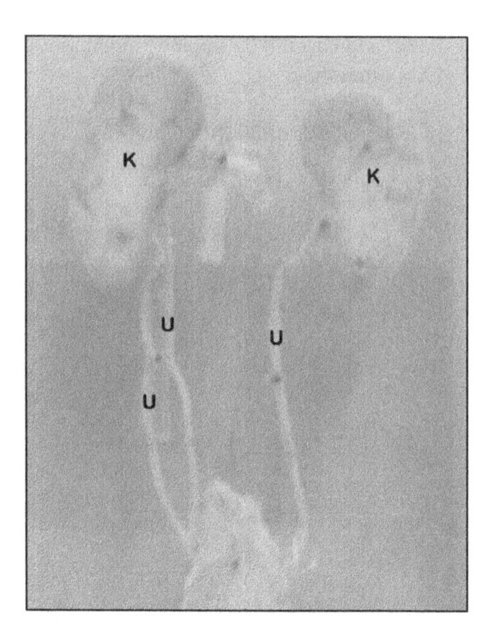

Plate 16 Dissected specimen of a fetus indicating double ureters on the right hand side. K = kidney; U = ureter.

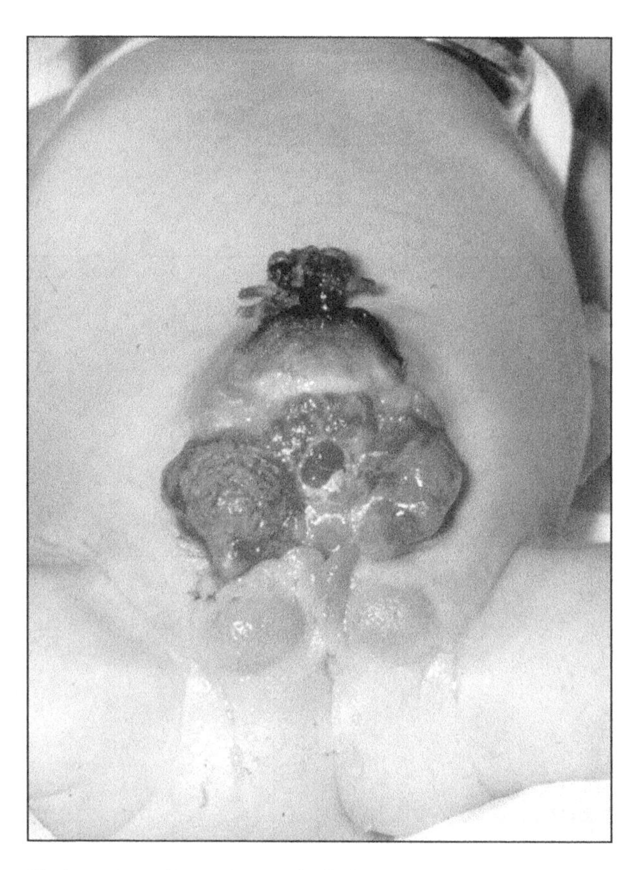

Plate 17 Ectopia vesica in which the posterior wall of the bladder is exposed through the deficient anterior abdominal wall. The degenerating umbilical cord is shown as a black area above the extruded bladder.

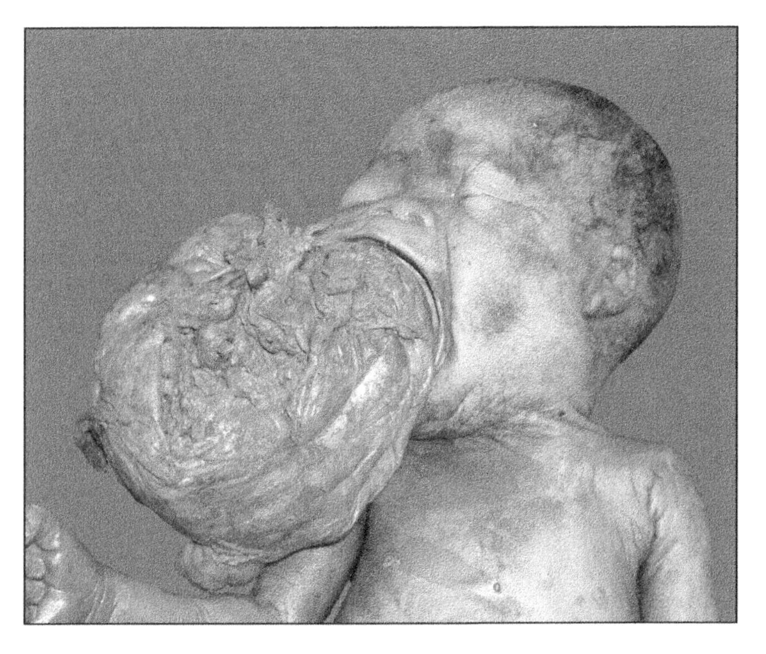

Plate 18 Fetus with oropharygeal teratoma. This is due to the abnormal migration of primordial germ cells to an ectopic site.

THE NERVOUS SYSTEM

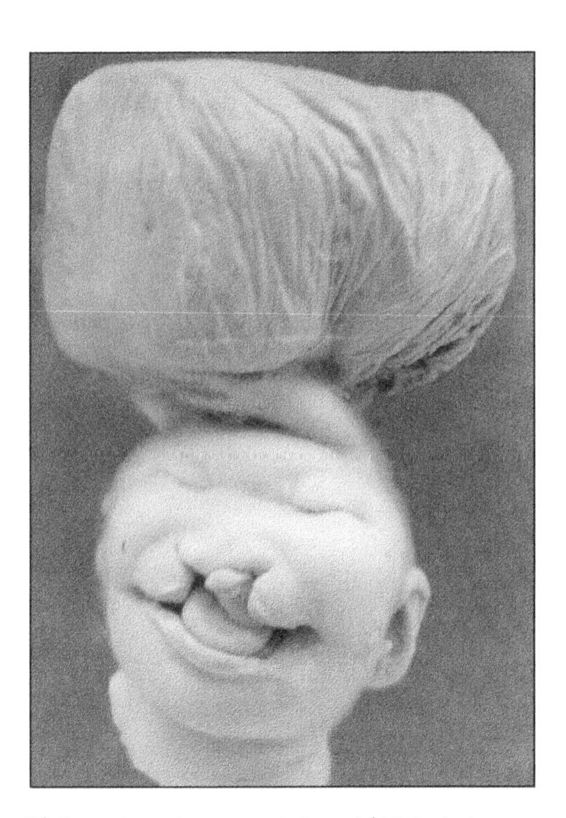

Plate 19 A meningoencephalocoel (this includes meninges and brain) which has herniated out of the skull.

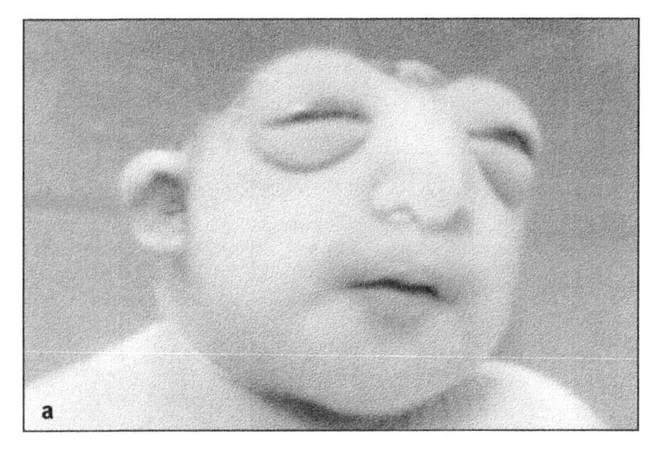

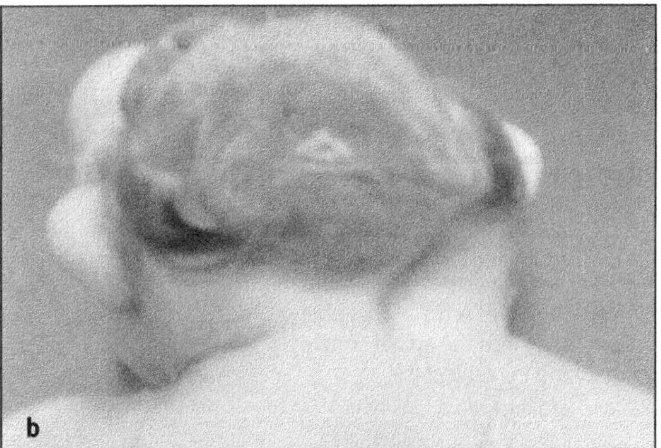

Plate 20 a and b This fetus (in anterior and posterior views) has mero-encephaly (formerly called anencephaly).

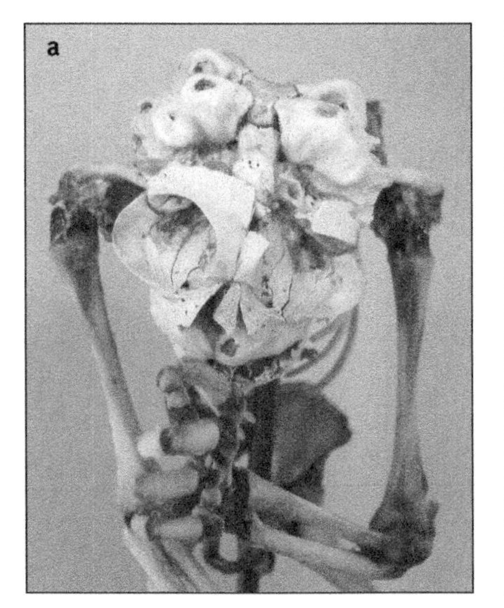

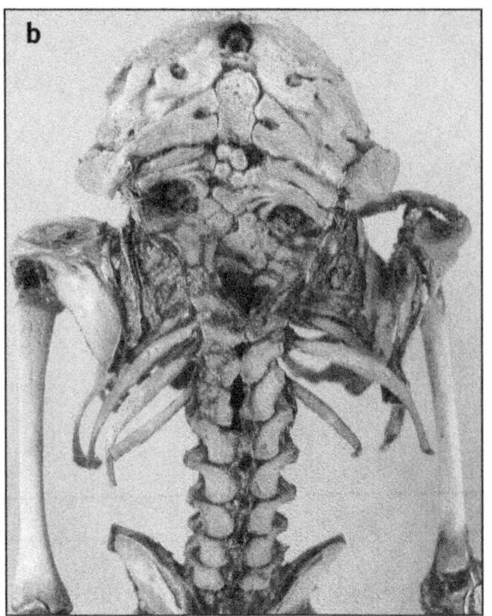

Plate 21a and b The skeleton of this fetus (in anterior (a) and posterior (b) views) shows mero-encephaly.

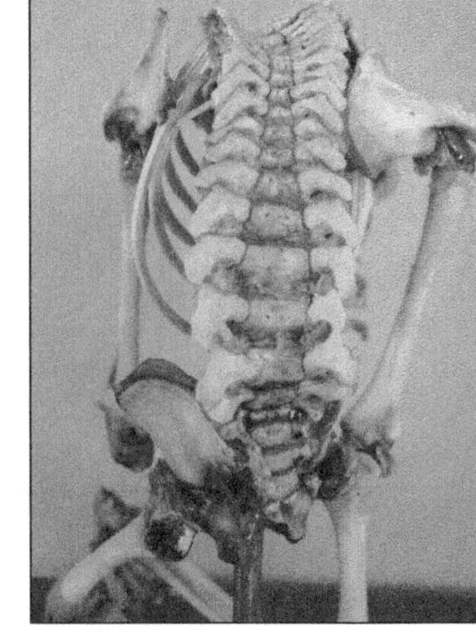

Plate 23 The skeleton of a fetus demonstrating rachischisis.

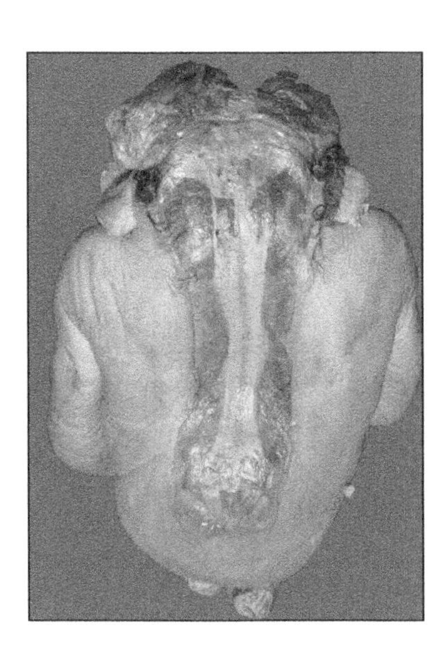

Plate 22 The posterior view of this fetus indicates a mero-encephaly extending into the spinal cord resulting in rachischisis.

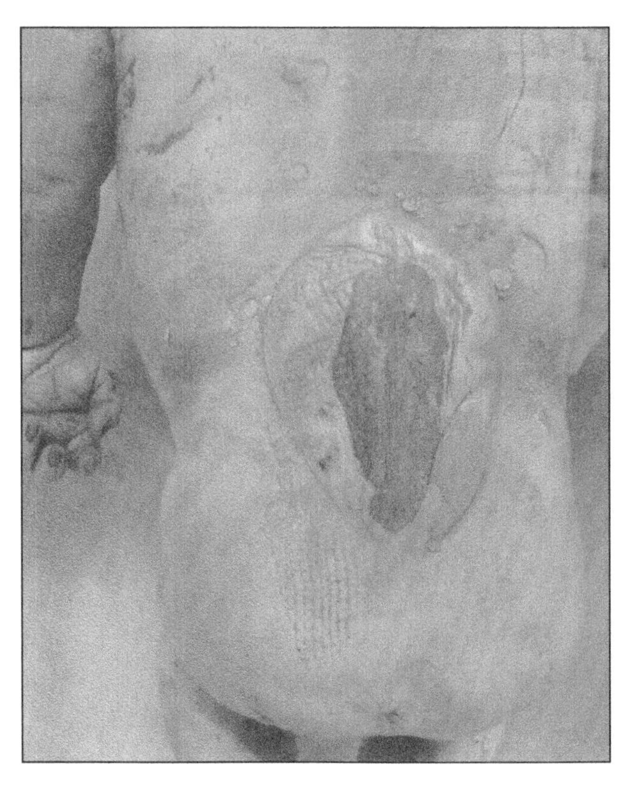

Plate 24 A localised spina bifida of the lumbar region.

Limbs

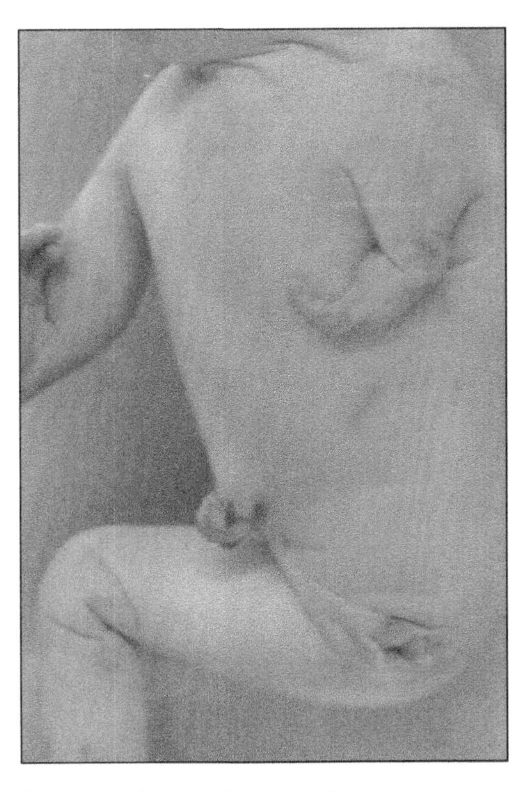

Plate 25 Deformity of upper limbs and absence of the left lower limb.

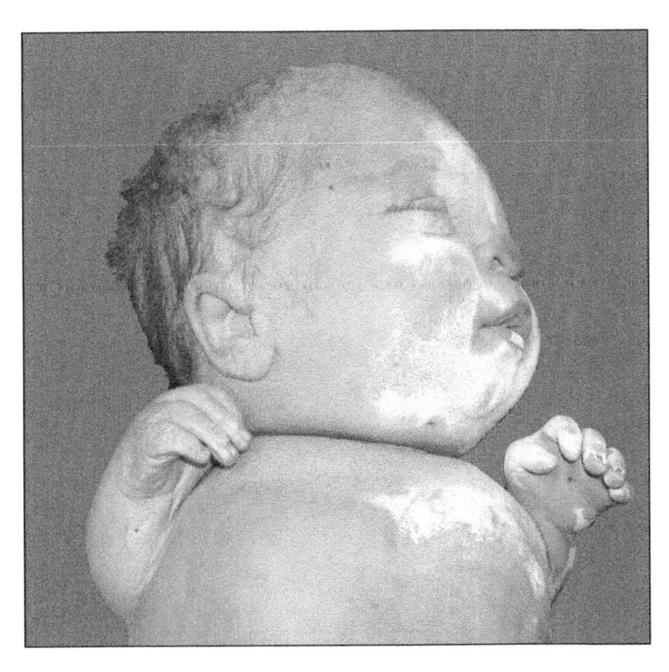

Plate 26 Diminutive upper limbs.

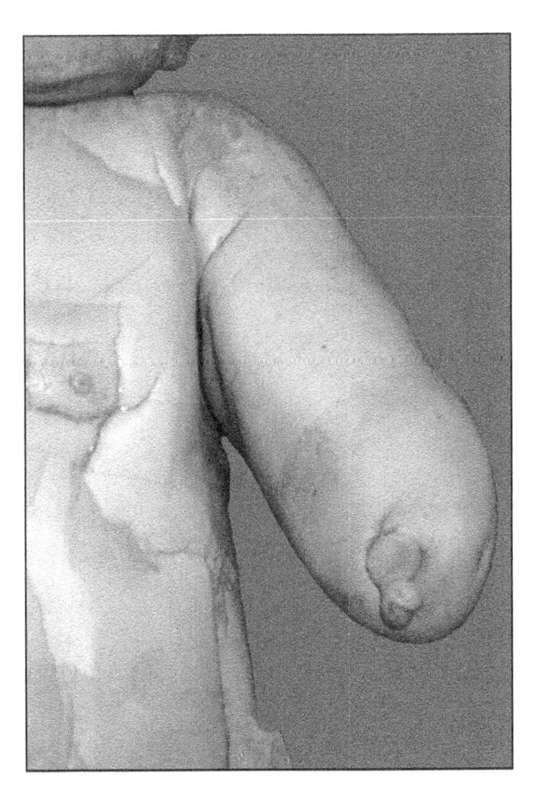

Plate 27 Absence of forearm and hand.

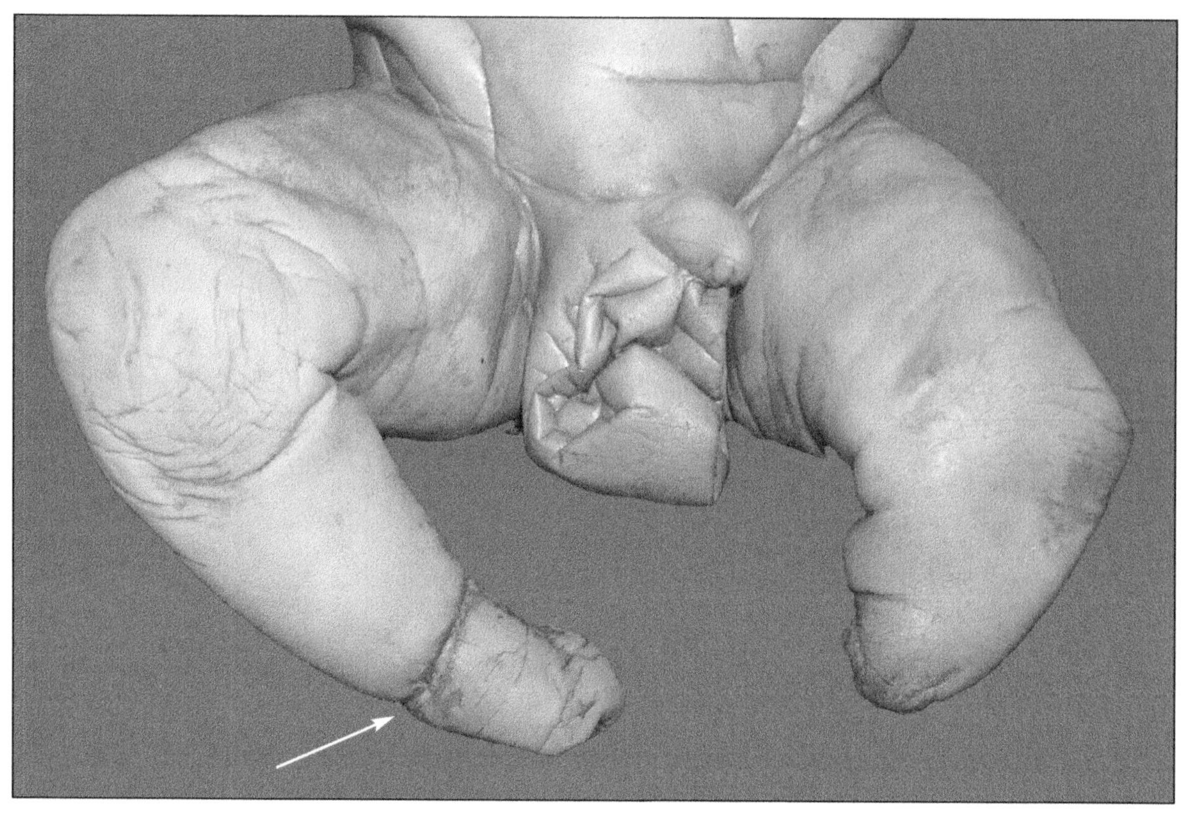

Plate 28 Evidence of the position of a band (◀——) which indicates the possibility of intra-uterine amputation of parts of the lower limbs.

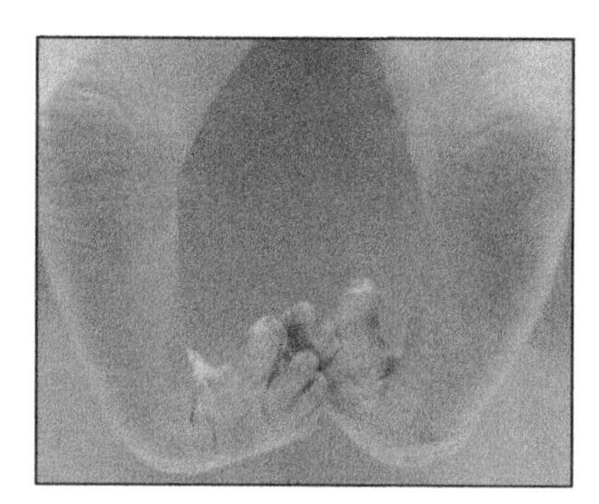

Plate 29 Bilateral club foot.

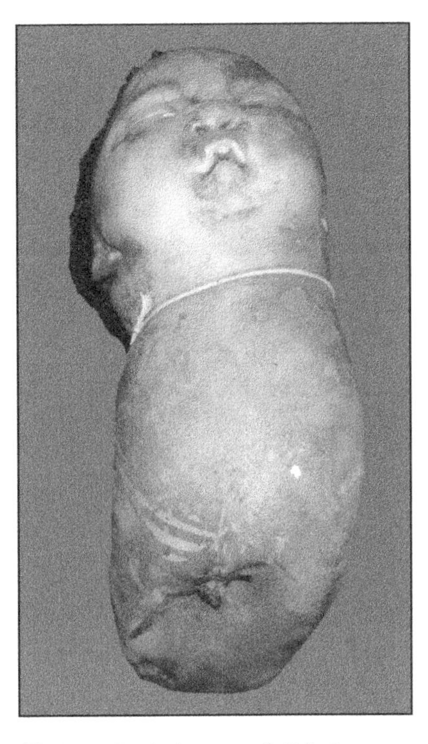

Plate 30 Total absence of all limbs.

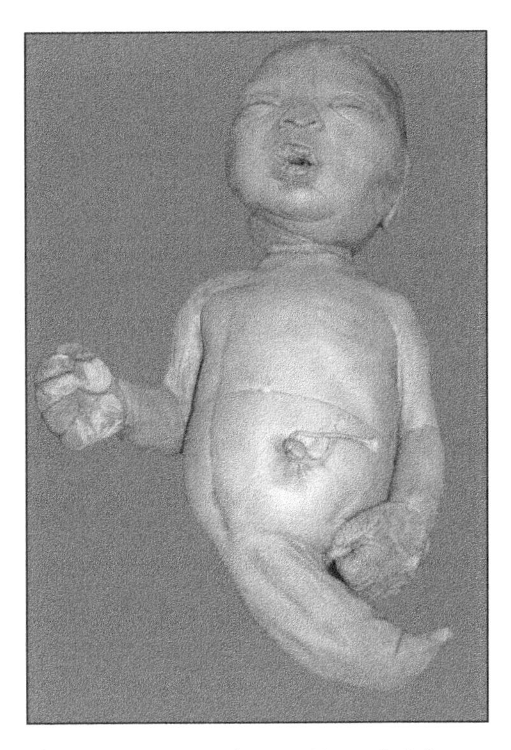

Plate 31 Sirenomelia (fusion of lower limbs).

THE INTEGUMENT

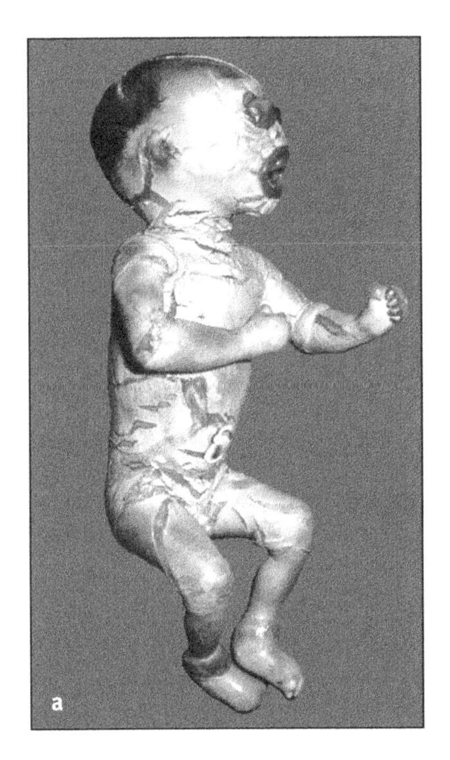

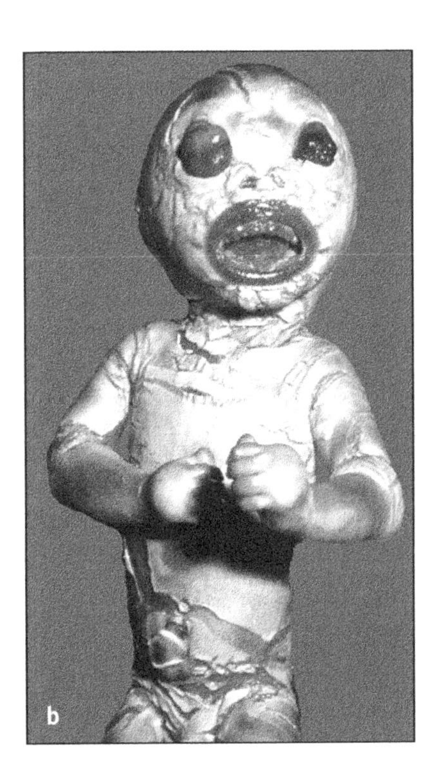

Plate 32a and b The condition known as icthyosis is shown in this fetus. This is a condition of the epidermis in which there is excessive cornification of the superficial layers of the skin.

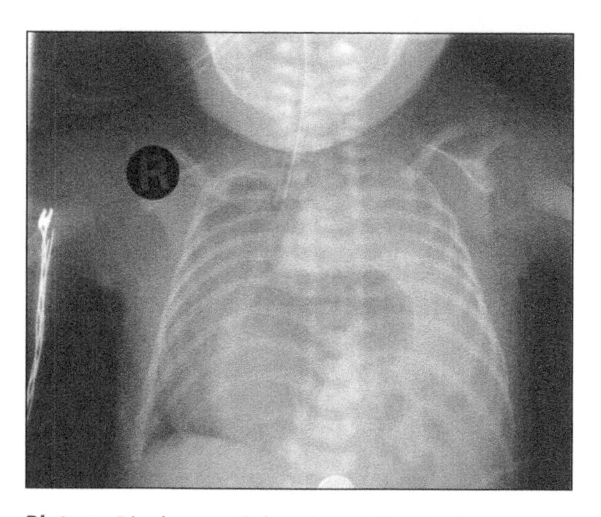

Plate 33 Diaphragmatic hernia on left side of neonate containing loops of intestine.

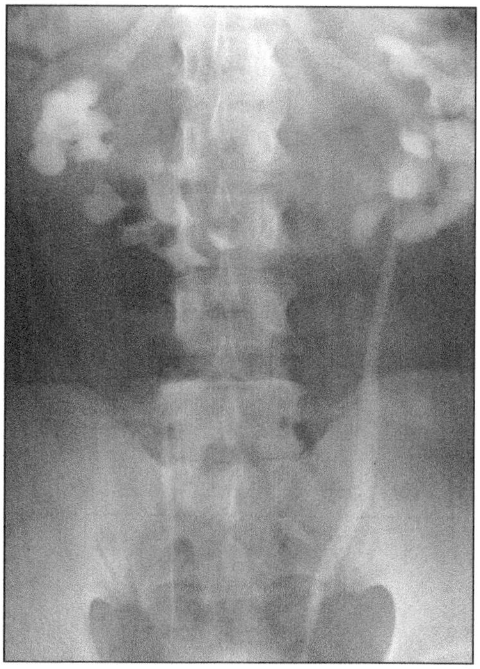

Plate 34 Horse-shoe kidney. The two kidneys are fused across the midline anterior to the vertebral column. Note the presence of a thin ureter on the right side.

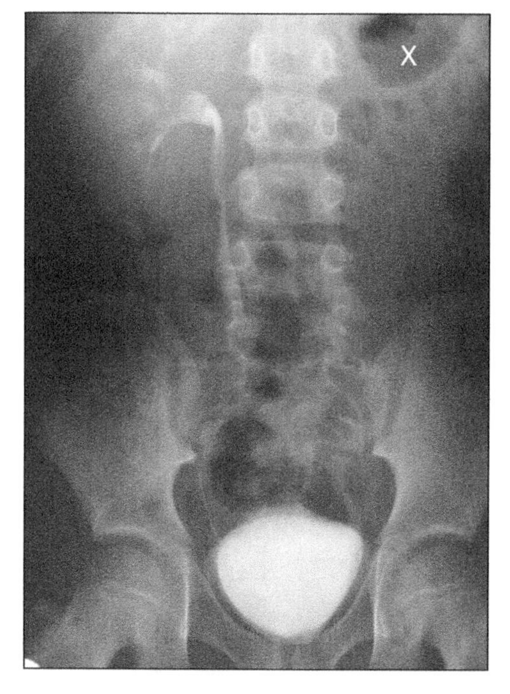

Plate 35 Unilateral kidney (on right side). Note the gas bubble on the upper left quadrant of the abdomen (X) indicating the stomach and the lack of a kidney shadow.

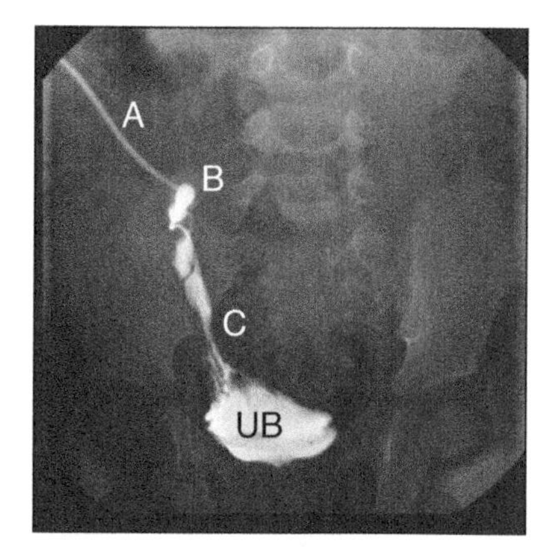

Plate 36 Patent urachus. A = catheter in umbilicus; B = umbilicus; C = urachus; UB = urinary bladder.

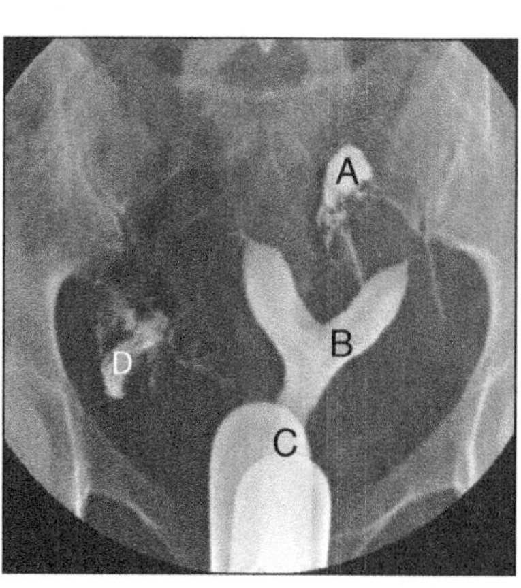

Plate 37 Bicornuate uterus.
A = fimbriated end of uterine tube on left side; B = bilateral cornua of uterus; C = vagina partially obscured by protective strip; D = fimbriated end of uterine tube on right side.

Printed and bound by CPI Group (UK) Ltd, Croydon, CR0 4YY

16/04/2025

14658448-0004